Environmental Science and Engineering

Environmental Engineering

Series editors

Ulrich Förstner, Hamburg, Germany
Wim H. Rulkens, Wageningen, The Netherlands
Wim Salomons, Haren, The Netherlands

More information about this series at http://www.springer.com/series/3172

Nilgün Balkaya • Sinan Guneysu
Editors

Recycling and Reuse Approaches for Better Sustainability

Springer

Editors
Nilgün Balkaya
Department of Environmental Engineering
Istanbul University-Cerrahpaşa
Istanbul, Turkey

Sinan Guneysu
Department of Environmental Engineering
Istanbul University-Cerrahpaşa
Istanbul, Turkey

ISSN 1431-2492
Environmental Engineering
ISSN 1863-5520 ISSN 1863-5539 (electronic)
Environmental Science and Engineering
ISBN 978-3-030-07110-3 ISBN 978-3-319-95888-0 (eBook)
https://doi.org/10.1007/978-3-319-95888-0

This Springer imprint is published by the registered company Springer Nature Switzerland AG
The registered company address is: Gewerbestrasse 11, 6330 Cham, Switzerland

Preface

This book covers the latest in recycling and reuse research focused toward greater sustainability and is inclusive of chapters authored by the world's leading thinkers and practitioners in the field. The authors' contributions are presented in five chapters. The first chapter, Recycling and Reuse, contains the results of the studies on soil improvement with reused tires and pets, domestic wastewater stream segregation, beneficial use of dredged materials, and university students' recycling behavior. The second chapter, Solid Waste Management, comprises of the results of the studies on life cycle environmental impact analysis, phosphorus management, recovery of phosphorus and nitrogen from sewage sludge, and pyrolysis of sludge and wastes. The third chapter, Renewable Energy, includes the results of the studies on energy production from excess pressure in municipal water supply systems, biofuel production from carbon dioxide gas, energy performance of residential buildings, and green roofs and urban life sustainability. The fourth chapter, Environmental Studies, deals with the results of the studies on evaluation of natural radioactivity of cements, ecological footprint, climate change and greenhouse gasses, and assessment of heavy metal pollution in urban soils. The final chapter, Wastewater Management, contains the results of the studies on anaerobic acidogenic bioreactor, heterogeneous photocatalytic treatment, gray water management, treatment of olive oil mill wastewater, and dye removal from wastewater.

Environmental issues with up-to-date literature and experimental focuses are contained making this text attractive to students, researchers, and professionals working in recycling and reuse, solid waste management, renewable energy technologies, environmental studies, and wastewater management issues within the geoscience, engineering, and chemistry fields.

We would like to thank Professor İbrahim Dincer, Christopher T. Coughlin (the Senior Editor of the Springer Nature), Ho Ying Fan (Associate Editor of Springer Physics), and Ms. Chandhini Kuppusamy (Production Editor). We would like to acknowledge the authors for their contribution to the book.

Istanbul, Turkey

Nilgün Balkaya
Sinan Guneysu

Contents

Contributors

Emine G. Abanozoğlu Corlu Engineering Faculty, Namık Kemal University, Tekirdağ, Turkey

Elanur Adar Department of Environmental Engineering, Yildiz Technical University, Istanbul, Turkey

Ahmet Adiller Department of Environmental Health, Uskudar University, Istanbul, Turkey

Semiha Arayıcı Department of Environmental Engineering, Istanbul University-Cerrahpaşa, Istanbul, Turkey

Serdar Aydın Department of Environmental Engineering, Istanbul University-Cerrahpaşa, Istanbul, Turkey

Mehmet Aydoğan Department of Environmental Engineering, Suleyman Demirel University, Isparta, Turkey

Coşkun Ayvaz Department of Environmental Engineering, Istanbul University-Cerrahpaşa, Istanbul, Turkey

Fatos Germirli Babuna Department of Environmental Engineering, Istanbul Technical University, Istanbul, Turkey

Nilgün Balkaya Department of Environmental Engineering, Istanbul University-Cerrahpaşa, Istanbul, Turkey

Müge Balkaya Department of Civil Engineering, Istanbul Technical University, Istanbul, Turkey

Bilsen Beler-Baykal Department of Environmental Engineering, Istanbul Technical University, Istanbul, Turkey

Perihan Biçer Corlu Engineering Faculty, Namık Kemal University, Tekirdağ, Turkey

Ayla Bilgin Department of Environmental Engineering, Artvin Çoruh University, Artvin, Turkey

Emine Çelik Department of Environmental Engineering, Suleyman Demirel University, Isparta, Turkey

Ahmet Demir Department of Environmental Engineering, Yildiz Technical University, Istanbul, Turkey

Hamit Erdal Institute of Social Sciences, Atatürk University, Erzurum, Turkey

Halil Ibrahim Erdal Turkish Cooperation and Coordination Agency, Republic of Turkey Prime Ministry, Ankara, Turkey

Sinan Guneysu Department of Environmental Engineering, Istanbul University-Cerrahpaşa, Istanbul, Turkey

Gokce Güney Department of Environmental Engineering, Dokuz Eylul University, Izmir, Turkey

Senem Yazici Güvenç Department of Environmental Engineering, Yildiz Technical University, Istanbul, Turkey

Selda Yiğit Hunce Department of Environmental Engineering, Istanbul University-Cerrahpaşa, Istanbul, Turkey

Hatice İnan Department of Environmental Engineering, Gebze Technical University, Kocaeli, Turkey

Nazlıcan Karabağ Department of Environmental Engineering, Istanbul University-Cerrahpaşa, Istanbul, Turkey

Bektas Karakelle Cekmece Nuclear Research, and Training Center, TAEK, Istanbul, Turkey

Caner Kepoğlu Department of Environmental Engineering, Namik Kemal University, Tekirdağ, Turkey

Süreyya Meriç Department of Environmental Engineering, Namik Kemal University, Tekirdağ, Turkey

Merve Mermertaş Department of Environmental Engineering, Istanbul Technical University, Istanbul, Turkey

Ayse Muhammetoğlu Department of Environmental Engineering, Akdeniz University, Antalya, Turkey

Habib Muhammetoğlu Department of Environmental Engineering, Akdeniz University, Antalya, Turkey

Ersin Namlı Department of Industrial Engineering, Istanbul University-Cerrahpaşa, Istanbul, Turkey

Semih Nemlioğlu Department of Environmental Engineering, Istanbul University-Cerrahpaşa, Istanbul, Turkey

Yalçın Aşkın Öktem Department of Environmental Engineering, Istanbul University-Cerrahpaşa, Istanbul, Turkey

Burcu Onat Department of Environmental Engineering, Istanbul University-Cerrahpaşa, Istanbul, Turkey

Atakan Öngen Department of Environmental Engineering, Istanbul University-Cerrahpaşa, Istanbul, Turkey

Ertuğrul Ordu Corlu Engineering Faculty, Namık Kemal University, Tekirdağ, Turkey

Şeyma Ordu Corlu Engineering Faculty, Namık Kemal University, Tekirdağ, Turkey

Emine Elmaslar Özbaş Department of Environmental Engineering, Istanbul University-Cerrahpaşa, Istanbul, Turkey

Hüseyin Kurtuluş Özcan Department of Environmental Engineering, Istanbul University-Cerrahpaşa, Istanbul, Turkey

Can Burak Özkal Department of Environmental Engineering, Namik Kemal University, Tekirdağ, Turkey

Ülkü Alver Şahin Department of Environmental Engineering, Istanbul University-Cerrahpaşa, Istanbul, Turkey

Demet Seyhan Istanbul Technical University, Istanbul, Turkey

Naim Sezgin Department of Environmental Engineering, Istanbul University-Cerrahpaşa, Istanbul, Turkey

Delia Teresa Sponza Department of Environmental Engineering, Dokuz Eylul University, Izmir, Turkey

Ugur Emre Temelli Department of Property Protection and Security, Istanbul University-Cerrahpaşa, Istanbul, Turkey

Ayla Uysal Department of Environmental Engineering, Suleyman Demirel University, Isparta, Turkey

Burcu Uzun Department of Environmental Engineering, Istanbul University-Cerrahpaşa, Istanbul, Turkey

Gamze Varank Department of Environmental Engineering, Yildiz Technical University, Istanbul, Turkey

Cansu Vural Fundamental and Industrial Microbiology Department, Ege University, Izmir, Turkey

Part I
Recycling and Reuse

Examination of Soil Improvement with Reused Tires and Pets

Perihan Biçer, Ertuğrul Ordu, Emine G. Abanozoğlu, and Şeyma Ordu

Abstract Various studies have been executed for the aim of obtaining new products via reused waste materials or to use such materials as admixture materials in already available products. The widespread, worldwide popularity of automobile traffic and increasing quantity of PET (polyethylene terephthalate) pose risks for human health, environmental safety and aesthetical concerns, etc. However, the reuse of waste tires and plastic materials in various applications has been encouraged and increased globally.

The objective of the present study is to analyze not only the usability of such wastes by adding waste tire and waste PET materials into the soil but also to determine their effects on the soil characteristics. Sieve analysis, hydrometer test, pycnometer, and compaction and CBR (California bearing ratio) tests have been conducted on a sample soil from Corlu region. In optimum water content, CBR tests have preliminarily been conducted on a natural soil sample, which was then ensued by adding into the soil sample 2% ratio of tire, 2% ratio of PET, 4% ratio of PET + 0.5% ratio of tire, and 0.5% ratio of PET + 4% ratio of tire (by weight), respectively. The results of the laboratory tests performed revealed that a significant change was not observed in the CBR values of the waste tire, while the PET admixture led to an increase in the CBR values.

Keywords Reuse · CBR values · Compaction · Waste tire · PET (polyethylene terephthalate)

P. Biçer · E. Ordu (✉) · E. G. Abanozoğlu · Ş. Ordu
Corlu Engineering Faculty, Namık Kemal University, Tekirdağ, Turkey
e-mail: eordu@nku.edu.tr

N. Balkaya, S. Guneysu (eds.), *Recycling and Reuse Approaches for Better Sustainability*, Environmental Science and Engineering,
https://doi.org/10.1007/978-3-319-95888-0_1

Introduction

Corresponding to population growth as well as the industrial developments in modern day which in turn boosted new demands, considerable quantities of slag, fly ash, plastic wastes, cullets, waste tires of vehicles, and many other waste materials emerge during production as well as usage process. Safe storage of the emerging waste materials has become an increasingly toilsome and costly process in addition to triggering huge problems such as environmental pollution. It is therefore essential to make the best of the industrial wastes in various usage areas for the betterment of national economy. As a consequence of the increased use of automobiles in the globe, there has been a parallel rise in the quantity of waste tires each year. Since these unified-part waste tires contain multiple cavities, they are hard to compress and cannot decompose in the short term; thus, it is not preferred to store them in solid-waste storage areas [1]. In civil engineering applications, waste tire parts are not used independently; they are used via mixing with soil at specific ratios thereby preventing potential geotechnical problems that might occur due to the low shear resistance of the waste tires [2].

It lasts more than a thousand of years for any given plastic materials to entirely disappear from the natural soil. Hence, one of the logical methods to mitigate plastic wastes collected in storage areas and the consequential environmental pollution is to employ such materials in relevant industries. In general sense, the largest area in the reuse of waste materials is the construction applications in which a remarkable portion of waste materials can be reused.

Laboratory compaction and CBR tests have been performed to measure the effects of waste tire and waste PET on the soil sample. Suggestions have been rendered upon analyzing the results of the experimental study.

Material and Methods

The sample from Corlu region, which then would be used for this analysis, had been taken to the laboratory. Sieve analysis, hydrometer analysis, and specific density tests were conducted to determine the physical characteristics of the said sample in the laboratory. To identify the effects of waste tire and PET admixture on the soil sample, compaction and CBR tests were repeated in the laboratory. Optimum water content and maximum dry unit weight of the natural soil sample were measured via implementing compaction test on the natural soil. CBR tests were conducted repetitively on the natural soil sample as well as by adding 2% ratio of tire, 2% ratio of PET, 4% ratio of PET + 0.5% ratio of tire, and 0.5% ratio of PET + 4% ratio of tire (by weight) to the natural soil, respectively.

For the purposes of the present study, granulometric characteristics (grain diameter distribution) of the soil were determined via sieve and hydrometer analysis tests. Figure 1 exhibits the granulometry curve obtained by evaluating the results of the sieve and the hydrometer analysis tests. At the end of these tests, soil classifica-

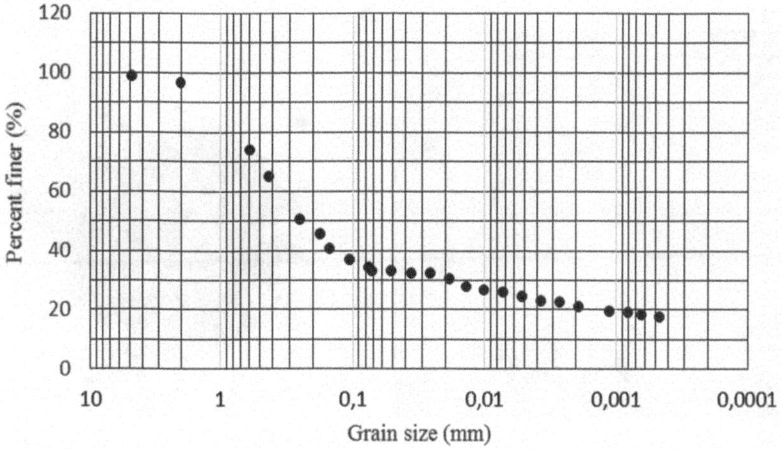

Fig. 1 Granulometry curve of the soil sample

Table 1 Gravel, sand, silt, and clay ratios of the soil sample

Gravel (%)	Sand (%)	Silt (%)	Clay (%)
1.25	64.62	9.8	24.3

Table 2 Uniformity coefficient and grading coefficient values of the soil sample

D_{10} (mm)	D_{30} (mm)	D_{60} (mm)	$C_u = \frac{D_{60}}{D_{30}}$	$C_c = \frac{(D_{30})^2}{(D_{10})(D_{60})}$
0.003	0.07	0.4	5.71	133.33

Table 3 Measuring the specific density of soil sample

Pycnometer weight (g) W_1	38.92	$G_s = \frac{(W_2 - W_1)}{(W_4 - W_1) - (W_3 - W_2)}$
Pycnometer + soil (g) W_2	48.93	
Water + pycnometer + soil (g) W_3	148.16	
Water + pycnometer (g) W_4	141.89	$G_s = 2.67$

tion was determined as clayey sand (SC) according to the Unified Soil Classification System (USCS).

In Table 1, gravel, sand, silt, and clay ratios of the soil obtained from soil's granulometry curve are presented. The uniformity coefficient (Cu) and the coefficient of gradation (Cc) for different diameter values are provided from the grain-size distribution curve, and they are given in Table 2.

Pycnometer test revealed that the grain unit weight of the soil sample was 2.67, and the results of specific density test were as displayed in Table 3.

Fractured PET bottle and shredded waste tire utilized in this study are as seen in Fig. 2.

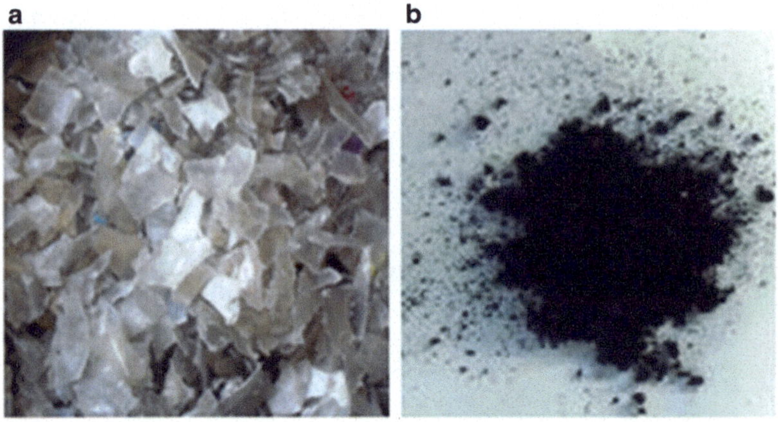

Fig. 2 Fractured PET bottle (**a**) and shredded waste tire (**b**)

Table 4 Parameters obtained from compaction test

Optimum moisture content, w_{opt} (%)	10.7
Maximum dry weight, γ_k (g/cm^3)	2.24

The most vital test is the compaction test in the aftermath of soil classification and determining unit weights. The compaction test used in this study is the modified Proctor test. The aim of the Proctor test is to determine the optimum water content and the maximum dry unit weight that could provide the best compaction in the field. The optimum water content in which the soil grains can compact the best is critical for the stabilization of the soils. The main principle of this test is to compress the material inside a certain pot by specific energies applied onto the definitive numbers of layers. At the end of the compaction test performed in the laboratory, the optimum water content of the soil was measured as 10.7%, and its maximum dry unit weight was measured as 2.036 g/cm^3. The results of the compaction test are given in Table 4, and the compaction test curve is shown in Fig. 3.

The CBR (California bearing ratio) is a test designed in 1929 by Highway Research Center stationed in the state of California (USA) to determine whether soils are fit to be used in the substructure of highways. California bearing ratio is expressed in percent, and it is obtained such that the resistance to penetration at a specified deflection is divided by the resistance offered by a standard crushed rock at the same penetration [3]. The CBR test is a penetration test that can be applied on a number of soil types ranging from clay to fine gravel. To perform the CBR test, materials were prepared in optimum water content predetermined in a modified proctor test and made ready upon compressing in a CBR mold. The soil samples, additive-free and containing 2% ratio of PET, 2% ratio of tire and 4% ratio of PET + 0.5% ratio of tire, and 0.5% ratio of PET + 4% ratio of tire, respectively, were prepared in optimum water content as seen in the CBR test graphic. The test results can be visualized in Fig. 4, and the CBR values can be further studied in Table 5.

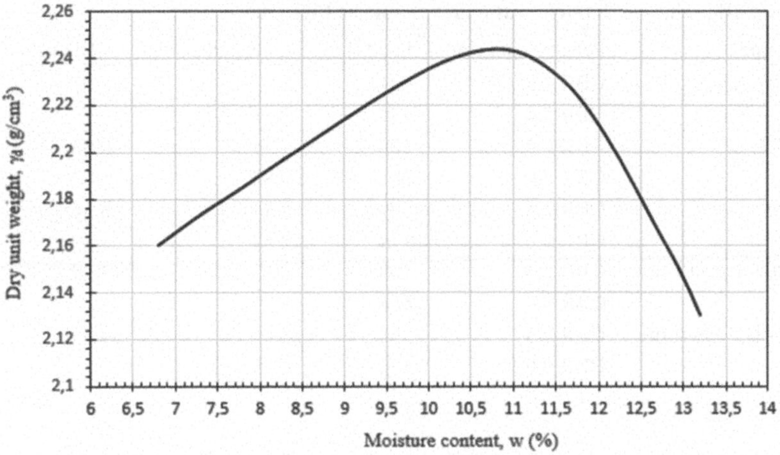

Fig. 3 Compaction test curve

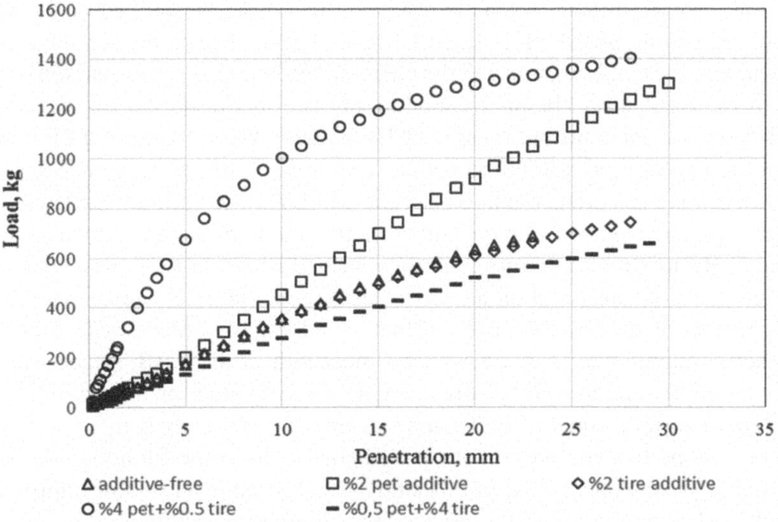

Fig. 4 Load-penetration curves in the CBR test

Results and Discussion

Reused waste material and by-products add financial value to waste materials by mitigating the use of already-limited natural resources and minimize potential environmental problems that would occur in the event of storing the mentioned materials. It is thus suggested that the industrial wastes be benefited by implementing in several usage areas for the improvement of the national economy. In the present

Table 5 CBR values in different waste tires and waste PET ratios

Test name	Penetration (mm)	Load (kg)	CBR values (%)	CBR values (%)
Additive-free	CBR 2.5	81	6	8.81
	CBR 5.0	174	8.81	
%2 PET	CBR 2.5	100	7.34	9.83
	CBR 5.0	201	9.83	
%2 tire	CBR 2.5	79	5.8	8.51
	CBR 5.0	174	8.51	
% 4 PET + % 0.5 tire	CBR 2.5	398	29.2	32.9
	CBR 5.0	674	32.9	
% 0.5 PET + % 4 tire	CBR 2.5	63	4.6	6.55
	CBR 5.0	134	6.55	

study, the usability of waste tire and plastic in soil improvement activities was analyzed. Besides, by adding shredded waste tire and admixture materials formed of waste PET pieces into the sandy soil, their effects on the CBR characteristics of the sandy soil were examined. When the results of the CBR tests applied on the natural soil and the results of the PET-added CBR tests were compared, it was seen that there was an observable increase in the CBR values, whereas a comparison with the result of the CBR tests obtained via admixing tire manifests that the tire did not provide a significant contribution to the CBR results. When 4% ratio of PET + 0.5% ratio of tire (by weight) admixture pattern was added to the soil, an increase in the CBR value was measured. When 0.5% ratio of PET + 4% ratio of tire (by weight) admixture pattern was added to the soil, a significant increase in CBR value was not measured. By mixing admixture materials formed of waste PET parts with sandy soil, the effects of sandy soil on the characteristics of the CBR were assessed.

The results of the tests manifested that by adding shredded waste tires to the sandy soil, a significant increase was not measured in the CBR values. However significant improvement could be obtained in the CBR values on a condition that the admixture material formed of PET parts was mixed with sandy soil.

It is our hope that the present study will shed light to the future studies on the same topic. As was underlined in this study, the desired level of soil improvement on sandy soil cannot possibly be achieved by adding tires solely.

For prospective studies focusing on the same topic analyzed hereby, it thus became crystal clear once again that there is an urgent need for conducting researches focusing on this process by using different recyclable materials.

Acknowledgments The authors would like to acknowledge the valued support of Namık Kemal University, Department of Scientific Researches Project for the NKUBAP.00.17.AR.13.04 no project titled as The Use of Waste Tires in Soil Improvement, Lokman Geri Kazanım A.Ş. for their assistance in obtaining waste tires, and Setaş Kimya Sanayi A.Ş. for their assistance in procuring waste PET materials.

References

1. Gönüllü MT (2004) Endüstriyel Kirlenme Kontrolü, 1st edn. Birsen Yayınevi, Turkey, pp 466–467
2. Zornberg JG, Alexandre RC, Viratjandr C (2004) Behaviour of tire shred-sand mixtures. Can Geotech J 41(2):227–241
3. Bowles JE (1988) Engineering properties of soils and their measurement. McGraw-Hill, Singapore

An Appraisal of Domestic Wastewater Segregation from the Perspective of Recovery, Recycling, and Reuse

Bilsen Beler-Baykal

Abstract One recent sustainable alternative for domestic wastewater management is stream segregation at the source to recover, recycle, and reuse valuable materials embedded in each stream. Gray water may be treated to remove organic matter and pathogens to return the reclaimed water to almost any point in the water cycle as an alternative source of water including toilet flushing and irrigation. Yellow water may be used as a significant source of nitrogen, phosphorus, and potassium as plant nutrients to be recycled as fertilizers through direct or indirect routes. Finally, organic matter in brown and black water may be used as a source of energy or as a soil conditioner after composting. Regardless of the form of segregation, each stream presents a new concept/route in terms of recycling and reuse to enable revaluation of this "waste" stream, i.e., domestic wastewater, as a beneficial source.

Keywords Stream segregation · Gray water · Yellow water · Brown water · Black water · Recovery · Alternative water source · Fertilizers from urine · Recycling and reuse

Introduction

In recent years, sustainable alternatives for conventional domestic wastewater management practices have been finding a considerable attention in the agenda. Several of those options also focus upon recovery of a number of different materials as well as energy to be recycled and reused. One of such alternatives is segregation of domestic wastewater into streams at the immediate source/point they are produced to be handled further to be valorized for various purposes. Common segregated domestic wastewater streams include black water, which refers to toilet wastewater, and gray water, all except wastewater coming from the toilet bowl, in addition to

B. Beler-Baykal (✉)
Department of Environmental Engineering, Istanbul Technical University, Istanbul, Turkey
e-mail: baykalb@itu.edu.tr

N. Balkaya, S. Guneysu (eds.), *Recycling and Reuse Approaches for Better Sustainability*, Environmental Science and Engineering,
https://doi.org/10.1007/978-3-319-95888-0_2

yellow and brown water, which are separately collected human urine and human feces, respectively.

Currently, the most common routes of segregation are separation of domestic wastewater in two components as gray water and black water or in three components as yellow water, brown water, and gray water. The latter is more commonly called as ECOSAN (ECOlogical SANitation) segregation which considers domestic wastewater as a source to be revaluated rather than a waste to be discarded. As such, segregated streams/ECOSAN approach offers a pathway to alleviate domestic wastewater-related water pollution problems while providing a new means for recycling and reuse of water, nutrients, and possibly energy embedded in wastewater from households. Within that context, gray water can be used as an alternative source of water upon proper treatment to recycle water itself, while yellow water is an effective source of plant nutrients to be used as fertilizers, and black/brown water are possible sources of organic matter for the production of biogas, especially methane, to be used as energy or compost to be used as a soil amendment [1–3].

Acknowledging its meaningful contribution to environmental pollution control and its respective benefits, this section will focus mainly upon the recycling/reuse aspects of stream segregation through recovering valuable material, and possibly energy, with an emphasis on gray water and yellow water. An appraisal of possible segregated domestic wastewater streams (gray, yellow, brown, and black water) will be presented from the perspective of recovery, recycling, and reuse of water and nutrients, together with a touch upon the potential for possible energy recovery as dictated by the contents of each stream. Emphasis will be placed on possible shortcuts and benefits of those streams in closed loops/circular operation in line with the suggested approach, and a comparison with current open loops/linear systems of conventional practice will be provided.

Segregation of Domestic Wastewater and Segregated Streams

Stream segregation is a recent domestic wastewater management concept which is based on strict source control in the form of separating streams at the very first point that they are generated, avoiding mixing of those different streams, and processing each one separately to match their characteristics for generating usable products to be forwarded to their next end uses. The main idea is to handle domestic wastewater as a source to be revaluated, not as a "waste" to be disposed of. The fundamental purpose is to create value out of those waste streams and use them as "inputs" or "raw materials" into some functional processes on one hand while taking care of the environment through water pollution control. The main actions/messages regarding this approach include segregating at the immediate source; keeping streams separate; avoiding mixing; closing material cycles, especially water and nutrients; creating value from this "waste" stream, i.e., domestic wastewater; and controlling pollution. This means collecting, recovering, recycling, and reusing the hidden value embedded in domestic wastewater to generate further benefits.

Table 1 Constituents and volumes of domestic wastewater [based on 1, 2 with minor revisions]

	Conventional domestic ww	Gray water	Yellow water	Brown water	Black water
Organic matter	100%	41%	12%	47%	59%
Nitrogen	100%	3%	87%	10%	97%
Phosphorus	100%	10%	50%	40%	90%
Potassium	100%	34%	54%	12%	66%
Pathogens	100%	Intermediate	Low	Very high	High
Volume	100%	75%	1%	0.1%	25%

Characteristics of the four possible streams, gray, black, yellow, and brown water, from domestic wastewater segregation available at this time are significantly different from each other in terms of constituents as indicated in Table 1. Conventional mixed domestic wastewater is also included in the table for comparison.

It may be observed from Table 1 that gray water which takes up ¾ of the volume is the mildest of all segregated streams as well as conventional domestic wastewater itself in terms of pollution potential with only organic matter to be removed and pathogens to be controlled before it can be reused in the water cycle for a number of different end uses. Table 1 also shows that the majority of the nutrients in domestic wastewater, i.e., over 80% of nitrogen and 50% of phosphorus, together with over 50% potassium, will reside in yellow water which constitutes only 1% of domestic wastewater volume [1], pointing at the very rich and concentrated nutrient content of this stream. In contrast to the low pathogenic content of yellow water, the greatest majority of pathogens and nearly half of the organic matter in conventional domestic wastewater will be encapsulated in brown water, which is equivalent to only 0.1% of the volume. In addition to the entire human metabolic wastes in yellow and brown water streams in ECOSAN segregation, black water would contain flush water, leading to relatively lower concentrations, and would typically be equivalent to about ¼ of the conventional domestic wastewater by volume. It is to be noted that black water and yellow and brown water cannot be obtained at the same time, that is, either black water or yellow and brown water will be produced in one given application, as opposed to gray water which is generated under all circumstances regardless of the type of segregation adopted.

An Overview of Recovery/Recycling/Reuse of Materials and Energy from Segregated Streams

It will be noted from the previous section that each segregated domestic wastewater stream is unique in its contents and can be matched/paired with relevant end uses depending upon its constituents. Additional characteristics and possible end uses

Table 2 Summary of characteristics of segregated domestic wastewater streams [2]

	Significant constituents	Constituents to be revaluated	Constituents to be eliminated prior to reuse	Valuable products	Final use
Gray water	Organic matter Pathogens	Water	Organic matter Pathogens	Water	Water cycle Flush water Service water Urban cleaning Irrigation
Yellow water	Nitrogen Phosphorus Potassium	Nitrogen Phosphorus Potassium	Pathogens Pharmaceuticals? Hormones?	Fertilizer	Agriculture Landscape Green areas
Brown water	Organic matter Phosphorus Pathogens	Organic matter Phosphorus	Pathogens	Biogas/energy Compost/soil conditioner	Energy Agriculture Landscape
Black water	Organic matter Nitrogen Phosphorus Potassium Pathogens	Organic matter Nitrogen? Phosphorus? Potassium?	Pathogens	Biogas/energy Compost/soil conditioner? Fertilizers?	Energy Agriculture? Landscape?

Table 3 Segregated domestic wastewater streams: processing options and possible end uses

Stream	Gray water	Yellow water	Brown water	Black water
Source	Wash water in the household, all domestic wastewater except toilet-bowl wastewater	Source separated human urine	Separately collected human feces, possibly flush water and toilet paper	Toilet wastewater containing human urine, feces, probably flush water and toilet paper
Processing options	Biological treatment Physicochemical treatment?	Direct use after storage/ dilution Indirect use after processing	Anaerobic processing Composting	Anaerobic processing, mostly in combination with vacuum toilets Composting
End uses	Water to be returned to almost any point in the water cycle after proper treatment	Fertilizer	Energy Agricultural use	Energy Agricultural use?

after proper processing of the four segregated streams are presented in Table 2, and possible processing alternatives are given in Table 3.

Gray water produced in several different washing functions in the household is an alternative source of water that may be returned to almost any point in the water cycle after proper treatment. It is rich in organic matter, which has to be removed from the liquid phase most commonly in a biological system before it can be reused for a line of final uses. Controlling pathogens is another priority action to be practiced before recycling reclaimed gray water [4].

Yellow water or source separated human urine presents an effective source of fertilizers as it contains the majority of macronutrients, nitrogen (N), phosphorus (P), and potassium (K), in domestic wastewater, which constitute the three main ingredients of fertilizers. Hence, the two major nutrients N and P which pose a threat to the water environment are significant constituents to be recovered for recycling in addition to K to support plant growth.

Urine may be applied either through direct or indirect routes. If the former is selected, storage for 6 months is recommended before application for hygienic safety [5, 6], and dilution is frequently practiced to alleviate the salinity hazard [7].

Processing of urine to remove and recover NPK from the liquid phase is needed for indirect use, and the most common methods seem to be struvite precipitation, stripping followed by absorption, or ion exchange/adsorption in combination with subsequent desorption, although others also appear in the literature [8–10]. Components which require special attention in this stream are pathogens [5, 6], although not as critical as in other segregated streams [11] together with pharmaceuticals and hormones, which are yet to be investigated further.

In case of brown water, organic matter is the most valuable component. Brown water may be used as a source of organics in anaerobic processes to produce biogas, which may in turn be used for generating energy. Digestate from biogas production is being investigated as a possible fertilizer as exemplified by Kocaturk-Schumacher et al. [12]. An alternative way of revaluating brown water is to make use of its organic content for producing compost to be returned to agricultural fields or landscape areas as a soil conditioner. In this case, a part of P may also be recycled.

Black water is composed of urine, feces, and flush water. It is rich in all major components listed for yellow and brown water, yet those components are considerably more dilute due to flush water which is almost ¼ of domestic wastewater by volume providing the liquid phase for dilution. Black water is rich in terms of all major constituents including NPK, organic matter, and pathogens. Since in addition to pathogens, concentrations are low and the volume is large, suitability of black water to be recycled for plant use has to be studied further. However, it contains organic matter, which may be converted to biogas, more significantly to methane. In several applications which revaluate this stream through energy recovery, vacuum toilets using minimal flush water were preferred to keep dilution at a minimum for more effective results. Composting organics from black water may also be potentially considered for recycling useful material.

One very important discourse of the ECOSAN concept is closing of material cycles. Figure 1 presents a comparison of the conventional cycle with the ECOSAN cycle. The former is an open cycle where fertilizers are produced from scratch using materials in gas, liquid, and solid phases as necessary to maintain the food chain mostly from limited sources and wastewater generated after consumption is treated and released into the atmosphere, water bodies, and soil. With ECOSAN, components of the wastewater cycle are returned back to the food cycle in a shortcut as a closed loop, eliminating some of the steps in the conventional one. As an example, in

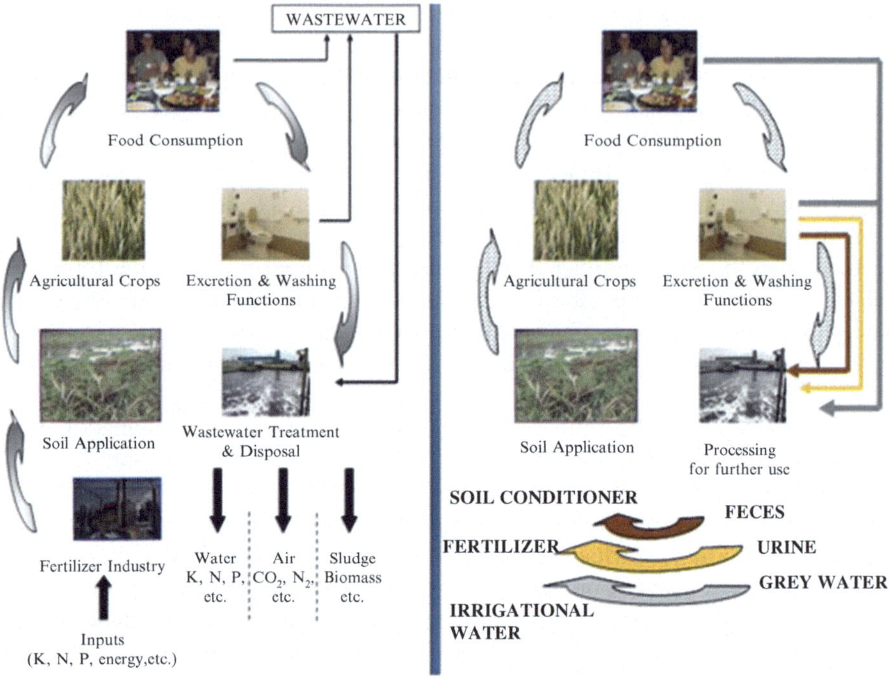

Fig. 1 Left, conventional (open loop/linear); right, ECOSAN (closed loop/circular) cycles

the conventional case ammonia, one of the most widely used nitrogenous fertilizers is synthesized from nitrogen gas and hydrogen through the Haber-Bosch process which is reported to use about 1% of the annual global energy consumption [13].

After wastewater generated as a result of excretion and washing functions is collected in one single pipeline and lead to domestic wastewater treatment plants, ammonia is converted into nitrogen gas through nitrification/denitrification processes which is released into the atmosphere. With the ECOSAN approach, however, urine is segregated in urine-diverting toilets as yellow water, which may either be applied directly or be used indirectly after processing to recycle it in the form of an ammonium-containing fertilizer, without converting it into nitrogen gas, which is subsequently synthesized into ammonia once again. As such, the ECOSAN approach provides a shortcut by bypassing conversion into nitrogen gas and subsequent need for the Haber-Bosch process.

It is also to be noted that "treatment" in the conventional cycle which basically aims to convert pollutants into acceptable final products to be released into various segments of the environment is replaced by "processing" in the ECOSAN cycle which refers to converting constituents of conventional wastewater into useful products to be recycle and reused.

An essential point in effective material recycling/reuse is to carry out material balances for material demand of the intended end use and material to be produced from respective streams to make sure that they match. Proximity between the final use area and the origin of materials to be recycled is a critical issue in terms of feasibility.

Recovering Water from Gray Water

Gray water which constitutes ¾ of the volume may either be collected as one entire entity in a combined/mixed stream or may be subdivided as light/dark or weak/strong gray water. As the name implies, light gray water has even a lower pollution potential as compared to mixed gray water and dark gray water.

Gray water presents a viable source of water, which may be returned to almost any point of the water cycle after pertinent treatment to meet the specific requirements of each respective end use. Due to having organic matter as its most predominant pollutant, gray water has to be treated to remove its organics content. This is generally done through the use of common biological treatment schemes, including land demanding ones like constructed wetlands and compact systems like rotating biological contactors (RBC) and membrane bioreactors (MBR). Monitoring and characterization in somewhat longtime spans are desirable to make the best selection as the characteristics of gray water may vary significantly [14]. Although not emphasized as clearly and strongly as organic matter in the literature, control of pathogens is essential in safeguarding hygienic well-being [4].

Toilet flushing and irrigation seem to be the most popular end uses although car washing, firefighting, urban cleaning, and groundwater recharge are some other areas of reuse. It is to be noted that the quantity of light gray water is just enough to cover the demand for flush water; hence only through recycling light gray water for flushing will secure 25% water savings and will avoid "wasting" water of drinking water quality for sweeping away human metabolic wastes. The contribution of gray water recycling especially in water stressed/water scarce areas should not be overlooked. More on the subject matter is presented in a later chapter of this book.

Recovering Plant Nutrients from Yellow Water

The annual quantity of NPK in human excreta on a global scale is equivalent to about 35–40% of the entire annual global fertilizer use, and the greatest portion of this comes from urine. It may be calculated from the information provided in [15] that urine from one person is equivalent to the amount of NPK fertilizers needed to produce about 200 kg of cereals on an annual basis or one loaf of bread per day. Obviously, recovery of nutrients from human urine is indeed a potential effective way of combating with hunger on a global scale, and its contribution to food security cannot be overlooked. This potential should be named as an indispensable source of fertilizers especially after demonstrating acceptable risks and finding the best route of recovery including elimination of pharmaceuticals and hormones if needed.

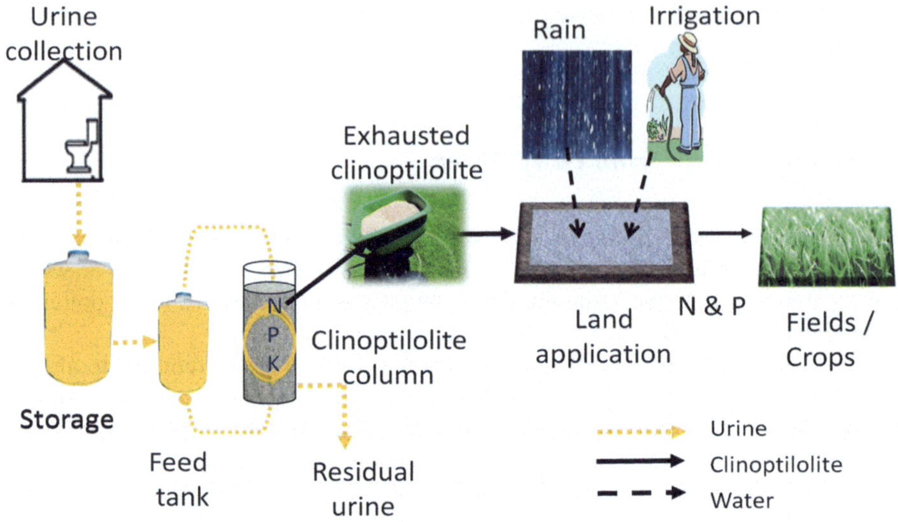

Fig. 2 Production and application of nutrient-enriched clinoptilolite as a urine-based fertilizer

Although there are also others suggested in the literature, the most widely used methods of nutrient recovery seem to be struvite precipitation, stripping followed by absorption, and ion exchange/adsorption followed by desorption. Mainly a P fertilizer is produced in struvite precipitation, struvite itself. N fertilizers are produced through stripping/absorption, ammonium sulfate or ammonium nitrate. Processing urine with the natural zeolite clinoptilolite, which is used as a soil conditioner itself, produces both N and P containing fertilizers. The predominating process in this case is ion exchange/adsorption, and the final product is nutrient-enriched clinoptilolite as illustrated in Fig. 2.

Of those three, struvite precipitation has received the greatest attention so far probably due to the hypothesis that P reserves in the globe are diminishing critically. Most of P recovery/struvite precipitation focused work is mainly concerned with P recovery that generally recommends nitrification/denitrification thereafter. This leads to vast amounts of N being released into the atmosphere without being recovered through the ECOSAN shortcut. While P depletion is a crucial fact, it is to be acknowledged that the amount of nitrogen in urine is several times more than phosphorus and on a global scale the requirement for nitrogen for agricultural purposes exceeds P similarly. This is one of the other indications which make elimination of N in this way a serious loss together with the vast amount of energy demand which has to be incorporated in the production of synthetic N fertilizers. While there is a substantial amount of work in the literature reporting each one of those processes in the literature, researchers have demonstrated production of respective fertilizers through struvite precipitation, stripping/absorption, and ion exchange/adsorption using exactly the same aliquot of urine to recover nutrients as struvite, ammonium sulfate, and nutrient-enriched clinoptilolite in lab scale experiments to enable comparisons of those three methods in single and multistage processing options [16]. The main conclusion drawn was that all processes were effective in the recovery of plant nutrients to be recycled as fertilizer, with the advantage of the first two as producing lower masses of the final product however with the disadvantage of recovering only one of the nutrients while "wasting" the other, especially when used in single stages. Upon processing with clinoptilolite, a commonly used soil conditioner, an NP fertilizer is produced, with the advantage of recovering both nutrients in a single stage simultaneously, however, with a considerably larger final product mass to be handled. Nevertheless, nutrients in yellow water may be recovered using the best option to be recycled back into the food cycle with high efficiencies based on the specific conditions of the cases to be handled.

Concluding Remarks

When recycling and reuse are considered as the focal point, stream segregation provides a more effective route as compared to conventional domestic wastewater. Gray water may be used as an alternative source of water to save pristine water for more worthwhile purposes in an effort to support water resources and aid with the problem of water stress/scarcity. Nutrient-rich yellow water is an effective source of fertilizers and may help in maintaining food security. Brown/black water are potential sources of energy and raw material for compost production.

Previous investigations have shown that stream segregation is technically doable, economically compatible, and socially acceptable to a reasonable extent and may be promoted by awareness raising and supplying pertinent information to all sectors of stakeholders including final consumers and decision-makers. An increase in the number of success stories will definitely help in the willingness to adopt this new domestic wastewater management approach. Stream segregation is an open field both in terms of academic research and practical applications which will benefit both mankind and environmental sustainability for a more prosperous future.

References

1. Otterpohl R, Braun U, Oldenburg M (2003) Innovative technologies for decentralized water-wastewater and biowaste management in urban and peri-urban areas. Water Sci Technol 48(11):23–31
2. Beler Baykal B (2015) Stream segregation in household use: a review of grey water as an alternative source of water and yellow water as an alternative source of fertilizers. Water Qual Expo Health 7(1):27–37
3. Simha P, Ganesapillai M (2017) Ecological Sanitation and nutrient recovery from human urine: how far have we come? A review. Sustainable Environ Res 27:107–116
4. Giresunlu E, Beler Baykal B (2015) An appraisal of microbiological indicators for grey water reuse as a priority issue, CEST 2015 14th international conference on environmental science and technology, Rhodes, Greece, 3–5 Sep 2015

5. Hoglund C (2001) Evaluation of microbial health risks associated with the reuse of source separated human urine. PhD Thesis. Department of Biotechnology, Royal Institute of Technology, Stockholm
6. WHO (World Health Organization) Guidelines (2006) Excreta and greywater use in agriculture. The safe use of wastewater, excreta and greywater, vol IV. WHO, Switzerland
7. Pinsem W, Sathreanranon K, Petpudpong K (2004) Human urine as plant fertilizer: trial on pot basil. In: Proceedings of international IWA conference on wastewater treatment for nutrient removal and reuse, 26–29 Jan 2004, Bangkok, Thailand, vol II, p 174–178
8. Beler Baykal B, Bayram S, Akkaymak E, Cinar S (2004) Removal of ammonium from human urine through ion exchange with clinoptilolite and its recovery for further reuse. Water Sci Technol 50(6):149–156
9. Ganrot ZS, Dave G, Nilsson E (2007) Recovery of N and P from human urine by freezing, struvite precipitation and adsorption to zeolite and active carbon. Bioresour Technol 98:3112–3121
10. Basakcilardan Kabakci S, Ipekoglu N, Talinli I (2007) Recovery of Ammonia from human urine by stripping and absorption. Environ Eng Sci 24(5):615–624
11. Giresunlu E, Beler Baykal B (2014) A preliminary investigation on pathogenic indicators in segregated domestic wastewater streams. Paper presented at IWA 12th specialized conference on small water and wastewater systems & 4th specialized conference on resources oriented sanitation, Muscat, Sultanate of Oman, 2–4 Nov 2014
12. Kocaturk-Schumacher P, Bruun S, Zwart K, Jensen L (2017) Nutrient recovery from the liquid fraction of digestate by clinoptilolite. CLEAN-Soil Air Water 45(6):1500153. https://doi.org/10.1002/clen.201500153
13. Wood S, Cowie A (2004) Review of greenhouse gas emission factors for fertiliser production. Research and development division, State Forests of New South Wales. Cooperative Research Centre for Greenhouse Accounting for IEA Bioenergy Task 38
14. Giresunlu E, Beler Baykal B, Afacan E (2016) A comparative study on light grey water from buildings with different functions. In: Paper presented at proceedings 8. International conference on environmental science and technology. American Academy of Sciences, Houston
15. Wach FG (2007) Sustainable water management and Ecological Sanitation. ZeroM lecture notes. TUBITAK MAM Research Center, Gebze
16. Beler Baykal B, Dogan G (2016) Two stage processing options for enhancing total nutrient recovery from source separated human urine to be used as fertilizer. Presented at WEF IWA nutrient removal and recovery NRR 2016, Denver, 10–13 Jul 2016

Beneficial Use of Dredged Materials in Geotechnical Engineering

Müge Balkaya

Abstract Beneficial use of sediments is known as a sustainable and environmentally friendly alternative to confined disposal facilities and open water disposal. Since most of the geotechnical engineering applications such as roads, backfills, berms, and in situ capping require a large amount of material in their construction, the beneficial use of dredged materials in geotechnical engineering applications is a promising implementation. In this study, sediment dredging and management are introduced. The beneficial use of dredged materials in geotechnical engineering applications is reviewed, and the necessary requirements for their beneficial use in these applications are addressed.

Keywords Dredged material · Sediment · Beneficial use · Geotechnical applications · Waste utilization

Introduction

Sediments are transported into oceans, lakes, streams, and rivers by the effect of various natural occurrences such as gravity, wind, water, and glaciers [1, 2]. They are the mixtures of miscellaneous materials including transported soil particles from surface erosion, shells and coverings of mollusks and other animals, organic matter from dead and rotting vegetation and animals, industrial wastes, sewage, other organic and inorganic materials, and chemicals that suspend in or accumulate on the bottom of the water body [3]. Due to the diversity in their occurrences, their origins, engineering properties, and geometries may significantly differ from one another. They can be found in various grain shapes and sizes that can range from coarse gravel to fine silt and clay [4].

Various construction activities such as dams, dikes, levees, man-made navigation channels, and land uses such as agriculture and mining may disrupt the balance

M. Balkaya (✉)
Department of Civil Engineering, Istanbul Technical University, Istanbul, Turkey
e-mail: balkayamu@itu.edu.tr

N. Balkaya, S. Guneysu (eds.), *Recycling and Reuse Approaches for Better Sustainability*, Environmental Science and Engineering,
https://doi.org/10.1007/978-3-319-95888-0_3

between the amount of sediment entering and exiting the system. The disruption of natural sediment cycle may cause acceleration of sedimentation which may consequently lead to decreased water depths, interruption of navigation, and flooding around waterways [5, 6].

Every year millions of cubic meters of materials are dredged from harbors and waterways to ensure the continuity of navigation and prevent flooding around rivers. A large amount of dredged material is either dumped into confined disposal facilities or disposed into the open water. However, with the rapid increase in the urbanization, the space required for confined disposal facilities becomes limited, and therefore the cost of dumping into confined disposal facilities increases. Besides, contaminated sediments dumped into the open water may lead to physical, chemical, and biological threat to the water quality and the ecosystem. On the other hand, dredged material can be a valuable resource [5–15].

The possibility that the dredged materials may be an environmental threat, along with the increasing environmental consciousness, leads to the application of more restrictive legislations. Consequently, sediment management techniques such as open water disposal and confined disposal became scarce. On the other hand, the beneficial use of dredged sediments in various fields such as civil engineering, manufacture, and agriculture has become increasingly encouraged. It is obvious that the proposed beneficial use application of dredged materials should be economical, environmentally safe, and technically sound. However, all types of sediments may not be applicable for beneficial use, and some beneficial use applications may not fulfil the economical criteria [16].

In the scope of this study, sediment dredging and management are overviewed. The beneficial use applications of dredged sediments and the necessary requirements for their use in geotechnical engineering applications are summarized.

Dredging of Sediments

As stated by Novak et al. [17], every year deposition of sediments leads to a loss of approximately 1% in the storage volume of worldwide reservoirs, which leads to a number of serious problems in and around the waterways [18]. In order to prevent the unwanted consequences of sedimentation and increase water depths, maintain navigation channels, construct and maintain port and harbor facilities, and develop reservoirs for drinking water or energy production, sediments are excavated or removed from the bottom of the water bodies such as rivers, lakes, estuaries, and marine locations. This procedure is termed as dredging [7, 9, 15, 19, 20].

Dredging is essential for the safe and efficient operation of navigable waterways and harbors. Therefore, every year millions of cubic meters of materials are dredged from harbors and waterways worldwide [5, 15].

The properties of dredged sediments are diverse, complex, and strongly dependent on their mineral constituents and the microstructural system of the constituent particles. Dredged sediments are generally fine materials containing between 5 and

10% organic matter [8, 21, 22]. As a result of their extremely high water content, dredged sediments are highly compressible and soft materials with low shear strength [14, 23]. Besides, these materials can also contain different types and levels of contaminants [24]. According to the Waste Framework Directive (2008/98/EC), the dredged materials are classified as waste with the waste codes of 17 05 05/ (polluted sediments) and 17 05 06 (other sediments) [7, 24, 25].

The dredging process consists of three phases including excavation, transportation, and utilization or disposal of the dredged sediments. In the excavation phase, the sediments are loosened or dislodged from the bottom of the water body by a dredger. In the transportation phase, the dredged sediments are transported from the dredging area to the placement area. In the utilization or disposal phase, the dredged sediments are either beneficially used in various applications or disposed in an approved location [26, 27].

There are three types of dredging known as capital, maintenance, and remediation dredging. Capital dredging is used for purposes such as land reclamation and deepening waterways. In capital dredging, generally old and uncontaminated sediments are dredged. Maintenance dredging is applied for keeping waterways at a defined depth to ensure safe navigation. Contamination especially leads to problems in maintenance dredging because the free disposal of sediments in the aquatic system is not allowed by the given standards or regulations. Remediation dredging is used for solving environmental problems related to contaminated sediments [20].

Dredge ability of sediments is affected by various geotechnical properties such as the in situ shear strength (relative consistency, compactness, or rock compressive strength), grain size distribution (coarse and fine fraction), angularity of coarse grains, plasticity of fine grains, organic matter content, and the presence of non-soil materials such as shells and debris [28].

Our environment is affected by dredging operations and the disposal of dredged materials. In a sustainable dredging application, the dredged sediments should be managed successfully, be beneficially used for various purposes, and support the conservation of natural resources [20, 29].

Dredged Sediment Management

In order to provide navigable access to many ports and harbors around the world, sediment dredging is a process that should be performed continuously. Therefore, sediment management is an important issue for the safe and sustainable running of dredging operations and waterways. However, as stated by Harrington et al. [29], the management of dredged sediments, especially for small ports, may be an expensive process. Dredging operations produce huge amounts of sediments which can be used as a new source of materials that is suitable for various applications [24]. Consequently, in a sustainable sediment management operation, the dredged sediments should be regarded as a valuable resource instead of waste that needs to be disposed of [20].

The suitable management option for the dredged sediments should be decided after determining the physical, chemical, and biological characteristics of the dredged material [30]. The management options generally involve disposal (unrestricted, open water deposit, confined aquatic deposit, or confined deposit facilities) or beneficial use (with or without treatment) [30, 31].

Disposal of Dredged Sediments

After the dredging operation is performed, the collected sediments need to be taken care of [32]. If the transport costs for relocation of the dredged material at sea are unreasonable, the material has the potential to be beneficially used in various applications, it is inert, or its contamination level is beyond the permitted limits for sea disposal but acceptable for controlled land disposal, then the dredged sediment is placed on land [33]. However, the majority of the dredged material is either dumped into confined disposal facilities or disposed into the open water [5, 6, 8, 11, 15, 30].

In confined disposal, the dredged material is placed into upland or near shore diked containment facilities and isolated from the environment. Issues such as the physical impacts (e.g., volatilization to air, surface runoff, leaching, uptake by plants and animals) and site requirements (e.g., soil characteristics, value of land, and accessibility to the confined disposal facilities) should be carefully evaluated before deciding for disposal in confined facilities [34].

In open water disposal, on the other hand, the dredged material is placed into water bodies such as oceans, rivers, and lakes [35, 36]. However, when the dredged material is considered for open water disposal, it should be kept in mind that the sediments may be contaminated due to various reasons. Therefore, in order to avoid its detrimental impacts on the environment, the quality of the sediments must be evaluated before disposing into the open water [16]. In order to decide whether the dredged material is suitable for open water disposal, its physical/chemical/biological properties should be determined beforehand. On the other hand, even though the material is considered acceptable for open water disposal, it should also be evaluated in terms of potential beneficial uses [37].

Beneficial Use of Dredged Sediments

With the rapid increase in urbanization, the space required for new or expanded confined disposal facilities becomes limited, and therefore the cost of dumping into confined disposal facilities increases. However, due to the ongoing dredging operations worldwide, the quantities of the dredged sediments increase steadily over the years [38]. Besides, contaminated sediments dumped into the open water may lead to physical, chemical, and biological threat to the water quality and the ecosystem [25, 38]. Therefore, dumping of dredged sediments into the open water should be considered as the last alternative in dredged sediment management [38–40].

On the other hand, since the dredged sediments are available in huge quantities and easily transportable through waterways, they can be used as a renewable and valuable resource [41]. As stated by Mir et al. [42], the beneficial use of dredged sediments has social, environmental, and economic benefits which play an important role in the global sustainability. Various studies have shown that their beneficial use is a sustainable and environmentally friendly alternative to the confined disposal facilities and open water disposal [5, 13, 19, 36, 43–50].

The term "beneficial uses of dredged materials" is known to be used in the USA before the mid-1980s. However, the first documented use of this term in Europe appeared in dredging conference proceedings in about mid-1980s [51]. Beneficial use involves the placement or use of dredged materials for some productive purposes where they can provide the most benefit environmentally, economically, and socially [36, 43, 49, 52–55].

The beneficial use of dredged sediments is a significant component of the sediment management. Depending on the physical and chemical properties of the dredged sediments, a wide variety of beneficial use options are available [37]. In beneficial use applications, the dredged sediment is either substituted for traditional land-based sources or used for the production of a new material [29, 37, 56, 57]. Nevertheless, before considering these materials for possible beneficial use, an appropriate engineering design along with regulatory acceptance and protection of the human health and the environment should be in the first place. Under some circumstances, the dredged sediments should be treated before beneficial use. For this purpose, there is a wide range of treatment techniques available that can be applied depending on the characteristics of the sediment and the proposed reuse objective [31].

In a dredging project, a large number of stakeholders with numerous interests are involved who try to reach an agreement regarding the costs and timing of the project. Since some problems may arise due to conflicts between the stakeholders, long-term planning of a dredging project is of utmost importance for providing a successful application of the beneficial use of the dredged sediments. In order to maintain a successful beneficial use project, issues such as economic requirements, social benefits, environmental considerations, and regulations should be identified in the first place. Besides, ecology, geology, and hydrogeology of the dredging site should also be considered for the appropriate beneficial use options [51].

Evaluation of the Suitability of Dredged Sediments for Beneficial Use

The type of dredged material, the dredging site, the applied dredging technique, and the suitability to the particular beneficial use are the major issues regarding the potential uses of dredged materials [42]. Examination of the physical, chemical, and biological characteristics of the dredged material and determination of the potential future behavior of the material after reuse are important aspects of a successful beneficial use application. By means of a careful preliminary study, the suitability of the dredged material for numerous beneficial use applications can easily be determined, the undesirable environmental impacts can be minimized, and the beneficial use opportunity can be maximized [51].

Some of the useful tools which can expedite the decision-making process of dredged sediments for beneficial use can be stated as sediment transport modeling, environmental impact assessment, and ecological models. According to the environmental legislations, the use of these tools is necessary for dredging and disposal projects especially when they are located in or in the vicinity of sensitive natural resources. Besides, timing of such studies in a dredging project is also an important issue for a successful beneficial use application by avoiding and/or minimizing the adverse environmental effects and maximizing the potential for mitigation measures [51].

The composition of the dredged sediment; its physical, chemical, and mechanical properties; grain size distribution; characteristics of the site for application; the potential bioaccumulation of contaminants; and the existence of invasive species in the sediment are the main factors affecting the projected beneficial use of the sediment. Before deciding whether the dredged sediment is acceptable for beneficial use, it is necessary to evaluate the contamination levels of the material by chemical analysis [10, 57–60]. Besides, the salinity of the sediment, acidity, and nutrient levels are also important issues [60]. A number of geotechnical tests are available for the characterization of the mechanical properties of dredged sediments for beneficial use applications. The necessary parameters for use in geotechnical engineering applications include the grain size distribution, water content, permeability, consistency limits, organic matter content, shear strength, and compressibility [61].

Possible Beneficial Use Options for Dredged Sediments According to Contamination Levels

The contamination level of dredged sediment is the primary limitation of its beneficial use [51]. Therefore, the determination of contamination level of dredged sediment is a critical issue for considering its suitability for various potential beneficial uses [60]. Dredged sediment can be beneficially used when concentration of contaminants are below the legislation limits. However, if the contamination level of the sediment is above the threshold levels, then it is considered as waste. Then treatment or landfilling will be the only option for this type of sediment [59].

There are various remediation technologies available for the treatment of contaminated sediments such as the electrokinetic remediation, bioremediation, solidification, and immobilization [62, 63]. Besides, soft remediation technologies such as phytoremediation, land farming, and co-composting, based on the use of plants and/or microorganisms, can also be used as sustainable and effective technologies for the remediation of contaminated sediments [64, 65].

Clean, uncontaminated sediments have a wide range of beneficial uses including agriculture, forestry, horticulture, aquaculture, beach nourishment, habitat restoration and development, parks and recreation, strip-mine reclamation, solid waste management, and construction and industrial development [60].

When working with slightly to moderately contaminated sediments, pre-treatment and/or covering with clean material may be required before beneficial use. This type of dredged sediments can be used for the creation of a soil berm around a dredge material dewatering facility, upland fill beneath a parking lot, or as daily cover material at a landfill [66].

Highly contaminated sediments are not suitable for beneficial use without treatment especially if its potential risk for biomagnification is high and wildlife habitat development projects are proposed. As expected, the more contaminated the dredged sediment, the greater the constraints on beneficial use are applied. The important issue when dealing with highly contaminated sediments is to determine whether the type and level of contamination are in accordance with the proposed beneficial use application [60].

Possible Beneficial Use Options for Coarse and Fine–Grained Materials

Dredged sediments can be found in various grain shapes and sizes that can range from rock and coarse gravel to fine silt and clay [4, 42, 67]. Silt and clay are the most common materials obtained from maintenance dredging in ports, canals, and rivers [51, 57]. The beneficial use of coarse and fine-grained materials is different from each other as a result of their diverse engineering properties and the required preparation work prior to their use.

Coarse-grained materials have higher mechanical resistance compared to the fine-grained soils. Therefore, the beneficial use of coarse-grained materials such as sand is suitable for a wide range of applications [60, 68]. Rock, which may range from soft marl like sandstone and coral to hard rock like granite and basalt, can be a valuable construction material [42, 57]. It can be beneficially used in strip mine reclamation, solid waste management, offshore berms, habitat restoration parks and recreation, agriculture, forestry, horticulture, and aquaculture [57, 60]. Gravel and sand are used for beach nourishment, habitat restoration and development of strip mine reclamation/solid waste management, agriculture, forestry, horticulture, and aquaculture, parks and recreation, and construction/industrial development [42, 57, 60].

As stated by Zentar et al. [16], fine sediments are difficult to reuse in comparison with sands because of their mechanical behavior, high water and organic matter content, and the potential presence of pollutants contained. For instance, due to their high water content, it is necessary to dewater the fine-grained silt and clay prior to beneficial use [51, 57]. Besides, fine-grained sediments containing contaminants require treatment before beneficially used [51, 57]. However, consolidated clay can find more engineering purposes than soft clay with a high water content [42, 57]. A beneficial use option for fine-grained dredged material is to use it in solid waste management as landfill daily cover material. However, in order to be used for landfill daily cover, the dredged sediment should be dewatered and dried prior to beneficial use. Moreover, it should also satisfy landfill cover regulations and fulfill the physical and chemical criteria required for this purpose [60]. Other beneficial uses of fine-grained silt and clay are parks and recreation, construction/industrial development, habitat restoration and development, agriculture, forestry, horticulture, and aquaculture [57, 60].

Beneficial Use of Sediments in Geotechnical Engineering Applications

Since most of the geotechnical engineering applications require a large amount of material in their construction, the beneficial use of dredged materials in geotechnical engineering applications is a promising implementation. Dredged sediments can be used in various geotechnical engineering applications including erosion control, shore protection, construction of levees and berms, road construction and pavement systems (as base, subbase, and subgrade material), land improvement/reclamation applications (such as brownfields redevelopment and strip mine reclamation), solid waste management as landfill daily covers and capping material, earthwork constructions (such as structural fills, backfills behind retaining walls, highway embankments, backlands fill for port expansion projects), and environmental enhancement applications (such as beach nourishment, wetland creation/enhancement, habitat restoration/creation/development). Dredged sediments can also be used for some other purposes including the production of various construction materials such as concrete, lightweight aggregates, asphalt, bricks, cement, blocks, tiles, and ceramics and in various other disciplines as agriculture, forestry, horticulture, and aquaculture [3, 10, 24, 44, 45, 49, 55, 56, 59, 69–93].

Road and Rail Embankments and Beds

With the increase in urbanization, the demand for natural aggregates increased. However, the increased demand is accompanied by the possible shortage of natural resources. Large quantities of natural aggregates are used for road construction, and this application may eventually lead to the depletion problem of natural resources [94]. Therefore, the reuse of waste materials in civil engineering as an alternative raw material and consequently the conservation of natural resources is highly encouraged. Dredged sediments are available in huge quantities and easily transportable through waterways [41]. Therefore, the consumption of dredged materials in road construction is one of the main opportunities for their beneficial use.

However, roads are subjected to high loads. Therefore, the materials used in road construction require a high level of stability. Dredged sediments generally do not have the required mechanical characteristics in their raw form. Due to high water content, presence of organic matter and salt, and the low strength of the dredged sediments, these materials usually cannot be directly used in road construction and may require treatment before use. In this case, mixing with appropriate stabilizing agents such as lime, cement, or hydraulic binders is the widely used application for providing the required mechanical stability of the dredged sediments [14, 24, 67, 95, 96].

Various researchers have investigated the use of dredged sediments in road construction [10, 16, 21, 24, 38, 58, 97–103]. However, before using in road construction, the specific geotechnical properties such as the water content, organic matter content, specific gravity, particle size distribution, plasticity index, compaction properties, permeability, capillarity, California bearing ratio (CBR), shear strength, resilient modulus, and freeze-thaw durability of the dredged material should be determined [58, 94, 104–108]. The salinity of the water included in the sediment should also be determined because the high salinity of marine water has a negative effect on the mechanical strength development of cement-based materials [109]. Besides, the water contents of the materials usually used in road construction should be less than 20%. The high water content of sediment prevents its use in road construction without dewatering. Therefore, the sediment to be used for road construction should dry naturally in a storage area until the water content is <20% before use [14].

Beach Nourishment

Beach nourishment is used for creating a new beach or widening an existing one, preventing erosion, stabilizing the shoreline, feeding the littoral zone, and moderating the wave action. In order to perform beach nourishment operation, material from another site is added to an eroding shoreline. If natural replacement of material moved along the shoreline by various reasons such as wave movement and tidal currents is not available, then this operation might be necessary. Since a large number of beaches need continued maintenance, beach nourishment is an important beneficial use option for dredged sediments [36, 43, 110].

Material used for beach nourishment must be clean and have comparable grain size and aesthetic characteristics to that of the beach under consideration. Usually only the coarse-grained part of the dredged material (gravel and sand) is suitable for beach nourishment. Therefore, separation and decontamination of the dredged material are required for beach nourishment purposes [36, 43, 110].

Construction and Raising of Coastal Lands

Considering the nature of the dredged sediments, their beneficial use appears to be a suitable option in places where restoration of salty areas for introducing plant and animal species typical of marshy habitats is needed. When the dredged sediment is chemically, microbiologically, and ecotoxicologically of good quality, treatment will not be necessary, and the sediments can be directly used for the reconstruction and rising of coastal lands. The use of desalinated sandy fraction of the dredged sediments for the restoration of a coastal land can increase permeability which helps in the cultivation of various crops [96].

Habitat Creation or Restoration

Habitat creation or restoration using dredged sediments can occur in aquatic, wetland and upland environments, and for the restoration of an existing island or the creation of a new one. By this means, the dredged sediments can be beneficially used, coastal defenses can be supported in a more sustainable way, and biodiversity can be maintained or increased [111]. Upland wildlife habitats can be created in pre-existing dredged material containment areas that are no longer used, as well as by placement of dredged material on degraded lands or habitats. Native vegetation is then re-established to provide food and cover for wildlife. In aquatic or wetland environments, dredged material can be used to nourish, restore, or improve habitats [36, 49, 53, 112].

The use of dredged materials as fill is another suitable beneficial use option for dredged sediments. For this purpose, huge amounts of dredged material can be dewatered in holding areas and sold or given to public or private interests to be used as fill material [112]. The dredged sediment to be used as fill material can be either pumped or transported overland directly to the fill site. Depending on the type and amount of loads to which it will be subjected, and the purpose of the fill, the geotechnical requirements and mechanical properties of the fill material vary [113].

Execution of Dams and Embankments

Dredged sediments can be used for the construction of dams and embankments near the coastlines. However, in order to provide structural and environmental safety, the suitability of the dredged sediment should be evaluated before the construction activity [96].

Strip Mine Reclamation

Dredged sediments can be used in mine tailings as tosoil substitute in order to provide a suitable growing medium for vegetation and a cover that reduces the infiltration of water. For best vegetation growth, dewatered dredged sediment having a loamy texture is the most suitable material. Fine-grained or sandy silt can also be used for the establishment of vegetation. However, in order to provide optimum plant growth, the dredged material should be tested for pH, organic content, and soluble salts [112].

Landfill Daily Covers and Capping Material

Dredged sediments can be used as daily cover, liner, gas vent, leachate drain, and gas barrier material in sanitary landfills [112]. Since handling of dredged sediments in the slurry state is difficult, dewatering is necessary before beneficial use. Dewatered sediment can be easily hauled, spread, and compacted. In order to be accepted as a suitable cover material, the dredged sediments should be easily workable, have high strength, and moderate cohesion. Therefore, highly organic materials and peat cannot be used as landfill daily cover material. As reported by USACE [112], well-graded gravel with 10–15% sand and 5% or more fines serves as an excellent landfill cover material.

Dredged sediments can also be used in confined aquatic disposal or capping projects. Capping is the placement of clean dredged sediment over contaminated sediment to isolate contaminants. Contaminated sediment can be either capped in-place or placed inside a submerged depression and then capped with clean sediment free of contaminants, which is generally comprised of sand [110]. This application prevents the harmful effects of the contaminated material to the human health and the environment [49].

Parks and Recreation

The recreational land use of dredged material containment sites is one of the most favorable and applicable beneficial uses of dredged sediments. Recreation sites require lightweight structures and lots of open space. This is particularly suitable for the weaker foundation conditions of fine-grained dredged sediments [53, 112].

Geotechnical Properties and Beneficial Use of Dredged Sediments

Geotechnical tests allow us to know how the sediments would behave when beneficially used in geotechnical engineering applications. For example, in order to be used as subgrade layer or construction fill, the engineering properties of dredged sediment such as the shear strength, compressibility, permeability, plasticity, water content, and organic matter content need to be determined [8]. The high water content, the soft fine-grained nature, and the organic matter levels of the dredged sediments generally make it unattractive for earthwork construction and represent undesirable mechanical properties in many geotechnical engineering applications [114].

In many cases, the majority of dredged sediments is classified as low-plasticity silt (ML), high-plasticity silt (MH), high-plasticity organic (OH), or high-plasticity clay (CH) by the Unified Soil Classification System [115], which are among the poorest earthwork materials [69, 70]. As stated by Chiu et al. [116], the fine dredged sediments are very soft soils which have natural water content higher than the liquid limit and very low shear strength ($Cu < 50$ kPa). Therefore, this type of raw sediments should be treated before beneficial use. In order to enhance its strength, modify drainage characteristics, reduce settlement, reduce free liquids, and immobilize possible contaminants, the sediment can be blended with other soils or materials such as cement, lime, fly ash, and slag [5, 7, 71, 117–121].

Conclusions

Within the scope of sustainable sediment management, it can be said that dredged sediment should be regarded as a natural resource rather than a waste. Evaluating the required physical, chemical, and biological properties of the sediment would lead to the selection of the most appropriate beneficial use option for the dredged material under consideration. If necessary, the appropriate treatment or stabilization techniques should be applied to ensure environmental safety and adequate mechanical properties required for the projected beneficial use application.

Considering every year millions of cubic meter sediments are dredged from the harbors and waterways around the world, it can be stated that instead of dumping into the open water or disposing into landfills without taking the advantage of this valuable resource, finding beneficial use applications provides an essential step toward the sustainable management of ports, harbors, and aquatic resources. Besides, the use of sediments in various beneficial uses would prevent the risk of shortage of some other natural resources as well. Therefore, after determining whether the properties (chemical, geotechnical, etc.) and contamination levels are within acceptable limits, the beneficial use of sediments in various beneficial use applications should be encouraged.

References

1. Bortone G (2007) Sediment and dredged material treatment. In: Bortone G, Palumbo L (eds) Sustainable management of sediment resources, sediment and dredged material treatment, 1st edn. Elsevier, Amsterdam, pp 1–10
2. Montgomery DR, Zabowski D, Ugolini FC, Hallberg RO, Spaltenstein H (2000) Soils, watershed processes, and marine sediments. Int Geophys 72:159–194
3. EPA (1993) Selecting remediation techniques for contaminated sediment. Office of Water (WH-585) EPA-823-B93–001, Washington, DC
4. Walling DE, Webb BW (1996) Erosion and sediment yield: global and regional perspectives (Proceedings of an international symposium held at Exeter, UK, from 15–19 July 1996). IAHS Publ. No. 236
5. Zhu W, Zhang CL, Chiu ACF (2007) Soil–water transfer mechanism for solidified dredged materials. J Geotech Geoenviron Eng 133(5):588–598
6. Welch M, Mogren ET, Beeney L (2016) A literature review of the beneficial use of dredged material and sediment management plans and strategies Portland State University. Hatfield School of Government, Center for Public Service, Center for Public Service Publications and Reports. http://pdxscholar.library.pdx.edu/publicservice_pub/34. Accessed 5 Feb 2018
7. Zentar R, Wang D, Abriak NE, Benzerzour M, Chen W (2012) Utilization of siliceous–aluminous fly ash and cement for solidification of marine sediments. Constr Build Mater 35:856–863
8. Rekik B, Boutouil M (2009) Geotechnical properties of dredged marine sediments treated at high water/cement ratio. Geo-Mar Lett 29:171–179
9. Dia M, Ramaroson J, Nzihou A, Zentar R, Abriak NE, Depelsenaire G, Germeau A (2014) Effect of chemical and thermal treatment on the geotechnical properties of dredged sediment. Procedia Eng 83:159–169
10. Kamali S, Bernard F, Abriak NE, Degrugilliers P (2008) Marine dredged sediments as new materials resource for road construction. Waste Manag 28:919–928
11. Yu H, Yin J, Soleimanbeigi A, Likos WJ (2017) Effects of curing time and fly ash content on properties of stabilized dredged material. J Mater Civ Eng 29(10):04017199
12. Banoune B, Melbouci B, Rosquoet F, Langlet T (2016) Treatment of river sediments by hydraulic binders for valorization in road construction. Bull Eng Geol Environ 75:1505–1517
13. Flora A, Gargano S, Lirer S, Mele L (2017) Experimental evidences of the strengthening of dredged sediments by electroosmotic consolidation. Geotech Geoenviron Eng 35:2879–2890
14. Saussaye L, van Veen E, Rollinson G, Boutouil M, Andersen J, Coggan J (2017) Geotechnical and mineralogical characterisations of marine-dredged sediments before and after stabilisation to optimise their use as a road material. Environ Technol 38(23):3034–3046

15. Chan CM (2016) Geo-parametric study of dredged marine clay with solidification for potential reuse as good engineering soil. Environ Earth Sci 75:941
16. Zentar R, Abriak NE, Dubois V, Miraoui M (2009) Beneficial use of dredged sediments in public works. Environ Technol 30(8):841–847
17. Novak P, Moffat AIB, Nalluri C, Narayanan R (2007) Hydraulic structures, 4th edn. Taylor & Francis, Abingdon
18. Zhang M, Yu G, La Rovere A, Ranzi R (2017) Erodibility of fluidized cohesive sediments in unidirectional open flows. Ocean Eng 130:523–530
19. Childs JL (2015) Dredged material management categories for tracking beneficial use. http://www.dtic.mil/get-tr-doc/pdf?AD=ADA619013. Accessed 10 Jan 2018
20. SEDNET (2014) Moving sediment management forward-the four SedNet messages. https://sednet.org/download/Moving-Sediment-Management-Forward.pdf . Accessed 10 Jan 2018
21. Colin D (2003) Valorisation des sédiments fins de dragage en technique routière. PhD Thesis. Caen University, Caen
22. Rekik B (2007) Propriétés géotechniques des sédiments de dragage traités au ciment. PhD Thesis. Caen University, Caen
23. Peng X, Zhou Y, Jia R, Wang W, Wu Y (2017) Preparation of non-sintered lightweight aggregates from dredged sediments and modification of their properties. Constr Build Mater 132:9–20
24. Dubois V, Abriak NE, Zentar R, Ballivy G (2009) The use of marine sediments as a pavement base material. Waste Manag 29:774–782
25. European Parliament and Council (2008) Directive 2008/98/EC of the European Parliament and of the Council of 19 November 2008 on waste and repealing certain directives (Waste framework), LexUriServ. Do. 3–30. 2008/98/EC. 32008L0098
26. Welp T (2018) Dredging and dredged material disposal overview, US Army Engineer Research and Development Center, Vicksburg. https://tr.scribd.com/document/358431793/2-OverviewofDredgingandDMD-Welp-pdf. Accessed 28 Mar 2018
27. Dredging: the facts. https://www.iadc-dredging.com/ul/cms/fck-uploaded/documents/PDF%20Publications/dredging-literature-dredging-the-facts.pdf. Accessed 28 Mar 2018
28. Spigolon SJ (1993) Geotechnical factors in the dredgeability of sediments; report 1, geotechnical descriptors for sediments to be dredged, contract report DRP-93-3. U.S. Army Engineer Waterways Experiment Station, Vicksburg
29. Harrington J, Murphy J, Coleman M, Jordan D, Szacsuri G (2016) Financial modelling and analysis of the management of dredged marine sediments – development of a decision support tool. J Shipping Trade 1(7):2–10
30. HELCOM guidelines for management of dredged material at Sea and HELCOM reporting format for management of dredged material at Sea, Adopted by HELCOM 36–2015 on 4 March 2015, Baltic Marine Environment Protection Commission
31. Harrington JR, Murphy J, Coleman M, Jordan D, Debuigne T, Szacsuri G (2016) Economic modelling of the management of dredged marine sediments. Geol Geophys Environ 42(3): 311–324
32. Brils J, de Boer P, Mulder J, de Boer E (2014) Reuse of dredged material as a way to tackle societal challenges. J Soils Sediments 14(9):1638–1641
33. Grandchamp F, Sutton G, Harrington J, Belhadj E, Mcheik A, Masson E, Lemière B, Brakni S (2014) CEAMaS project: civil engineering applications for marine sediments. Fourth international symposium on sediment management (I2SM), Ferrara, Italy
34. National Research Council (1997) Committee on contaminated marine sediments: contaminated sediments in ports and waterways: cleanup strategies and technologies. National Academy Press, Washington, DC
35. Wes Haus N (2011) Beneficial reuse of dredged materials in upland environments MSc Thesis. Virginia Polytechnic Institute and State University, Virginia
36. Millrath K, Kozlova S, Shimanovich S, Meyer C (2001) Beneficial reuse of dredged material, progress report. Colombia University, New York

37. OSPAR guidelines for the management of dredged material (Reference Number: 1998–20), Ministerial meeting of the OSPAR Commission Summary Record OSPAR 98/14/1-E, Annex 43
38. Sheehan C, Harrington J (2012) Management of dredge material in the Republic of Ireland – a review. Waste Manag 32:1031–1044
39. Krause PR, Mcdonnell KA (2000) The beneficial reuse of dredged material for upland disposal. Harding Lawson Associates, Novato
40. Cappuyns V, Deweirt V, Rousseau S (2015) Dredged sediments as a resource for brick production: possibilities and barriers from a consumers' perspective. Waste Manag 38:372–380
41. Aouad G, Laboudigue A, Gineys N, Abriak NE (2012) Dredged sediments used as novel supply of raw material to produce Portland cement clinker. Cem Concr Compos 34:788–793
42. Mir BA, Kumar P, Kumar A, Kumar D, Kumar M, Mazumder M (2011) Geotechnical evaluation of dredged material from Dal Lake as a construction material, Proceedings of Indian Geotechnical Conference, p 725–728
43. Nightingale B, Simenstad C (2001) Executive summary: dredging activities: marine issues. http://www.google.com/url?sa=t&rct=j&q=&esrc=s&source=web&cd=1&ved=0ahUKEwiUh9WNqfDJAhVK22MKHeN8A3cQFggcMAA&url=http%3A%2F%2Fwww.wsdot.wa.gov%2Fresearch%2Freports%2Ffullreports%2F508.2.pdf&usg=AFQjCNE9KfxLqE97TAO0DUgScyqvuZKvzQ&sig2=j_Ou5IyxUAB8vBmxgIeyw&bvm=bv.110151844,d.cGc. Accessed 7 Feb 2018
44. PBC (Palm Beach County). Lakes, Estuaries and Lagoons. http://www.pbcgov.com/erm/lakes/estuarine. Accessed 12 Mar 2018
45. Vogt Inc C (2010) Beneficially using dredge materials to create/restore habitat and restore brownfields, and team collaborative efforts that have achieved success. http://www.google.com/url?sa=t&rct=j&q=&esrc=s&source=web&cd=1&ved=0ahUKEwim59alpLbJAhUBwmMKHZooARAQFggcMAA&url=http%3A%2F%2Fgreatlakesdredging.net%2Ffiles%2Fpdf%2FFinal-report-Beneficial-use-of-dredged-material-andcollaboration.pdf&usg=AFQjCNE2r5I1qPKh_VSqzF9YsnvEqICPsA&sig2=AbutiDbm-T-Ig25Qa5y6kg. Accessed 10 Jan 2018
46. SSJDC (Sacramento–San Joaquin Delta Conservancy) (2018). Restoration projects in the delta and Suisun Marsh. http://deltaconservancy.ca.gov/restoration-projects-delta-and-suisun-marsh. Accessed 7 Feb 2018
47. GreenVest (2014) Thin layer March. http://greenvestus.com/thin-layer-marsh-restoration. Accessed 7 Feb 2018
48. NWRD (National Wildlife Refuge Delaware) (2015) http://www.fws.gov/refuge/Prime_Hook/what_we_do/marshrestoration.html. Accessed 15 Mar 2018
49. Clark G, Knight DL (2013) Beneficial use of dredged material in the Great Lakes. http://greatlakesdredging.net/files/pdf/2013BeneficialUse-online-FINAL.pdf. Accessed 15 Mar 2018
50. Levacher D, Colin D, Perroni AC, Duan Z, Sun L (2006) Recycling and valorization of fine dredged sediments for their reuse as road material. IXe'me Journe'es Nationales Ge'nie Civil-Ge'nie Co^tier, Brest, pp 603–612
51. Paipai E (2003) Beneficial uses of dredged material, beneficial use: yesterday, today and tomorrow. Terra et Aqua 92:3–12
52. Reed D (2018) Dredging materials and environmental restoration: a win-win story. http://www.google.com/url?sa=t&rct=j&q=&esrc=s&source=web&cd=2&ved=0ahUKEwiblIad9fzJAhUS2mMKHdnXDokQFggiMAE&url=http%3A%2F%2Fmassbay.mit.edu%2Fresources%2Fpdf%2FDredgedPDF%2FAbstracts7.pdf&usg=AFQjCNFj6-ccx8mjqBwrS2ZMvUHXHGulQ&sig2=BCOsc4KjZvTPymnHZwL6Cw. Accessed 10 Jan 2018
53. EPA and USACE (United States Environmental Protection Agency and the US Army Corps of Engineers) (2007) Identifying, planning, and financing beneficial use projects using dredge materials. http://www.google.com/url?sa=t&rct=j&q=&esrc=s&source=web&cd=1&ved=0ahUKEwiSvZHdveHJAhUS1WMKHaUHBnwQFggcMAA&url=http%3A%2F%2Fwww.

epa.gov%2Fsites%2Fproduction%2Ffiles%2F2015-08%2Fdocuments%2Fidentifying_planning_and_financing_beneficial_use_projects.pdf&usg=AFQjCNFA3Se4PwZ6BvtjGHFeKjDc9XPS6w&sig2=PkAz73UtuJh--ZI9G5eyNw&bvm=bv.110151844,dcGc. Accessed 10 Jan 2018
54. Ozer-Erdogan P, Basar HM, Erden I, Tolun L (2016) Beneficial use of marine dredged materials as a fine aggregate in ready-mixed concrete: Turkey example. Constr Build Mater 124:690–704
55. Limeira J, Agullo L, Etxeberria M (2012) Dredged marine sand as construction material. Eur J Environ Civil Eng 16(8):906–918
56. Mchergui C, Aubert M, Buatois B, Akpa-Vinceslas M, Langlois E, Bertolone C, Lafite R, Samson S, Bureaua F (2014) Use of dredged sediments for soil creation in the seine estuary (France): importance of a soil functioning survey to assess the success of wetland restoration in floodplains. Ecol Eng 71:628–638
57. Deibel I, Lampe C, Ulbricht JP, Cnudde T, van Dessel G (2007) In: Bortone G, Palumbo L (eds) Beneficial use, sustainable management of sediment resources: sediment and dredged material treatment. Elsevier, New York, pp 119–132
58. Miraoui M, Zentar R, Abriak NE (2012) Road material basis in dredged sediment and basic oxygen furnace steel slag. Constr Build Mater 30:309–319
59. Mattei P, Cincinelli A, Martellini T, Natalini R, Pascale E, Renella G (2016) Reclamation of river dredged sediments polluted by PAHs by co-composting with green waste. Sci Total Environ 566–567:567–574
60. EPA (2007) Identifying, planning, and financing beneficial use projects using dredged material, beneficial use planning manual. US Environmental Protection Agency and US Army Corps of Engineers., EPA842-B-07-001, Washington, DC
61. Dredged material screening tests for beneficial use suitability, Technical Note DOER-C2 Feb 1998
62. Netzband A, Hakstege AL, Hamer K (2002) Treatment and confined disposal of dredged material. Dutch-German Exchange on Dredged Material, Part Two
63. Rulkens W (2005) Introduction to the treatment of polluted sediments. Rev Environ Sci Biotechnol 4:213–221
64. Bert V, Seuntjens P, Dejonghe W, Lacherez S, Thuy HT, Vandecasteele B (2009) Phytoremediation as a management option for contaminated sediments in tidal marshes, flood control areas and dredged sediment landfill sites. Environ Sci Pollut Res Int 16:745–764
65. Masciandaro G, Di Biase A, Macci C, Peruzzi E, Iannelli R, Doni S (2014) Phytoremediation of dredged marine sediment: monitoring of chemical and biochemical processes contributing to sediment reclamation. J Environ Manag 134:166–174
66. Childs JL, Abney A, Young M (2011) Beneficial reuse of dredged material-the regulatory approach. http://www.sludgesandsediments.com/downloads/dredging/1-_CH2M_HILL_Beneficial_Reuse_Childs.pdf. Accessed 7 Feb 2018
67. WP2-A8 requirements for sediment-based civil engineering formulations, CEAMaS http://ceamas.sediments.fr/sites/default/files/documents/wp2-a8_3-report_on_reuse_of_Sediments_in_concrete_and_geotechnical_applications_final.pdf. Accessed 3 Mar 2018
68. WP2-A6 definition of characterization methods, CAEMaS. http://ceamas.Sediments.fr/sites/default/files/documents/wp2_a6_definition_of_characterization_methods-october_2015.pdf. Accessed 12 Mar 2018
69. USBR (US Department of the Interior, Bureau of Reclamation) (1998) Earth manual, part 1, 3rd edn. US Department of the Interior, Bureau of Reclamation, Denver
70. Grubb DG, Davis A, Sands SC, Carnivale M, Wartman J, Gallagher PM (2006) Field evaluation of crushed glass-dredged material blends. J Geotech Geoenviron Eng 132(5):577–590
71. Malasavage NE, Jagupilla S, Grubb DG, Wazne M, Coon WP (2012) Geotechnical performance of dredged material–steel slag fines blends: laboratory and field evaluation. J Geotech Geoenviron Eng 138(8):981–991

72. Rekik B, Boutouil M (2011) Geotechnical properties of dredged marine sediments treated at high water/cement ratio. Geo-Mar Lett 29:171–179
73. LCEP (Lower Columbia Estuary Partnership) (2010) LCREP Draft Aug 2010
74. LCEP (Lower Columbia Estuary Partnership) (2011) Lower Columbia River Estuary Plan Comprehensive Conservation and Management Plan 2011 Update. http://www.Estuarypartnership.org/who-we-are/ourmanagement-plan. Accessed 21 Jan 2018
75. Roegner C, Fields S (2014) Mouth of the Columbia River: beneficial sediment deposition project benthic impact study 2014. http://www.google.com/url?sa=t&rct=j&q=&esrc=s&source=web&cd=1&ved=0ahUKEwietMTv4LTJAhWFK4gKHeQFAbQQFggcMAA&url=http%3A%2F%2Fwww.nwfsc.noaa.gov%2Fassets%2F26%2F8494_09082015_155342_Beneficial-dredge-MCR%2520.pdf&usg=AFQjCNE1fT-mBCwsCmOTLEK1GMpRarPOPA&sig2=e2Jpg7zdgHkZ3WozReGAhw. Accessed 12 Mar 2018
76. Cappuyns V, Deweirt V, Rousseau S (2015) Dredged sediments as a resource for brick production: possibilities and barriers from a consumers' perspective. Waste Manag 38:372–380
77. Murphy JP, Riordan JD, Harrington J, Sheehan C (2008) Construction and demolition waste and dredge material as landfill liner in Ireland. Geophys Res Abstr 10:EGU2008–A-11749 EGU General Assembly
78. Sheehan C, Harrington J, Murphy JD (2010) An environmental and economic assessment of topsoil production from dredge material. Resour Conserv Recycl 55:209–220
79. Yozzo DJ, Wilber P, Will RJ (2004) Beneficial use of dredged material for habitat creation, enhancement, and restoration in New York-New Jersey Harbor. J Environ Manag 73:39–52
80. Reine K, Clarke D, Ray G, Dickerson C (2013) Fishery resource utilization of a restored estuarine borrow pit: a beneficial use of dredged material case study. Mar Pollut Bull 73:115–128
81. Limeira J, Etxeberria M, Agulló L, Molina D (2011) Mechanical and durability properties of concrete made with dredged marine sand. Constr Build Mater 25:4165–4174
82. Said I, Missaoui A, Lafhaj Z (2015) Reuse of Tunisian marine sediments in paving blocks: factory scale experiment. J Clean Prod 102:66–77
83. Kang G, Tsuchida T, Athapaththu AMRG (2016) Engineering behavior of cement treated marine dredged clay during early and later stages of curing. Eng Geol 209:163–174
84. Kazi Aoual-Benslafa F, Kerdal D, Mekerta B, Semcha A (2014) The use of dredged sediments as sand in the mortars for tunnel lining and for environmental protection. Arab J Sci Eng 39:2483–2493
85. Liu W, Cui H, Dong Z, Xing F, Zhang H, Lo TY (2016) Carbonation of concrete made with dredged marine sand and its effect on chloride binding. Constr Build Mater 120:1–9
86. Marmin S, Dauvin J, Lesueur P (2014) Ocean & coastal management collaborative approach for the management of harbour-dredged sediment in the bay of seine (France). Ocean Coast Manage 102:328–339
87. Aziz M, Ghorbel E, Bibi M (2016) The possibility of using dredging sludge in manufacturing cements: optimization of heat treatment cycle and ratio replacement. Constr Build Mater 106:330–341
88. Cantré S, Saathoff F (2013) Investigation of dredged materials in combination with geosynthetics used in dike construction. Procedia Eng 57:213–221
89. Tang YX, Miyazaki Y, Tsuchida T (2001) Practices of reused dredging by cement treatment. Soils Found 41(5):129–143
90. Watabe Y, Saegusa H, Shinsha H, Tsuchida T (2011) Ten year follow-up study of airfoam treated lightweight soil. Ground Improv Inst Civ Eng 164(G13):189–200
91. Seng S, Tanaka H (2011) Properties of cement-treated soils during initial curing stages. Soils Found 51(5):775–784
92. Watabe Y, Noguchi T (2011) Site-investigation and geotechnical design of D-runway construction in Tokyo Haneda airport. Soils Found 51(6):1003–1018
93. Lee CR (1999) Case studies: characterization tests to determine dredged material suitability for beneficial uses. In: DOER technical notes collection, ERDC TN-DOER C-7. US Army Engineer Research and Development Center, Vicksburg

94. Bernard F, Abriak NE, Damidot D (2002) Recycling of sea sands in the civil engineering field. Dredging symposium. Dunkirk, France
95. Maherzi W, Benzerzour M, Mamindy-Pajany Y, van Veen E, Boutouil M, Abriak NE (2018) Beneficial reuse of Brest-Harbor (France)-dredged sediment as alternative material in road building: laboratory investigations. Environ Technol 39(5):566–580
96. Sedi.Port.Sil (2009) Sediment management guidelines, recovery of dredged SEDIments of the PORT of Ravenna and Silicon extraction, Project LIFE09 ENV/IT/000158 LIFE+ Environment Policy and Governance
97. Achour R, Abriak NE, Zentar R, Rivard P, Gregoire P (2014) Valorization of unauthorized sea disposal dredged sediments as a road foundation material. Environ Technol 35(16):1997–2007
98. Pinto PX, Al-Abed SR, Barth E, Loftspring C, Voit J, Clark P, Ioannides AM (2011) Environmental impact of the use of contaminated sediments as partial replacement of the aggregate used in road construction. J Hazard Mater 189(1–2):546–555
99. Silitonga E, Levacher D, Mezazigh S (2009) Effects of the use of fly ash as a binder on the mechanical behaviour of treated dredged sediments. Environ Technol 30(8):799–807
100. Wang DX, Abriak NE, Zentar R, Xu W (2012) Solidification/stabilization of dredged marine sediments for road construction. Environ Technol 33(1–3):95–101
101. Zentar R, Dubois V, Abriak NE (2008) Mechanical behaviour and environmental impacts of a test road built with marine dredged sediments. Resour Conserv Recycl 52(6):947–954
102. Zdiri M, Abriak NE, Ben Ouezdoua M, Neji J (2009) The use of fluvial and marine sediments in the formulation of roller compacted concrete for use in pavements. Environ Technol 30(8):809–815
103. Tribout C, Husson B, Nzihou A (2011) Use of treated dredged sediments as road base material: environmental assessment. Waste Biomass Valori 2:337–346
104. Abayazeed SD, El-Hinnawi E (2011) Characterization of Egyptian smectitic clay deposits by methylene blue adsorption. Am J Appl Sci 8(12):282–1286
105. Aprile F, Lorandi R (2012) Evaluation of cation exchange capacity (CEC) in tropical soils using four different analytical methods. J Agric Sci 4(6):278–289
106. Hang PT, Brindley GW (1970) Methylene blue absorption by clay minerals. Determination of surface areas and cationexchange capacities (clay-organic studies XVII). Clay Clay Miner 18(4):203–212
107. Sarkar D, Haldar A (2009) Cation exchange capacity determination of soils with methylene blue exchange. Agropedology 19(1):63–67
108. Mackiewicz SM, Ferguson EG (2005) Stabilization of soil with self-cementing coal ashes. World of coal ash (WOCA). American Coal Ash Association, Farmington Hills, pp 1–7
109. Kaushik SK, Islam S (1995) Suitability of sea water for mixing structural concrete exposed to marine environment. Cem Concr Compos 17:177–185
110. Sediment Management Handbook for Dredge and Fill Projects (2011) Prepared for Port of Long Beach, Anchor QEA LP and Thomas Johnson Consultant LLC, 100343–02
111. Fletcher CA, Dearnaley MP, Nottage AS, Stevenson JR, Feates NG, Burt TN (2002) Habitat creation and the beneficial use of muddy dredged material in the United Kingdom. In: Pederson J, Adams EE (eds) Dredged material management options and environmental considerations, proceedings of a conference. Massachusetts Institute of Technology, Cambridge, pp 94–100
112. USACE (2015) Dredging and dredged material management, engineer manual no: 1110–2-5025, EM 1110–2-5025. US Army Corps of Engineers (USACE) publications, Washington, DC
113. SKM (2013) Improved dredge material management for the great barrier reef region. Great Barrier Reef Marine Park Authority, Townsville https://www.Environment.gov.au/system/files/resources/7beff474-15b0-4033-84d7-193d984fabb2/files/gbr-dredge-material-management.pdf. Accessed 12 Mar 2018
114. Rekik B, Boutouil M (2008) Organic matter and cement contents effects on geotechnical properties of cement treated sediments. In: Abriak NE, Damidot D, Zentar R Proceeding of

international symposium sediment management, 9–11 July 2008, Lille, France. Ecole des mines de Douai/I2SM, Lille, p 225–233
115. ASTM D2487–17 Standard practice for classification of soils for engineering purposes (Unified Soil Classification System)
116. Chiu CF, Zhu W, Zhang CL (2008) Yielding and shear behaviour of cement-treated dredged materials. Eng Geol 103:1–12
117. Mitchell JK (1981) Soil improvement state of the art report. Proceedings of 10th International Conference on Soil Mechanics Found Engineering, vol 4, p 509–565
118. Meegoda JN, Member ASCE, Partymiller K, Richards MK, Kamolpornwijit W, Librizzi W, Tate T, Noval BA, Mueller RT, Santora S (2000) Remediation of chromium contaminated soils-pilot scale investigation. Practice Period Hazard Toxic Radioactive Waste Manage 7:7–15
119. Watabe Y, Tsuchida T, Furuno T, Yuasa H (2000) Mechanical characteristics of a cement treated dredged soil utilized for waste reclamation landfill. Proceedings of International Symposium on Coastal Geotechnical Engineering Practice, Yokohama, Japan, p 739–745
120. Maher A, Bennert T, Jafari F, Douglas WS, Gucunski N (2004) Geotechnical properties of stabilized dredged material from New York–New Jersey Harbor. Geol Prop Earth Mater 1874(1):86–96
121. Grubb DG, Chrysochoou M, Smith CJ, Malasavage NE (2010) Stabilized dredged material. I: parametric study. J Geotech Geoenviron Eng 136(8):1011–1024

University Students' Recycling Behavior and Attitudes Toward the Disposal of Solid Wastes

Nilgün Balkaya and Ayla Bilgin

Abstract This study aims to determine the methods used by the university students for the disposal of potentially recyclable wastes (paper and cardboard, glass, metal cans, etc.) and assess university students' recycling behavior. For this purpose, a survey was conducted on students at Istanbul University, Turkey. Depending on the survey results, suggestions on how to improve recycling of package wastes, spent batteries, and waste electrical and electronic equipments in the university campus are presented. The effect of gender on recycling behavior of university students and attitudes toward the disposal of potentially recyclable wastes were also discussed.

Keywords Recycling activities · University students · Solid waste · Survey

Introduction

Universities are like small cities which have large size and crowded population, and the campuses host various activities. They have serious effects on the environment both directly and indirectly [1]. Universities also play important roles in the development of our society. Therefore, they have a particular societal responsibility, especially, relating to the sustainable protection of the environment and the use of resources [2]. As stated by Gallardo et al. [3], universities are kind of institutions where a large number of people go to work or to study every day. Besides, university communities use different services such as cafeterias, canteens, and photocopies which have environmental impacts and need a cleaning service as well. It is known that an effective way of reducing environmental damage is to reduce the waste generation. Therefore, many universities have taken considerable steps toward reducing

N. Balkaya (✉)
Department of Environmental Engineering,
Istanbul University-Cerrahpaşa, Istanbul, Turkey
e-mail: nbalkaya@istanbul.edu.tr

A. Bilgin
Department of Environmental Engineering, Artvin Çoruh University, Artvin, Turkey

N. Balkaya, S. Guneysu (eds.), *Recycling and Reuse Approaches for Better Sustainability*, Environmental Science and Engineering,
https://doi.org/10.1007/978-3-319-95888-0_4

waste generation in order to reduce the environmental impacts caused by their activities.

In order to make provisions for waste management activities at a university, in the first place, it is necessary to know the composition, amount, and distribution of the waste generated in its facilities. The obtained information regarding the generated waste will help to organize and ensure an efficient recycling and reuse of the wastes [3]. However, the success of recycling programs does not solely depend on technology. It must be remembered that it also depends on the involvement of people. Therefore, the improvement and maintenance of environmentally responsible behavior have major impacts on the success of recycling programs [4]. As stated by Zhu [5], universities supply recycling infrastructures on campus for students. However, in order to increase the participation of students to recycling activities, their recycling behaviors should be understood beforehand.

There are numerous research studies on the environmental attitudes and recycling. However, the environmental attitudes and recycling of university students have not been taken into account in most of these studies. On the other hand, the recycling behavior of this population is of utmost importance in terms of environment, since today's university students will be the decision-makers of recycling programs in their communities in the future. If the attitudes and behaviors of university students are understood, the authorities designing recycling programs can develop and modify programs to suit the students' needs, and these individuals can be informed by the organized education programs [6]. Philippsen [7] stated that understanding the recycling behavior of students would be important for policy-makers or authorities responsible for the implementation of recycling programs on campus. The obtained information can be used to implement specific measures to encourage the participation of students in recycling programs.

In this study, the preliminary findings obtained during the first phase of a research project were presented. University students' recycling behavior and attitudes toward the disposal of potentially recyclable wastes such as paper/cardboard, glass, and metal cans were assessed. In order to determine the effects of gender, it was also discussed whether there are significant differences between the behaviors of the female and male university students toward the disposal of potentially recyclable wastes. Besides, some suggestions were given for improving the campus recycling rates and increasing the participation for recycling activities.

Material and Methods

This study was conducted over a 5-month period beginning in January 1, 2014, and running through May 31, 2014, in Avcilar Campus of Istanbul University, Istanbul, Turkey.

The survey was conducted on 446 respondents ($n = 446$) from faculties of engineering, management, veterinary, school of physical education and sports, and vocational school of technical sciences in the campus. The data were collected using

Table 1 Gender, age range, and faculty/college of the students surveyed

	Frequency	Percent (%)
Gender		
Female	256	57.4
Male	190	42.6
Age		
18–25	423	94.8
26–35	21	4.7
36–45	1	0.2
46>	1	0.2
Faculty/college		
Engineering	140	31.4
Management	66	14.8
Veterinary	59	13.2
School of physical education and sports	61	13.7
Vocational school of technical sciences	60	13.5
Total	446	100

a face-to-face survey (on-site method), which has strong randomness and high representativeness [8]. The respondents were read questions and asked to mark their responses on the survey. Most of the respondents (94.8%) were between the ages of 18 and 25. 57.4% of the respondents who completed the questionnaire were females, and 42.6% were males. The general information such as the gender and age of the respondents are given in Table 1.

The survey contained questions on the respondents' general information and their attitudes toward the disposal of potentially recyclable wastes. The general information included questions regarding the gender, age, and the faculty or school of the respondents. In the other part, the respondents were asked how they disposed of their waste papers and cardboards, waste glasses, waste metal cans, plastic bags, spent batteries, and waste electrical and electronic equipments. It was also asked if they knew where the nearest authorized collection point (e.g., community bring center belonging to the municipalities) for the waste electrical and electronic equipments is. The respondents were also asked whether they preferred rechargeable battery when purchasing battery. The recycling of plastic bottles was not in the scope of this survey.

The analyses were performed using SPSS-19 program, and the relationships between the answers were determined using the frequency and Crosstab methods and ANOVA analysis. All data were presented at a significance level of 0.05 with a confidence level of 95%.

Results

Recycling and University Students: How Do University Students Dispose of Potentially Recyclable Wastes (Paper and Cardboard, Glass, Metal Can, Etc.)?

The responses provided by the respondents from various faculties and colleges to the question "How do you dispose of your waste papers and cardboards?" are shown in Fig. 1. As can be seen from Fig. 1, 43% of the respondents stated that they put them in paper and cardboard recycling bin. According to the survey results, this response was followed by the response "I reuse them at home" (31.3%). The survey results indicated that, although the rate (43%) of the response "I put them in a paper and cardboard recycling bin" was not very high, the respondents seemed to have environmental awareness about waste paper and cardboard recycle.

The respondents were asked how they disposed of their plastic bags. The distribution of responses of the respondents from various faculties and colleges to the question "How do you dispose of your plastic bags?" is shown in Fig. 2. Majority (69.4%) of the respondents stated that they reused them. Sixteen percent of the respondents stated that they put them in trash. The survey results indicated that the most common behavior was the reuse of plastic bags (Fig. 2). The reuse activities seemed to be more common than recycling.

The responses of the students participating in the survey to the question "How do you dispose of your waste glasses?" are shown in Fig. 3. 40.8% of the respondents stated that they put their waste glasses in trash, and 32.1% of them stated that they put their waste glasses in glass recycling bin (Fig. 3). The survey results showed that most of the respondents put most of their waste glasses in trash.

More than half (52%) of the respondents answered the question "How do you dispose of your waste metal cans?" as "I put my waste metal cans in trash." According to the questionnaire results, this response was followed by the response "I put my waste metal cans in metal recycling bin" (32%) (Fig. 4). The survey results showed that most of the respondents put their waste metal cans in trash instead of putting them in metal recycling bin.

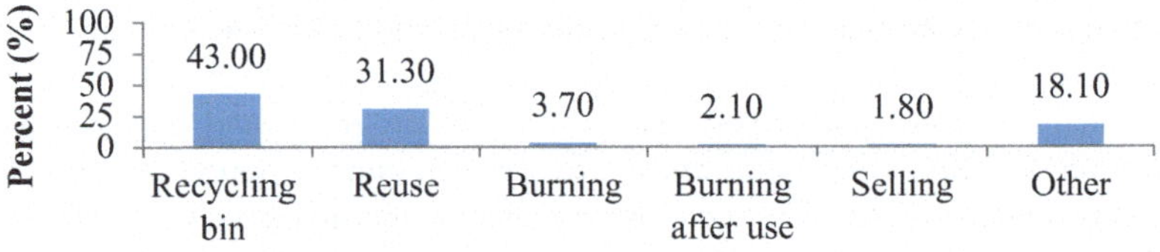

Fig. 1 Responses of the respondents from various faculties and colleges to the question "How do you dispose of your waste papers and cardboards?"

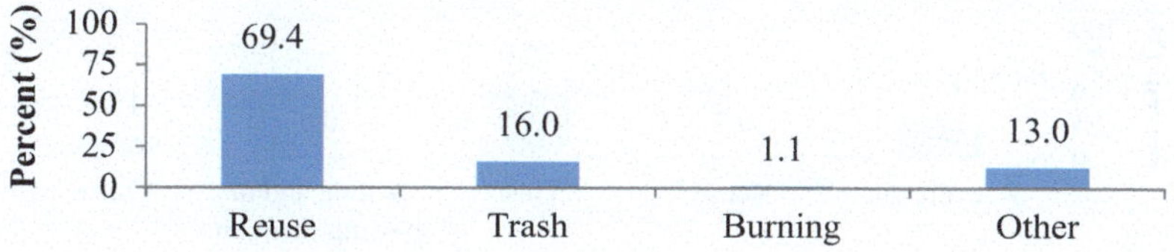

Fig. 2 Responses of the respondents from various faculties and colleges to the question "How do you dispose of your plastic bags?"

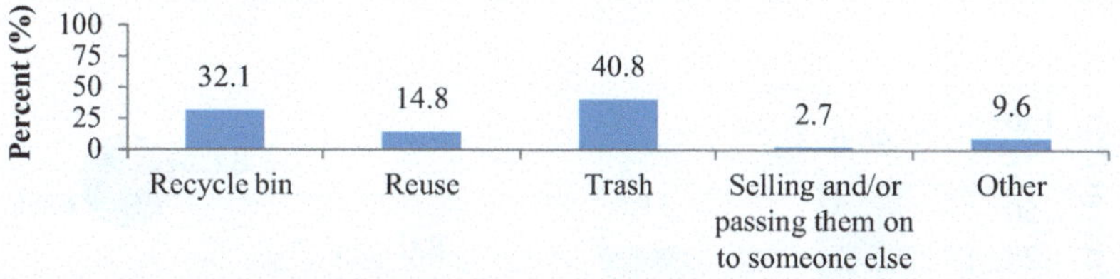

Fig. 3 Responses of the respondents from various faculties and colleges to the question "How do you dispose of your waste glasses?"

The respondents were asked "How do you dispose of your spent batteries" and "Do you prefer rechargeable battery when purchasing battery?" The responses were "I put my spent batteries in battery recycling bin" (66%) and "I put my spent batteries in trash" (31%). The majority (63%) of the respondents replied the question "Do you prefer rechargeable battery when purchasing battery?" as "Yes" (Fig. 5). The survey results indicated that most of the respondents had environmental awareness about spent batteries. The rates of preference for purchasing rechargeable battery and recycling of waste batteries were high among the respondents.

In the survey study, the following questions were asked to the respondents from various faculties and colleges: "Do you have any information about the concept of waste electrical and electronic equipment?" "Is there any authorised collection point for waste electrical and electronic equipments that you know?" "How do you dispose of your waste electrical and electronic equipment?" The responses provided by the students to these questions are presented in Figs. 6 and 7. It can be seen from Fig. 6 that more than half (56%) of the respondents had information about the concept of waste electrical and electronic equipment (WEEE). However, 68% of them had no information about the existence of authorized collection points (e.g., community bring centers belonging to the municipalities) for WEEE. As can be seen from Fig. 7, 50% of the respondents sell their WEEE to get economic benefits. Nineteen percent of the respondents throw their WEEE with household's solid waste. Only 18% of the respondents give their WEEE to the authorized collection points.

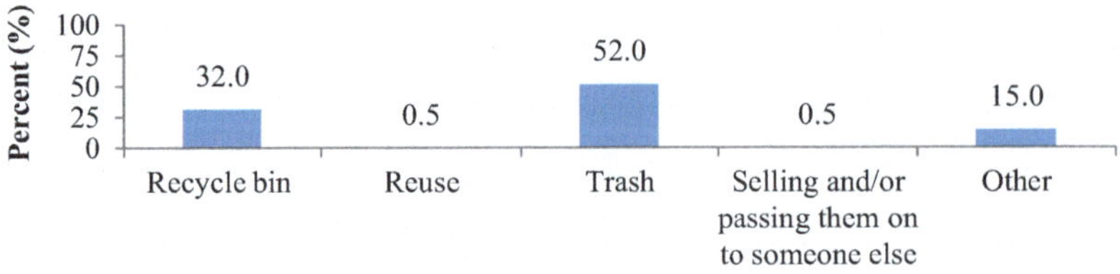

Fig. 4 Responses of the respondents from various faculties and colleges to the question "How do you dispose of your waste metal cans?"

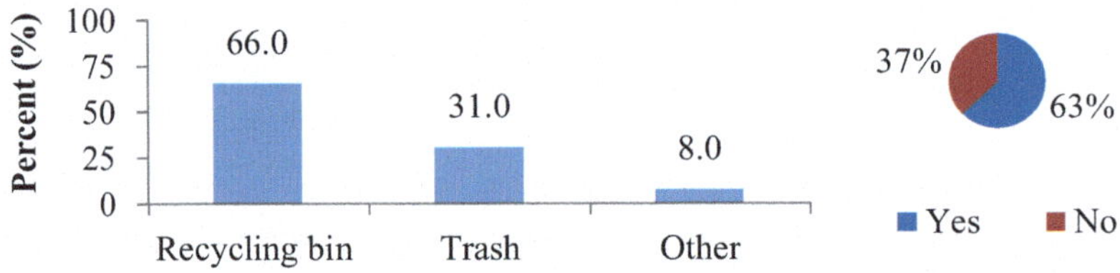

Fig. 5 Responses of the respondents from various faculties and colleges to the question "How do you dispose of your spent batteries?" and "Do you prefer rechargeable battery when purchasing battery?"

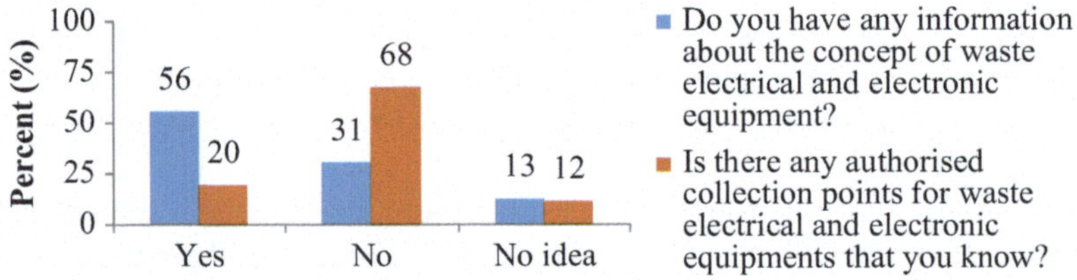

Fig. 6 Responses of the respondents from various faculties and colleges to the questions regarding "Do you have any information about the concept of waste electrical and electronic equipment?" and "Is there any authorised collection point for waste electrical and electronic equipments that you know?"

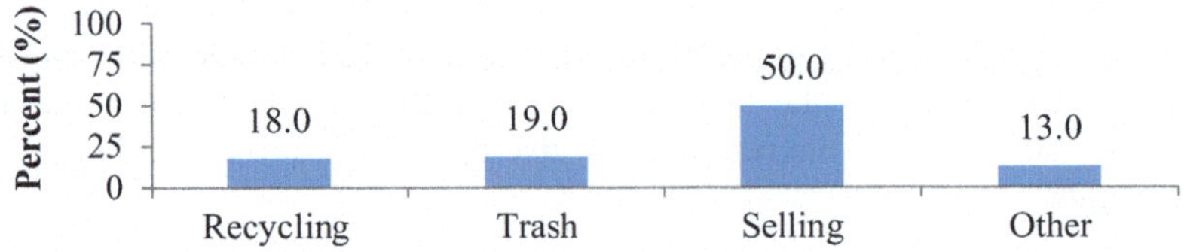

Fig. 7 Responses of the respondents from various faculties and colleges to the question "How do you dispose of your waste electrical and electronic equipments?"

Recycling and Gender: Is There a Difference Between the Behaviors of Male and Female University Students Regarding Their Recycling Activities?

The relationship between the genders, which is a demographic variable, and the university students' recycling behavior were examined using the ANOVA analysis and discussed below. If the "P" value was <0.05 at a 95% confidence level, it was concluded that a statistically significant difference was observed among the answers.

In Fig. 8, the responses of the female and male respondents to the question "How do you dispose of your waste papers and cardboards?" are shown. According to Fig. 8, it can be said that 45% of the female students and 40% of the male students contributed to paper and cardboard recycling. When ANOVA analysis was performed at 95% confidence level for determining whether there is a significant difference between the behaviors of the female and male university students regarding the disposal of waste papers and cardboards, no significant difference ($P = 0.75 > 0.05$) was observed.

The responses provided by the female and male respondents to the question "How do you dispose of your plastic bags?" are shown in Fig. 9. When the responses were assessed according to the gender, it was seen that the plastic bags were reused by 68.6% of the female students and 75% of the male students. Besides, it was seen that 16.5% of the female students and 15.3% of the male students put their plastic bags in trash (Fig. 9). When ANOVA analysis was performed at 95% confidence level for determining whether there is a significant difference between the behaviors

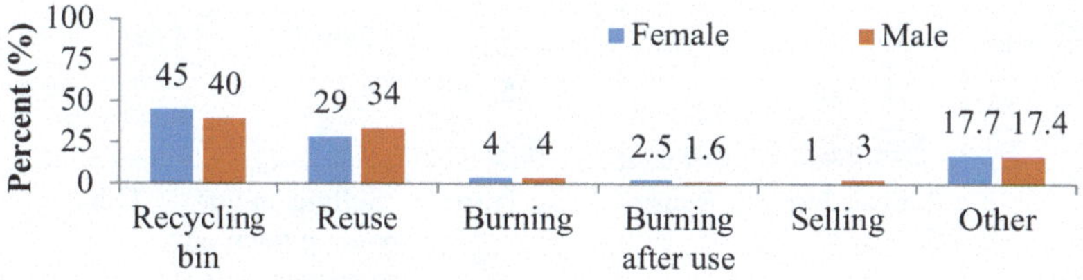

Fig. 8 Responses provided by the female and male respondents to the question "How do you dispose of your waste papers and cardboards?"

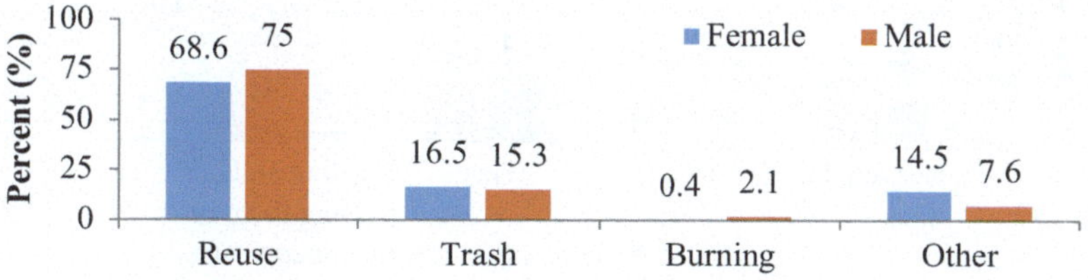

Fig. 9 Responses provided by the female and male respondents to the question "How do you dispose of your plastic bags?"

of the female and male students regarding the disposal of plastic bags, a significant difference was not observed ($P = 0.35 > 0.05$).

The responses of the female and male students participating in the survey to the question "How do you dispose of your waste glasses?" are shown in Fig. 10. The survey results showed that 34% of the female respondents and 30% of the male respondents put their waste glasses in glass recycling bin. Seventeen percent of the female respondents and 12% of the male respondents reused their waste glasses. Thirty-five percent of the female respondents and 48% of the male respondents put their waste glasses in trash (Fig. 10). These results showed that the response of the female respondents as "I put my waste glasses in trash" was lower than that of the male respondents. When ANOVA analysis was performed at 95% confidence level, no significant difference was found between the behaviors of the female and male respondents regarding the disposal of their waste glasses ($P = 0.28 > 0.05$).

Figure 11 shows the responses of the female and male respondents to the question "How do you dispose of your waste metal cans?" When the survey results are evaluated according to the gender, it can be seen that 33% of the female respondents and 32% of the male respondents put their waste metal cans in metal recycling bin. Fifty-four percent of the female respondents and 50% of the male respondents put their waste metal cans in trash. According to these results, it can be said that the male and female respondents had similar behaviors in the disposal of waste metal cans (Fig. 11). The results of the ANOVA analysis (at 95% confidence level) showed that a significant difference was not observed ($P = 0.29 > 0.05$) between the responses of the female and male respondents to the question: "How do you dispose of your waste metal cans?"

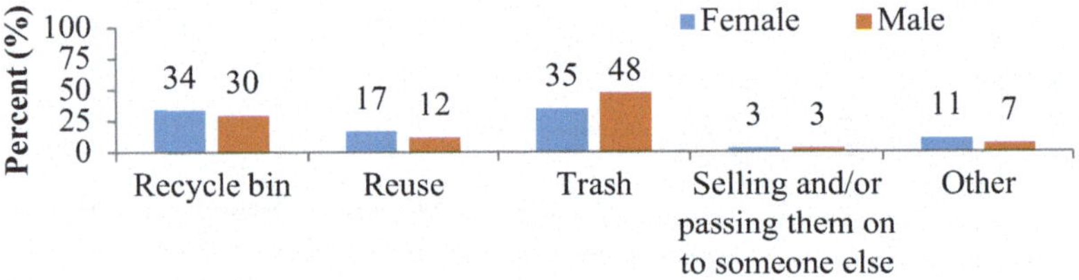

Fig. 10 Responses provided by the female and male respondents to the question "How do you dispose of your waste glasses?"

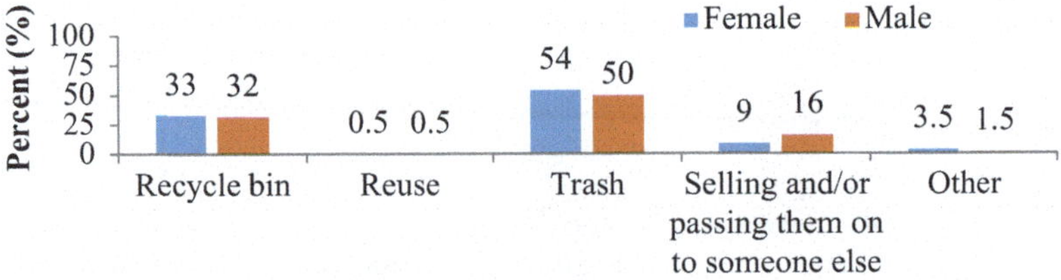

Fig. 11 Responses provided by the female and male respondents to the question "How do you dispose of your waste metal cans?"

The responses of the female and male respondents to the questions "How do you dispose of your spent batteries?" and "Do you prefer rechargeable battery when purchasing battery?" are shown in Fig. 12. As can be seen from Fig. 12, 70% of the female respondents and 64% of the male respondents put their spent batteries in battery recycling bin. Thirty percent of the female respondents and 36% of the male respondents put their spent batteries in trash. The female (65%) and male (64%) students prefer buying rechargeable battery during battery selection. These results showed that the female and male respondents had similar behaviors in the disposal of their spent batteries and in purchasing rechargeable battery (Fig. 12). When the ANOVA analysis was performed at 95% confidence level in order to determine whether there was a significant difference between the female and male students' behaviors in the disposal of spent battery, no significant difference was found ($P = 0.848 > 0.05$).

In Fig. 13, the responses of the female and male students to questions "Do you have any information about the concept of waste electrical and electronic equipment?" and "Is there any authorised collection point for waste electrical and electronic equipments that you know?" are shown. Fifty-four percent of the female respondents and 58% of the male respondents stated that they had information about the concept of *WEEE*. Sixty-three percent of the female respondents and 76% of the male respondents said that they had no information about the existence of authorized collection points for WEEE (Fig. 13). The results of the ANOVA analysis

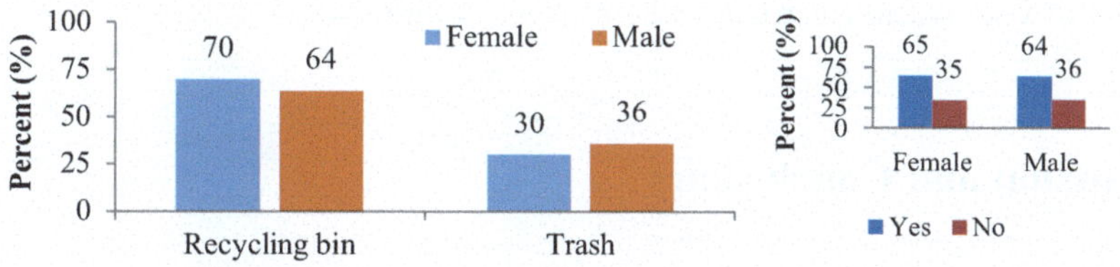

Fig. 12 Responses provided by the female and male respondents to the question "How do you dispose of your spent batteries?" and "Do you prefer rechargeable battery when purchasing battery?"

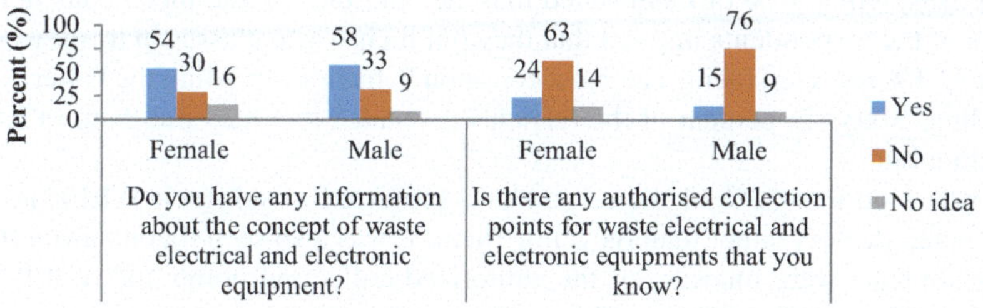

Fig. 13 Responses provided by the female and male respondents to the questions "Do you have any information about the concept of waste electrical and electronic equipment?" and "Is there any authorised collection point for waste electrical and electronic equipments that you know?"

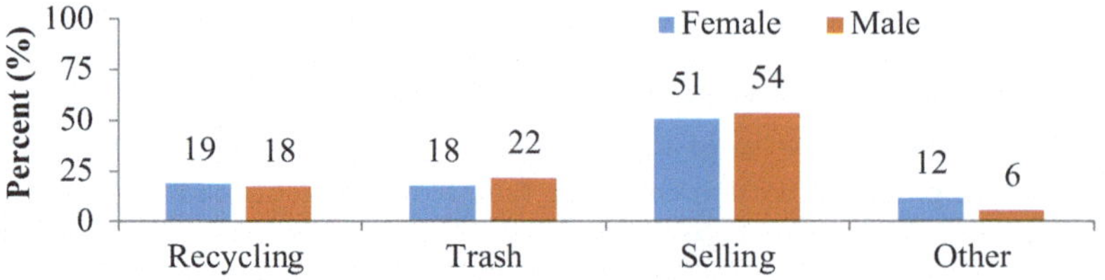

Fig. 14 Responses provided by the female and male respondents to the question "How do you dispose of your waste electrical and electronic equipments?"

indicated that a significant difference was not observed ($P = 0.470 > 0.05$) between the responses of the female and male students to the questions "Do you have any information about the concept of waste electrical and electronic equipment?" and "Is there any authorised collection point for waste electrical and electronic equipments that you know?"

The answers given by the female and male students participating in the survey to the question "How do you dispose of your waste electrical and electronic equipments?" are presented in Fig. 14. As can be seen from Fig. 14, 51% of the female respondents and 54% of the male respondents sell their WEEE, and 18% of the female respondents and 22% of the male respondents throw their WEEE with household's solid waste. The results of the ANOVA analysis (at the 95% confidence level) indicated that a significant difference was not observed ($P = 0.098 > 0.05$) between the responses of the female and male students to the question "How do you dispose of your waste electrical and electronic equipments?"

Discussion and Conclusion

The survey results showed that paper and cardboard (43%) were recycled more frequently than glass (32.1%) and metal (cans) (32%). The survey results also indicated that many respondents (69.4%) kept their plastic bags at home and used them again. Thirty-two percent of the respondents stated that they recycled their waste metal cans, while 52% of them stated that they put their waste metal cans in trash. 40.8% of the respondents implied that they put their waste glasses in trash, whereas about 32.1% recycled them. The most common behavior regarding the batteries was recycling. Sixty-six percent of the respondents stated that they put them in battery recycling bin.

It was seen that 50% of the respondents preferred to sell their WEEE to local small scrap dealers rather than recycling them. It was also seen that majority (68%) of the students were unaware of the authorized collection points for WEEE. This situation also indicated the absence/lack of collection/recycle systems for WEEE. The main reasons for the small amount (18%) of WEEE recycle may be due to the fact that most of the university students (68%) were either unaware of the authorized collection points for WEEE or preferred selling their WEEE to a small fee for extra income (50%), rather than recycling it. According to the survey results, it can be estimated that approximately 19% of the WEEE ended up in landfills. Indeed, selling of WEEE by students' means giving the item away for reuse. The obtained results are similar with the earlier studies carried out with households by Islam et al. [9] and Darby and Obara [10]. Islam et al. [9] revealed that 30% of the households were selling the WEEE to local small scrap dealers to get economic benefit. Around 20% of the households were throwing the WEEE with the household's solid waste. In a study carried out in the UK, Darby and Obara [10] stated that 97% of the households were not recycling small WEEE and the majority of their disposal of small WEEE items were via civic amenity sites (33%) or in the household refuse (26%).

Our survey results also indicated that paper/cardboard (31.3%), plastic bags (69.4%), glass (14.8%), and metal (cans) (~1%) were reused. Besides, the reuse for plastic bags was more common than recycling.

In our study, the recycling rates of paper and cardboard (43%) were found to be lower than the ones at various universities and colleges around the world. For example, recycling rates in Wartburg College, University of Northern Iowa, and Hawkeye Community College in the USA were reported as 100% for paper and 91% for cardboard, 76% for paper and 75% for cardboard, and 62% for paper and 68% for cardboard, respectively. However, the recycling rates of the students at Avcilar Campus for glass (32.1%) were higher than the ones of Wartburg College (15%), University of Northern Iowa (10%), and Hawkeye Community College (8%) [6]. The recycling rates for paper and glass in Rhodes University in South Africa were reported as 61% and 42%, respectively [11]. It can be stated that the recycling rates of the students at Rhodes University for paper and glass were higher than those of the students at Avcilar Campus, Istanbul University. It can be said that indeed, this was an expected result. For example, in a study by Izagirre-Olaizola et al. [12], it was stated that the recycling behaviors of university students from the USA and Spain were quite different. Seventy-five percent of the university students in Spain and 33.5% of the university students in the USA usually recycled. The difference between the university students' recycling behaviors in different countries was explained by these authors as the structural and cultural differences across countries.

According to our survey results, it can be said that some strategies are required to encourage the recycle of the solid wastes at Avcilar Campus. Since paper and cardboard were being recycled a little more frequently than glass, metal (cans), and WEEE, in particular, the strategies to improve the levels of participation toward the recycle of glass, metal (cans), and WEEE should be developed. Educational and promotional programs on the benefits and importance of recycling activities [13], and workshops and optional courses on solid waste management and recycling [14], should be organized for both the students and the academic staff. For example, for a more promoting recycling, the university can organize a seminar on the benefits and importance of recycling at the beginning of every semester [15]. Remarkable campaigns should also be organized on the campus. In a study by Tangwanichagapong et al. [16], it was stated that the recycling rate was increased from 1.8% to 12% by the presence of recycling facilities and remarkable campaigns at the university. Besides, collaboration with student services could be helpful for encouraging recycling activities [15].

To increase the awareness about recycling, information should be given to the campus population through brochures and posters [14, 15]. The usage of signs for recycling and instructions given on the university website can be efficient ways of increasing the awareness about recycling. Posters should be hanged on the notice boards in the most frequented places such as the cafeterias, library, and hallways of the campus buildings [15].

As pointed out by Bailey et al. [15], bin location also has an effect on the recycling levels. Therefore, the recycle bin locations at the campus should be taken into account, and recycling bins should be placed in highly frequented places (cafeterias, library, recreational areas, departmental offices/faculties/schools, etc.) and closer to the point of resource consumption. More bins should be placed in the campus. In the study by Kelly et al. [4], most respondents (83%) who were students in Massey University, New Zealand, stated that they would recycle more if there were more bins around the campus. Mtutu and Thondhlana [11] pointed out in their study that the provision of more recycling bins that are strategically located on campus could increase the tendency to recycle. The placement of metal can/bottle recycling bins next to trash can be useful in increasing the recycling level. The recycling bins should be easily accessed. As pointed out by Vicente-Molina [17], people's participation in recycling activities might be prevented due to the lack of easy access to the recycling bins. Recycling signs should be posted near recycling bins. In a study carried out by Katzev and Mishima [18], it was reported that the paper recycling in a college mail room increased after recycling signs were posted near the waste bins.

According to the statistical analysis, it was seen that there was no significant difference between the female and male students regarding the awareness and levels of participation into recycling activities at Avcilar Campus. In a study by Clay [19], it was revealed that the males were more likely to recycle than the females both at home and in the university. Oztekin et al. [20] stated that the percent of variance in explaining the recycling behavior of the males from Turkish university community was higher than that of the females. On the other hand, in a case study conducted at a Chilean university for the environmental attitudes and the behaviors of college students, it was found that the females had more positive environmental attitudes than the males among the first year students [21]. Vicente-Molina et al. [17] revealed that the gender affected environmental behavior and the females were more likely to carry out environmentally friendly activities in both advanced and emerging countries. Zelezny et al. [22] and Hunter et al. [23] stated that the females performed more pro-environmental behavior than the males. Plavsic [24] also revealed that the gender influenced the recycling behavior and the females were more willing to recycle.

Kelly et al. [4] stated that there was no significant relationship between on-campus recycling behavior and gender, which is consistent with our study. In a case study on white paper recycling at Oklahoma State University, it was found that no relationship was observed regarding gender on the recycling behavior and attitudes of students [25]. Similar observations were also reported by Oskamp et al. [26], Schultz et al. [27], Vencatasawmy et al. [28], Lorgunpai and Lertchaiworakul [29], Ugurlu [30], and Bailey et al. [15]. A recent study conducted at Basque Country University by Vicente-Molina et al. [31] also revealed that there was no significant difference between the genders regarding recycling.

As stated before, in this study, the preliminary findings obtained from the first phase/stage of a research project were discussed. Although a further research is needed, it can be stated that these preliminary findings may provide insight for implementing recycling schemes on campus.

Acknowledgments This work was financially supported with funding grant No BAP-38831 by the Research Project Unit of Istanbul University. The authors are grateful to the Research Project Unit of Istanbul University for the financial support.

References

1. Alshuwaikhat HM, Abubakar I (2008) An integrated approach to achieving campus sustainability: assessment of the current campus environmental management practices. J Clean Prod 16:1777–1785
2. Alshuwaikhat H, Abubakar I (2008) An integrated approach to achieving campus sustainability: assessment of the current campus environmental management practices. In: Proceedings of the 4th international Barcelona conference on higher education, Higher education for sustainable development, vol 7. GUNI, Barcelona
3. Gallardo A, Edo-Alcón N, Carlos M, Renau M (2016) The determination of waste generation and composition as an essential tool to improve the waste management plan of a university. Waste Manag 53:3–11
4. Kelly TC, Mason IG, Leiss MW, Ganesh S (2006) University community responses to on-campus resource recycling. Resour Conserv Recycl 47:42–55
5. Zhu N (2016) Impact of communication appeals on recycling behaviors among undergraduate students. M.Sc. Thesis, Purdue University, Open Access Theses. 1025. http://docs.lib.purdue.edu/open_access_theses/1025
6. Wilcox MA (2014) A study of college student attitudes and behaviors related to recycling. PhD Thesis, University of Northern Iowa, Electronic Theses and Dissertations. 2. http://scholarworks.uni.edu/etd/2
7. Philippsen Y (2015) Factors influencing students' intention to recycle. M.Sc. Thesis, University of Twente
8. Yin J, Gao Y, Xu H (2014) Survey and analysis of consumers' behaviour of waste mobile phone recycling in China. J Clean Prod 65:517–525
9. Islam T, Abdullah AB, Shahir SA, Kalam MA, Masjuki HH, Shumon R, Rashid H (2016) A public survey on knowledge, awareness, attitude and willingness to pay for WEEE management: case study in Bangladesh. J Clean Prod 137:728–740
10. Darby L, Obara L (2005) Household recycling behaviour and attitudes towards the disposal of small electrical and electronic equipment. Resour Conserv Recycl 44:17–35
11. Mtutu P, Thondhlana G (2016) Encouraging pro-environmental behaviour: energy use and recycling at Rhodes University, South Africa. Habitat Int 53:142–150
12. Izagirre-Olaizola J, Fernández-Sainz A, Vicente-Molina MA (2015) Internal determinants of recycling behaviour by university students: a cross-country comparative analysis. Int J Consum Stud 39:25–34

13. Wan C, Cheung R, Shen GQ (2012) Recycling attitude and behaviour in university campus: a case study in Hong Kong. Facilities 30(13):630–646
14. Tehrani AM, Hosseindoost G, Miranzadeh MB (2016) Knowledge and attitude level of students about solid waste recycling; Kashan University of Medical Sciences. Int Arch Health Sci 3(1):13–17
15. Bailey J, Pena M, Tudor T (2015) Strategies for improving recycling at a higher education institution: a case study of the University of the West Indies, Cave Hill campus, Barbados. Open Waste Manage J 8:1–11
16. Tangwanichagapong S, Nitivattananon V, Mohanty B, Visvanathan C (2017) Greening of a campus through waste management initiatives: experience from a higher education institution in Thailand. Int J Sustain High Educ 18(2):203–217
17. Vicente-Molina MA, Fernández-Sáinz A, Izagirre-Olaizola J (2013) Environmental knowledge and other variables affecting pro-environmental behaviour: comparison of university students from emerging and advanced countries. J Clean Prod 61:130–138
18. Katzev RD, Mishima H (1992) The use of posted feedback to promote recycling. Psychol Rep 71:259–264
19. Clay S (2005) Increasing university recycling: factors influencing recycling behaviour among students at Leeds University. Earth Environ 1:186–228
20. Oztekin C, Teksöz G, Pamuk S, Sahin E, Kilic DS (2017) Gender perspective on the factors predicting recycling behavior: implications from the theory of planned behavior. Waste Manag 62:290–302
21. Heyl M, Díazy EM, Cifuentes L (2013) Environmental attitudes and behaviors of college students: a case study conducted at a Chilean University. Rev Latinoam Psicol 45(3):487–500
22. Zelezny LC, Chua PP, Aldrich C (2000) Elaborating on gender differences in environmentalism. J Soc Issues 56:443–457
23. Hunter LM, Hatch A, Johnson A (2004) Cross-national gender variation in environmental behaviors. Soc Sci Q 85:677–694
24. Plavsic S (2013) An investigation of gender differences in pro-environmental attitudes and behaviors. Honors Scholar Theses, University of Connecticut. http://digitalcommons.uconn.edu/srhonors_theses/404
25. Brown PL (2007) Case study on white paper recycling at Oklahoma State University. M.Sc. Thesis, Oklahoma State University
26. Oskamp S, Harrington MJ, Edwards TC, Sherwood DL, Okuda SM, Swanson DC (1991) Factors influencing household recycling behavior. Environ Behav 23:494–519
27. Schultz PW, Oskamp S, Mainieri T (1995) Who recycles and when? A review of personal and situational factors. J Environ Psychol 15:105–121
28. Vencatasawmy CP, Öhman M, Brännström T (2000) A survey of recycling behaviour in households in Kiruna, Sweden. Waste Manage Res 18:545–556
29. Lorgunpai S, Lertchaiworakul J (2010) Understanding the attitude of Thai students in Sweden toward recycling system: a study of container deposit system. Master Dissertation, Malardalen University, Sweden
30. Ugulu I (2015) A quantitative investigation on recycling attitudes of gifted/talented students. Biotechnol Biotechnol Equip 29(S1):20–26
31. Vicente-Molina MA, Fernandez-Sainz A, Izagirre-Olaizola J (2018) Does gender make a difference in pro-environmental behavior? The case of the Basque Country University students. J Clean Prod 176:89–98

Part II
Solid Waste Management

Life Cycle Environmental Impact Analysis of HDPE Packaging Materials for Different Disposal Options

Merve Mermertaş and Fatos Germirli Babuna

Abstract The aim of this study is to evaluate the environmental impacts of high-density polyethylene (HDPE) packaging materials with the help of life cycle methodology. The production of HDPE plastics has two consecutive steps of granulate production and blow molding. When production of HDPE and transportation stages are considered, highest shares in global warming potential (GWP), photochemical ozone creation potential (POCP), acidification potential (AP), and eutrophication potential (EP) are due to granulate production. On the other hand, blow molding is the main contributor to ozone depletion potential (ODP) with regard to production and transportation stages. Incineration is observed to exert positive impacts on POCP, AP, and EP. Recycling has positive effect on all impact categories except EP. Besides, as landfilling elevates environmental impacts, it is not recommended as an end-of-life fate for HDPE packaging wastes.

Keywords Life cycle assessment · Plastics · High-density polyethylene · Environmental impacts · Sustainability

Introduction

Life cycle assessment (LCA) can be used as a thriving tool to envisage environmental burdens and gains for products and services. Environmental hot spots are put forth; various alternatives are evaluated and compared in a holistic perspective via LCA. The picture obtained in LCA aids decision-makers, producers, and even environmentally aware consumers.

An annual growth by 5.5% is expected in Turkish plastic sector that has a capacity usage of approximately 73% [1]. Plastic and rubber product industry is among

M. Mermertaş · F. Germirli Babuna (✉)
Department of Environmental Engineering, Istanbul Technical University, Istanbul, Turkey
e-mail: germirliba@itu.edu.tr

N. Balkaya, S. Guneysu (eds.), *Recycling and Reuse Approaches for Better Sustainability*, Environmental Science and Engineering,
https://doi.org/10.1007/978-3-319-95888-0_5

the booming sectors [2]. The highest share (by 41%) in total production of plastics in Turkey is allocated for production of plastic packaging [1].

Packaging materials made up of plastics have widespread usage ranging from food and cleaning agents to beverages. Ever-increasing amount of waste plastics is the expected outcome of such applications. It is stated by Brandt and Pilz [3] that the energy consumption and greenhouse gas emissions will be elevated when plastic packaging is substituted by other materials. Therefore, besides developing alternative packaging materials, ways to reduce the environmental impacts of plastics are of importance. High-density polyethylene (HDPE) is among the three most required plastics together with polypropylene (PP) and low-density polyethylene (LDPE) [4]. Furthermore, between various types of plastics, the most commonly used ones in packaging are made up of HDPE [5].

In this context, the aim of this study is to provide a comprehensive environmental profile of HDPE used as a packaging material via life cycle approach in a cradle-to-grave scope. A clear holistic picture is obtained for incineration, landfilling, and recycling as end-of-life alternatives.

Materials and Methods

The functional unit used in this study is 1 kg HDPE material. System flowchart is illustrated in Fig. 1.

The scope of this study is defined as from cradle to grave. The scope covers HDPE production, transportation, and end-of-life management. HDPE production consists of HDPE granulate production followed by blow molding. The main raw material of polyethylene production is crude oil. As illustrated in Fig. 1, ethylene is obtained from gas oil or naphtha in a steam-cracker. Ethylene is then converted into HDPE under low-pressure conditions in a fluidized bed reactor where the reaction takes place in gas phase. The generated HDPE is then subjected to blow molding, where after it is shaped. There are no omitted manufacturing processes in the conducted LCA.

Transportation is related to raw materials, product distribution, and conveyance to alternative end-of-life site. Transportation of HDPE pellets from refinery to manufacturing plant (470 km road) is performed with trucks.

In terms of energy requirements, average European electricity mix is used for grid mix.

Incineration, landfilling, and recycling are alternative end-of-life destinations for HDPE packaging materials. For transportation to end-of-life facilities, an average road of 32 km is assumed to be covered by trucks as specified by US EPA Waste Reduction WARM Model [6].

Global warming potential (GWP), acidification potential (AP), eutrophication potential (EP), photochemical ozone creation potential (smog) (POCP), and ozone depletion potential (ODP) are the environmental impact categories considered.

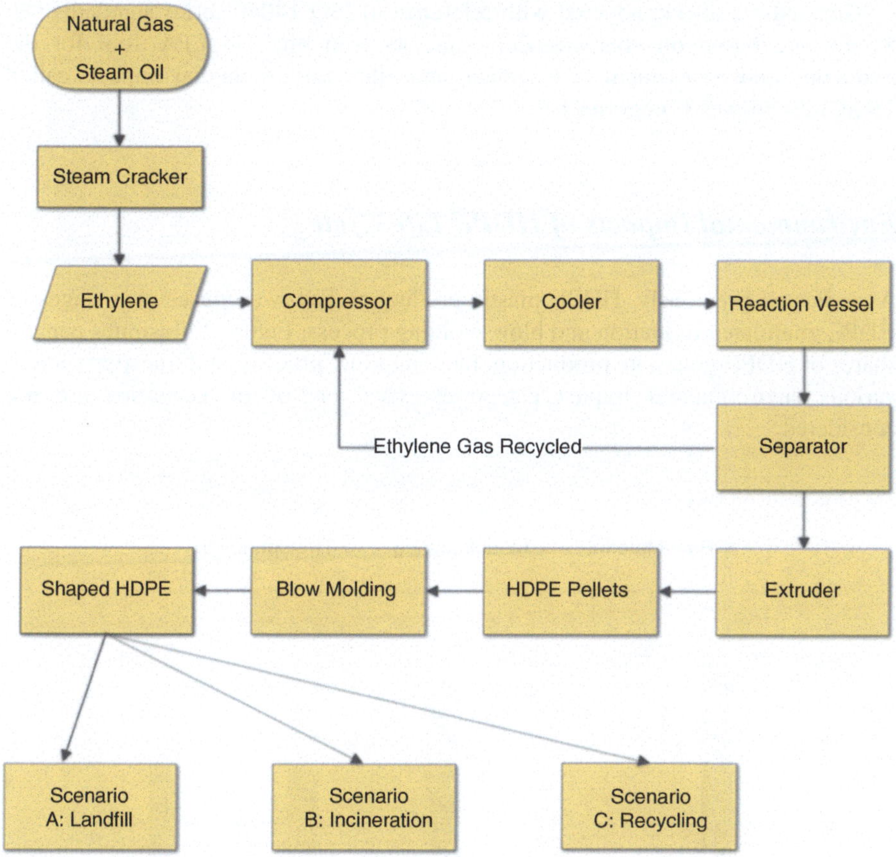

Fig. 1 System flow chart

Table 1 Inputs and output

Inputs	
High-density polyethylene granulate (kg)	1.03
Electricity (MJ)	4.795
Compressed air 7 bar (Nm^3)	0.08
Diesel (kg)	0.018
Output	
High-density polyethylene (kg)	1

The LCA model was created using GaBi DB Version 6.115 which provides the life cycle inventory database for life cycle engineering. Primary inputs and outputs given in Table 1 are obtained from GaBi database [7].

The study is also conducted with reference to ISO 14040 series standards [8]. Results are based on characterization factors from the US EPA Tool for the Reduction and Assessment of Chemical and other Environmental Impacts called TRACI 2.1 Impact Categories [9].

Environmental Impacts of HDPE Life Cycle

As indicated previously, HDPE plastic production has two consecutive stages of HDPE granulate production and blow molding process. Figure 2 illustrates percent shares of HDPE granulate production, blow molding process, and transportation in various environmental impact categories when end-of-life scenarios are not considered.

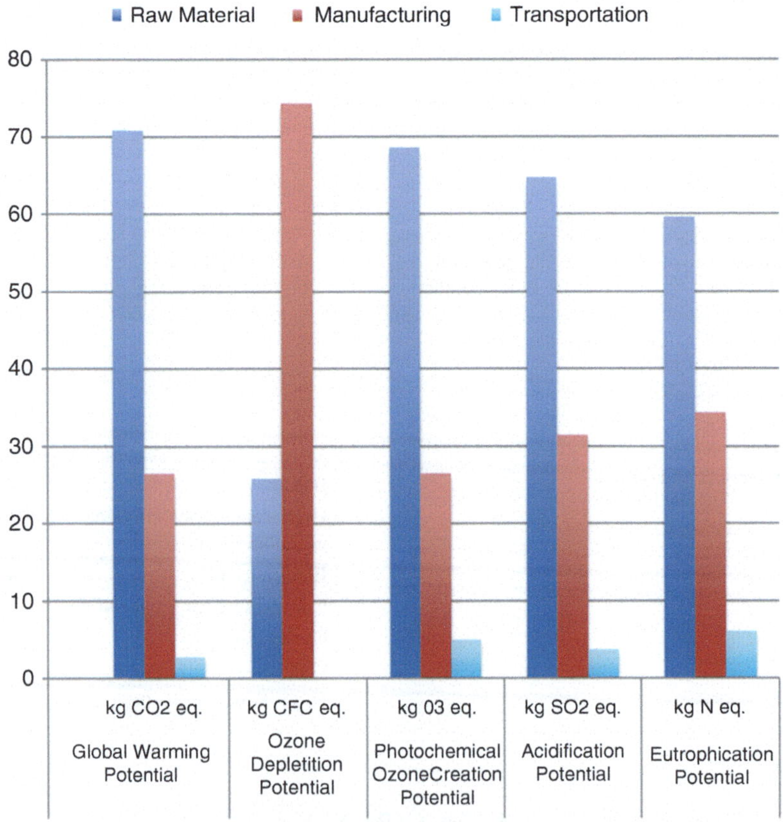

Fig. 2 Percent shares of HDPE granulate production (raw material), blow molding process (manufacturing), and transportation in various environmental impact categories

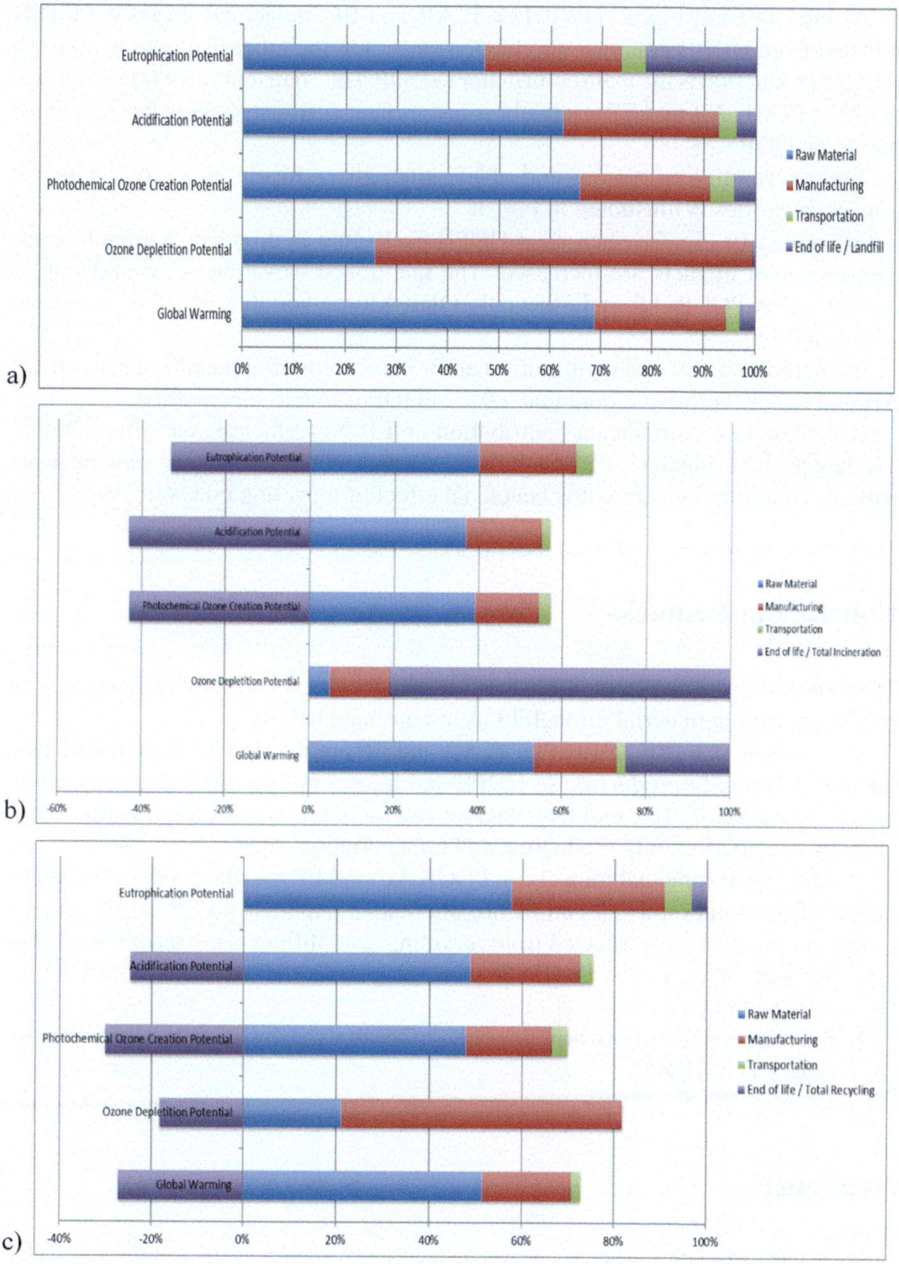

Fig. 3 Percent shares in impact categories. (**a**) Landfill. (**b**) Incineration. (**c**) Recycling

As evident from Fig. 2, GWP, POCP, AP, and EP impact categories are highly affected from HDPE granulate production stage. On the other hand, blow molding stage of production is the main contributor to ODP. The contribution of transportation to GWP, POCP, AP, and EP relatively is insignificant. Transportation has no contribution on ODP.

The contribution of various end-of-life alternatives for different environmental impact categories is illustrated in Fig. 3.

According to Fig. 3, when used HDPE packaging is directed to a landfill, the environmental impacts are increased. The mentioned elevation is insignificant in case of GWP, POCP, AP, and especially ODP. More than 20% of EP is associated with landfilling.

Incineration as an end-of-life destination lowers environmental impacts on EP, AP, and POCP. However, more than 80% of ODP is due to incineration.

Recycling has insignificant contribution on EP. Nevertheless, recycling alternative has positive impact on POCP, GWP, AP, and ODP. The results are consistent with the literature indicating the beneficial effect of recycling on GWP [10].

Concluding Remarks

The concluding remarks of this study appraising the environmental impacts of HDPE packaging material through LCA are outlined below.

The highest shares in GWP, POCP, AP, and EP are due to HDPE granulate production, when only production of HDPE and transportation stages are considered. On the other hand, blow molding stage of production is the leading contributor to ODP by considering only production and transportation stages.

Positive environmental impacts on POCP, AP, and EP are observed for the application of incineration as an end-of-life alternative. Apart from EP, all the impact categories are positively affected from recycling. Landfilling is not recommended as a proper way of dealing with HDPE packaging wastes due to its contribution to environmental impacts.

It is recommended to perform the same LCA analysis by using the local data on the production of HDPE.

References

1. TPIF (2017) Turkish Plastics Sector Monitoring Report, Turkish Plastic Industry Foundation PAGEV, p 33
2. SPA (2015) Chemical Industry Working Group Report, State Planning Agency, 10th Development Plan, Ankara, p 89
3. Brandt B, Pilz H (2011) The Impact of plastic packaging on life cycle energy consumption and greenhouse gas emissions in Europe. http://denkstatt-group.com/files/the_impact_of_plastic_

packaging_on_life_cycle_energy_consumption_and_greenhouse_gas_emissions_in_europe.pdf. Accessed 25 Mar 2018
4. Elduque A, Elduque D, Javierre C, Fernandez A, Santolaria J (2015) Environmental impact analysis of the injection molding process: analysis of the processing of high-density polyethylene parts. J Clean Prod 108:80–89
5. Ebottles (2018) Plastic bottle resin materials. https://www.ebottles.com/resins.html. Accessed 01 Jan 2018
6. US EPA (2016) Waste reduction WARM model. https://www.epa.gov/warm. Accessed 05 Jan 2018
7. GaBi (2016) GaBi DB Version 6.115. Software and database for life cycle engineering. LBP, University of Stuttgart and PE International AG, Leinfelden-Echterdingen
8. ISO (2004) ISO 14001:2004 environmental management system requirements standard
9. US EPA (2012) TRACI 2.1. Tool for the reduction and assessment of chemical and other environmental impact. EPA/600/R-12/554, p 24
10. Brogaard LK, Damgaard A, Jensen MB, Barlaz M, Chistensen TH (2014) Evaluation of life cycle inventory data for recycling systems. Resour Conserv Recycl 88:30–45

Evaluation of Phosphorus Substance Flows for a Long-Term Resource Management

Demet Seyhan

Abstract Resource management, akin to hazardous waste management, must adopt a cradle-to-grave perspective for key resources. One good reason is that substance flows go unnoticed in the regional metabolism and the global trade of all goods (im Güterfluss). This in turn conceals causality in depletion and pollution, or hinders an effective conservation of resources and the environment, leaving its management symptomatic. Phosphorus (P) is a key nonrenewable resource, which must be imported to almost all countries, and its use as fertilizer cannot be substituted posing a constraint on the global food production and for the long term. This paper shows the P-flows used and lost abroad (the Hinterland) to produce the goods imported into one country, as a continuum of our earlier study and for the cases of Turkey and Austria. These Hinterland flows represent the actual and total raw material consumption of a country and can in some cases dominate the overall system belittling front-end or country-wide recycling and conservation efforts. In particular, losses in global agriculture as well as the magnitude of mining wastes must be considered for effective decisions on P-management. Taking the Hinterland into account will link those global and regional P-flows and help in setting the right priorities for P-management. Implication for recycling and the circular economy is the potential shown for recovery, which shall only take place alongside with reducing big losses and the inefficiencies within one country and elsewhere.

Keywords Resource management · Phosphorus · Material flow analysis · Evaluation method · Hinterland

---

D. Seyhan (✉)
Istanbul Technical University, Istanbul, Turkey

N. Balkaya, S. Guneysu (eds.), *Recycling and Reuse Approaches for Better Sustainability*, Environmental Science and Engineering,
https://doi.org/10.1007/978-3-319-95888-0_6

Introduction

The economy is run by resources. Some of them are extractive and finite, and especially for these, we strive for circular economy designs. One such resource, also essential to human survival, is managed unsustainably and demanded each year in ever-growing amounts: phosphorus (P) is the subject of this work on resource management. In order to effectively conserve this resource, flows and stocks of it must be known at national and global scales. Such information was gathered, and a method, SFA for P (substance flow analysis for P), was presented by Seyhan [1] based on Brunner et al. [2]. In it not only flows and stocks of P consumed within one country are estimated but also those incurred elsewhere due to this consumption, i.e., indirectly.

The paradigm shift from waste management into resources management is reflected well in the case of phosphorus: in the beginning was P, the pollutant in wastewater, to be removed from it in a tertiary treatment step, from the perspective of sanitation and environmental engineering. Then, as early as in the 1980s [3], there was a talk of phosphorus recovery from sewage sludge in order to reduce P raw material imports to a country, which would also economically justify the advanced treatment of wastewater. Although enhanced biological phosphorus removal (EBPR) had begun with a serendipitous discovery in the late 1960s, there was a peak in research in the 1990s, and attempts have been made to recover P from EBPR sludge starting from the 2000s [4]. By 2012, the amount of P in sewage was said to be able to cover for or substitute 20% of the worldwide P-rock demand and that EBPR was a mature technology; however, subsequent P recovery technologies still needed to be worked on [5]. We can narrate the case of struvite in a parallel manner, the precipitation of which was also a major line of research and development (R&D) for nutrient recovery from wastewater. This was the first tier of resource management for P as depicted in Fig. 1: resource recovery from wastewater. It has all started with advanced wastewater treatment technologies, at the end of the pipe.

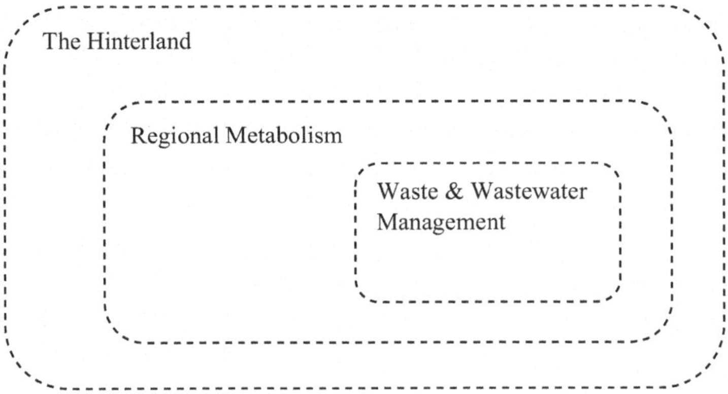

Fig. 1 Resource management going back to the cradle or the mine, one tier at a time (ca. one decade apart), starting from the grave or the end of pipe

Material flow analysis (MFA) is a well-suited tool for decision support on regional nutrient management. Derived from MFA, substance flow analysis (SFA) depicts the substance flows both in raw materials and in finished goods. Substance flows of P from the 1990s reflect the central European experience: the anthropogenic use of the resource was inefficient and burdened with high losses, especially in agriculture [6], leading to accumulation of P in the soil and eutrophic surface waters. A comprehensive SFA involves all relevant flows and stocks of P, and in the 2000s, it was used for the resource management of P in Turkey and in Austria [7]. Flows and stocks were identified and quantified at three levels. Ranking the six main processes in importance with their efficiencies and their sub-processes, P-metabolisms of these regions were observed. Measures and technologies in the agricultural sector seemed indeed urgent, and recycling at the end of pipe still made sense if the tertiary treatment was all complete and exhaustive. This was the second tier (Fig. 1): resource management of a region. A resource accounting of those substance flows enabled this new perspective.

Global environmental and resource issues have a larger scale in space and time requiring a wider perspective. For depleting resources and especially for those resources essential to human survival, an outlook involving distant times and places becomes vitally important. Such long perspectives relate to basic human rights when it comes to key resources and are shown to be urgent in providing long-term distributional benefits for the globe and for future generations [8, 9]. Like in these economic analyses, material flows and their long-term evaluation can extend beyond regional boundaries too: tracing them back to the mine, we can see the wider repercussions of a regional metabolism on the global, overall system. This notion of Hinterland flows was introduced already by 2006 [1], where the Hinterland and the Region were depicted as two concentric, rounded rectangles depicting regional flows and their impacts inflicted abroad. In this comparative study, a third, subsystem-level, sectoral analysis was also made. It was not until 2014 that system boundaries in the form of concentric rectangles were displayed again, this time in a review paper by Chowdhury et al. [10]. Their aim was to order and categorize all P-studies until then into a cascade of analyses, much as we quantified in our multilayer analysis for Turkey and Austria [2], followed a decade later by another multilayer study on China with the national borders as the outermost. Although these later studies proposed sketches similar to our notion, they miss the point of having production processes beyond country borders, such as agriculture, and suffice at most with the mining process out there.[1] Other approaches were developed after 2010 that too widen the scope of analysis, with or without process thinking: "Virtual phosphorus ore" requirement for Japan [11] or "P footprints" for the diets of 170 countries [12] were calculated, which do track flows back to the mine. Compared to these studies that do not propose comprehensive methodologies nor define boundaries, we present a widely applicable systematic approach based on SFA, with

[1] Referring also to our study [7], Chowdhury mentions that no recycling was reported by us, while recycling for both countries in the form of manure, sludge, and even compost was shown or mentioned in the said study, which is less overloaded with data than the original work [1].

Table 1 Suggested priorities for P-management at various levels

Global priorities	National priorities	User priorities
Resource allocation Mining wastes Agricultural technologies Food security and safety	Efficient resource use Waste reduction/avoidance (vermeidung) Recycling and recovery	Sustainable consumption Sustainable production

complete scope. It builds on our previous work [7] and follows the approach published by Seyhan and Brunner [13], which deliberates and explains the choice on the system boundaries: while involving the Hinterland links those global and regional P-flows arising from a metabolism, the former can render front-end measures become obsolete in some cases. Thus, juxtaposing the two will help in reconciling short-term (regional) and long-term (global) goals and issues, encouraging more conscious decision-making in this realm, and also informing us toward making the background goals and those front-end solutions more congruent, harmonious, and in tune with one another. Table 1 offers a possible prioritization of this sort, one that can emerge in an informed society.

In this study, we aimed at the long-term, sustainable management of phosphorus resources by studying the Hinterland flows of countries. The concept of "Hinterland use" refers to the total amount of P used and lost abroad for the production of all goods imported into one country. For this, we visualized an additional system boundary beyond national borders and incorporated those upstream (initial; prior to import) Hinterland processes. The objectives of this paper are to show:

1. A country's consumption of the resource P and its total share in primary resource use including those used in the Hinterland
2. The pollution potential generated in the Hinterland due to the consumption patterns within one country
3. A comparative analysis of cases demonstrating the wider application of the method

Method

The framework used is explained in Seyhan and Brunner [13]. A simple system (Fig. 2) is proposed there, reducing the flows and the stocks to the essentials for our purposes, displaying at this level of analysis only the average values, and discussing the long-term management. According to this, P-management should deal not only with the short-term concerns of efficiency increases and regional pollution but also with the long-term concern of resource depletion and stock changes.

A comparison to the well-known indicators like total material requirement (TMR) and hidden flows (HF—the ecological rucksack) of the Wuppertal school shall be made at this point [14]. Bringezu et al. carried out their MFA at the goods level. They estimated the total mass of all goods used by a national economy,

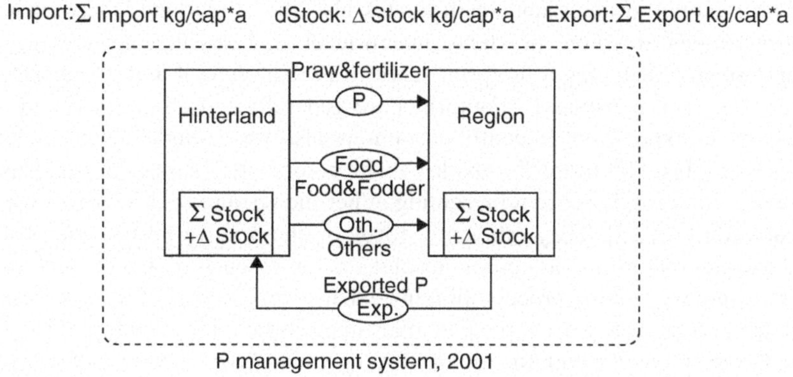

Fig. 2 The phosphorus management system for a Region together with its Hinterland [13]

together with their "ecological rucksack." The latter refers to the hidden flows—hidden because not registered, not having a price nor being traded: "Total quantity (in kg) of natural material moved (physically displaced) by humans in order to generate a good." These do normally not show up in those MFA works based on conventional statistics or in economic analyses. TMR, per definition, covers all the goods used (domestic + import) together with their hidden flows. The Hinterland consumption of P then can be seen as an allusion to TMR, in the sense that it includes both imported amounts and their ecological rucksack (the hidden flows involved in their production) but at a substance level. Beside the level of aggregation, another difference is that TMR also involves inland hidden flows. In our work [1], those inland hidden flows that are "not accounted for" in statistics (like erosion, manure, losses, soil accumulation) were estimated within the regional SFA. Thus, they are not hidden and do not relate to the Hinterland analysis presented here. Again contrary to TMR, inland mining of the resource—P drawn from their own soil stock in our cases—is not included in the parameter defined here. One weakness of our approach lies in its purpose: it is not intended to determine a total environmental impact like overall material use of a society; it just shows the share of an importing country in the global P-resource use, as well as the potential of P pollution created in its Hinterland. Hinterland mining would not directly show the total consumption of those few countries holding their own reserves (like the USA or China), from a global stock. However, the original work [1] incorporates a national stock to represent such cases.

SFA for P [1] described the entire regional metabolism using the mass balance principle and so differs from recent footprint and embodied P analyses. It estimates and balances substance flows with data double-checked at input and output ends of the processes from various sources; and it provides additional cross-check that relies on the systems' inner validation. These country-wide balances were carried out in gross amounts, and once the work was done on two levels including the uncertainties, the evaluation of the system follows: at a third level (and this time in averaged per capita flows), this proxy system is designed for the assessment of those balanced regional systems and incorporates also the Hinterland. In comparison, footprint studies are more practical, macro-level analyses that are service- and product-oriented deducing a footprint score from one sectoral and global data set, such as "the food consumed." Footprints are generally loosely defined and often lack clarity in expression. Recently accounting also for a single substance across regions as in SFA, "footprints" provide us with comparative studies of performance summarised in a single score representing either the resource extraction or the carbon emission [12, 15]. They efficiently focus on the relative performances in the World League rather than assigning absolute values to each flow and stock in one country, while we employ process thinking exhaustively: stages of resource use and losses depict redundancies and the potentials in the system for meaningful improvements. For example, through the use of SFA, upstream flows or the input side of the system depicts pollution potentials there besides the resource use, while the output side points at the resource potential there besides accounting for emissions alone. As such, mining waste flows or an estimation of the amounts extracted from the ground are provided here.

Turkey and Austria are chosen to be the case studies; time- and cost-efficient collection of direct data was made possible due to residence in these countries as well as due to previous work done. They also represent a 1:10 ratio in terms of size and population providing two complete different cases geographically, culturally, and economically. In fact, adapting the same SFA for both countries having different systems of data collection aspired for a model widely applicable for most countries and in reality anticipated hard-to-compare varieties to follow. We now intend to study the USA and China for their Hinterland flows, due to their sizes and the speed of change in P-flows over time, while data will most probably be derived from secondary sources, which seem available by now. Also, the former two countries were 100% importers of the resource, whereas the latter two belong to those few countries holding P-deposits and mining them.

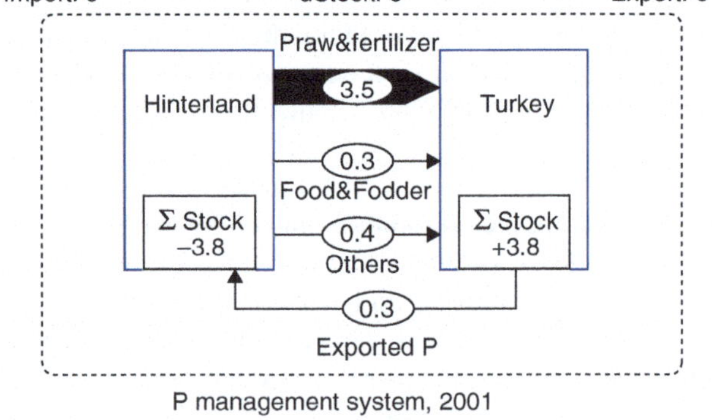

Fig. 3 Phosphorus imports, exports, and stocks of Turkey, kg/cap*a [13]

Results and Discussion

Below, the P import, export, and stock of Turkey and Austria are presented in Figs. 3 and 4, and then the corresponding P-flows in the Hinterland required to produce those imports are shown (Figs. 5 and 6).

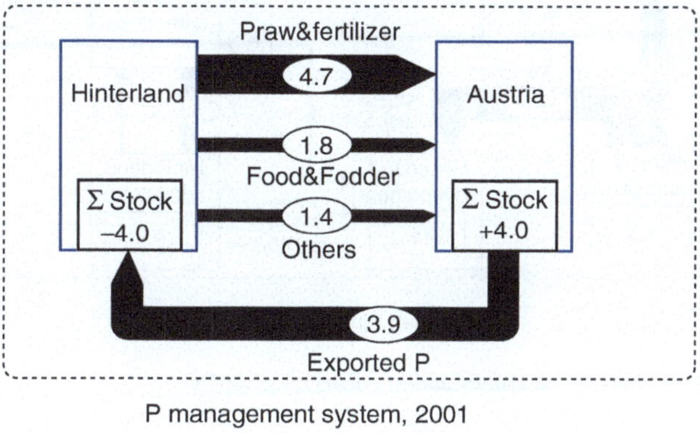

Fig. 4 Phosphorus imports, exports, and stocks of Austria, kg/cap*a

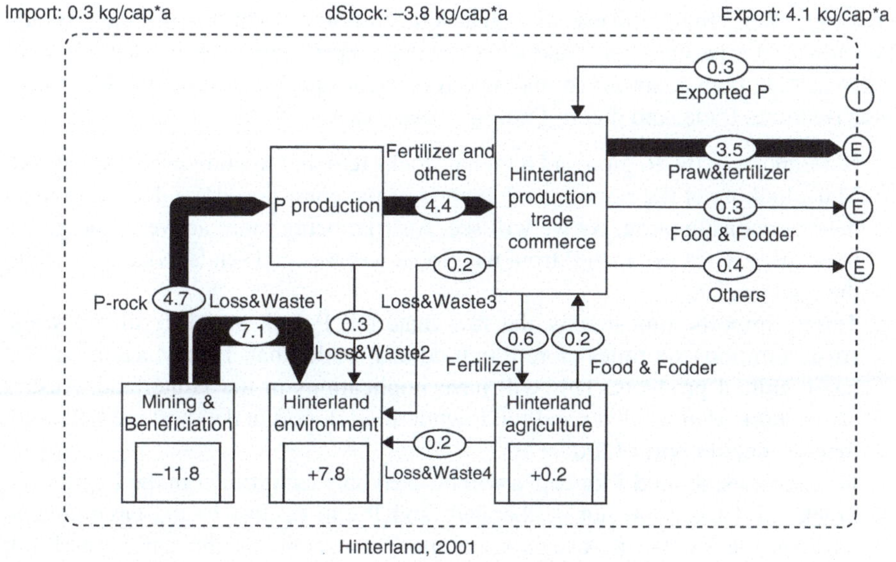

Fig. 5 Phosphorus flows in the Hinterland of Turkey, kg/cap*a [13]

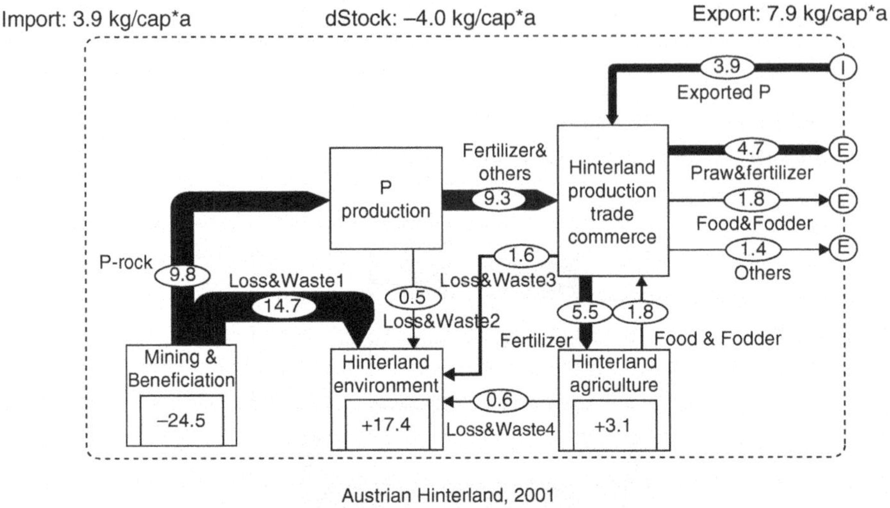

Fig. 6 Phosphorus flows in the Hinterland of Austria, kg/cap*a

The Overall System Including the Hinterland

For a long-term and sustainable management of P, the total raw material use of countries, their Hinterland use, as well as total emissions to the environment should be known. Comparing Figs. 3 and 4, we see that surprisingly there is a net import of 4 kg/cap*a in both countries (rounding errors and averages), despite big differences in agriculture, trade, and diet of P in these two cases.

1. Net imports to the Region and regional stocks remain the same in both countries. Thus, looking at the imports and exports of the countries alone does not reflect their ore consumption. As we will see, Austria, being more active in the global trade, uses twice as much P from the global reserves and sends twice as much to the total stocks.
2. Turkey imports almost only the raw material, P-rock and P-fertilizer, while Austria imports six times more P in food and fodder than Turkey, a quarter of it being animal products. This will have implications in the Hinterland. Austria also exports P in fertilizer and food: while food import and export are balanced, there is a net import of fodder-P.
3. In both cases, around 3 kg/cap*a will be used only as fertilizer in the Region [1].
4. Stocks in Turkey are not deliberated, and P can be lost to the environment, whereas the stocks in Austria seem more conscious as the agricultural soil (enriched soil), the landfills (protected environment), and only then the environment [1, 7, 13].

The Hinterland

Previously we showed that a net import of 4 kg/year remains in both countries. Now looking at the Hinterland, we observe that the corresponding resource used abroad is much larger and can differ in various cases (Figs. 5 and 6). We know that P consumed in diet is 0.6 kg per person in both countries [1]. To reach that end, around 12 and 25 kg/person and year are extracted from the global reserves in case of Turkey and Austria, respectively. The difference arises from the fact that Austria is more involved in animal agriculture and such diet as well as in the exports of P-involving goods such as food and fodder.

The efficiencies of use in the Hinterland are chosen to be similar to those in the region with the exception of mining: mining and beneficiation have the highest inefficiency in terms of P. Otherwise, the biggest inefficiency is in agriculture. P applied on soils is not immediately available to plants, and animal agriculture leads to manure-bound P that is not properly recycled to the soil. Similar to Seyhan [1], Ma et al. [16] assessed P-management scenarios with indicators defined for this purpose and reported massive amounts of manure being wasted into waterways in China. The authors suggest importing of animal products and fodder to reduce inefficiencies and the losses to the environment within the country. However, as they too recognize, imports just transfer the inefficiencies of production to the exporting countries. Clearly, new agricultural technologies and practices are needed. There are other regions in the world with predominantly and increasingly animal-based diets, having problems with manure surpluses, which require solutions toward a clean recovery of P. To that end, we currently develop a solution to recover P from manure that shall be used firstly in the rural areas of Turkey [17].

Concerning the emissions to the environment, the first concern has been the eutrophication of surface waters in a Region. Still in Turkey and in Austria, this type of pollution control will be completed for point and diffuse sources, respectively. In this study, we aimed to take into account also those emissions in the Hinterland. Agricultural losses to the Hinterland environment arise mainly due to the imports of food and fodder, thus being larger for Austria than for Turkey (Fig. 6). Yet, the largest loss in the Hinterland of an importing country is caused due to mining and beneficiation. For Turkey, around 7 kg P/cap*a is lost to the environment from mining alone, which is around 60% of all the losses to the environment, point, or diffuse alike, both in the Region and the Hinterland.

As for the secondary resources of P in the region, and resource recovery, we showed previously [7] that recoverable amounts of P in these countries come mainly in manure and sewage sludge, and half a kilogram per person or more is already recovered in both cases. This alone will fall short of replacing the 10- or 20-fold higher raw material extraction in the Hinterland so as to overcome global resource scarcity. Necessary efficiency increase of the processes will be reducing losses, and new, better ways of recovery shall be developed to have a more sustainable P-management over the entire space and time.

Conclusions

Resources do fire the economy, and so the recognition of a depletable resource or a sudden price spike of it leads to reactive research and recovery. Nutrient management is a classical problem of human history and will need longer perspectives. Once the initial haste is over, we can ponder and reevaluate how efficient the overall system is and what that implies for each country. After spotting those inefficiencies, we can make a deliberate choice also to push measures further upstream toward the cradle, which will be one major advantage of systems thinking.

With the inclusion of the Hinterland, a much higher per capita consumption of P is observed in raw material terms. A SFA with wider scope shows both the consumption abroad for the production of all imports and the losses there. These can be much larger than the pollution or losses in the Region and should be given precedence in that case. Responsibility falls on all countries, especially concerning non-renewable resources like P, to support sustainable production and consumption as well as R&D toward recovery. Both would serve a lesser depletion and a better distribution of these resources. To be able to track such improvement, we also aim to further our model toward a new anthropogenic definition of "depletion": communicating resource depletion and consumption of a country in a clear way that involves the Hinterland use and those recovered amounts, we shall differentiate between the two outputs as emissions to the environment and those recoverable forms of P, ordering them into separate stocks, within the regional model.

Acknowledgments The author wishes to thank her supervisor, Prof. Paul H. Brunner, for his time, for his method, and for his faith, with great respect. Thanks also to the Austrian Exchange Service, ÖAD, for the scholarship.

References

1. Seyhan D (2006) Development of a method for the regional management and long-term use of non-renewable resources: the case for the essential resource phosphorus. Dissertation. TU Wien
2. Brunner PH, Daxbeck H, Henseler G, von Steiger B, Beer B, Piepke G (1990) Der regionale Stoffhaushalt im Unteren Bünztal: Entwicklung einer Methodik zur Erfassung des regionalen Stoffhaushaltes. Eidgenössische Anstalt für Wasserversorgung, Abwasserreinigung und Gewässerschutz EAWAG, CH-8600 Dübendorf; Switzerland
3. Regitze L, Jens T, Jens H (1984) Phosphorus recovery from sewage for agriculture. Waste Manage Res 2(4):369–378
4. Seyhan D, Erdincler A (2003) Effect of lime stabilisation of enhanced biological phosphorus removal sludges on the phosphorus availability to plants. Water Sci Technol 48(1):155–162
5. Yuan Z, Pratt S, Batstone DJ (2012) Phosphorus recovery from wastewater through microbial processes. Curr Opin Biotechnol 23(6):878–883
6. Baccini P, Brunner PH (2012) Metabolism of the anthroposphere, analysis, evaluation, design, 2nd edn. MIT Press, Cambridge

7. Seyhan D (2009) Country-scale phosphorus balancing as a base for resources conservation. Resour Conserv Recycl 53(12):698–709
8. Seyhan D, Weikard HP, Ierland E (2012) An economic model of long term phosphorus extraction and recycling. Resour Conserv Recycl 61:103–108
9. Weikard HP, Seyhan D (2009) Distribution of phosphorus resources between rich and poor countries: the effect of recycling. Ecol Econ 68(6):1749–1755
10. Chowdhury RB, Moore GA, Weatherley AJ, Arora M (2014) A review of recent substance flow analyses of phosphorus to identify priority management areas at different geographical scales. Resour Conserv Recycl 83:213–228
11. Matsubae K, Kajiyama J, Hiraki T, Nagasaka T (2011) Virtual phosphorus ore requirement of Japanese economy. Chemosphere 84(6):767–772
12. Metson GS, Bennett EM, Elser JJ (2012) The role of diet in phosphorus demand. Environ Res Lett 7(4):043044
13. Seyhan D, Brunner PH (2018) The choice of system boundaries for effective long-term management of phosphorus. J Mater Cycles Waste Manage 20(1):345–352
14. Bringezu S, Schmütz H, Moll S (2003) Rationale for and interpretation of economy- wide materials flow analysis and derived indicators. J Ind Ecol 7(2):43–64
15. Hertwich EG, Peters GP (2009) Carbon footprint of nations: a global, trade-linked analysis. Environ Sci Technol 43(16):6414–6420
16. Ma L, Wang F, Zhang W, Ma W, Velthof G, Qin W, Oenema O, Zhang F (2013) Environmental assessment of management options for nutrient flows in the food chain in China. Environ Sci Technol 47(13):7260–7268
17. Seyhan D, Canli G, Akduman I, Yilmaz T (2015) Heating stove to effectively burn manure, with recycling. Pending patent. TR2015/11113, Turkey

Recovery of Phosphorus and Nitrogen from Sewage Sludge as Struvite Using a Combined Alkali Hydrolysis and Thermal Treatment Process

Ayla Uysal, Mehmet Aydoğan, and Emine Çelik

Abstract Alkali (NaOH) hydrolysis and low-temperature thermal treatment were applied to digested sludge for the recovery and release of PO_4^{3-} and NH_4^+. The Box-Behnken design was applied in the hydrolysis using 0.5M NaOH for the optimization of the conditions that affect the nutrients and metal release from digested sludge. PO_4^{3-} release was positively associated with decreasing liquid/solid ratio and increasing temperature. Moreover, Ca, Fe, and Zn releases also increase with increasing temperature. However, the temperature did not have any effect on NH_4^+ release. An optimal condition for the release of nutrients and metals was obtained at a liquid/solid ratio of 10/1 (mL/g), a temperature of 40 °C, and a reaction time of 40 min. In this optimal condition, the concentrations of PO_4^{3-} and NH_4^+ released were 921.00 and 819.15 mg/L, respectively. The removal rates of PO_4^{3-} and NH_4^+ from hydrolyzed sludge liquid by struvite crystallization were 95.27% and 77.95% in this condition. Struvite obtained had low Ca, Fe, and Al content. The produced struvite meets the legal limits for fertilizer use in terms of Cd, Cu, Ni, Pb, Zn, Hg, and Cr content specified by Turkish regulations.

Keywords Alkali hydrolysis · Metal release · Nutrients recovery · Sewage sludge · Struvite · Thermal treatment

Introduction

The amount of sludge produced by biological sewage treatment plants around the world has been constantly increasing. Thus, the treatment and disposal of excess sewage sludge pose a challenge for wastewater treatment plants. The phosphorus and nitrogen contents of sludge, due to the application of biological nutrient removal technologies, range between 4% and 9% [1, 2]. Sludge is the main channel for phosphorus

A. Uysal (✉) · M. Aydoğan · E. Çelik
Department of Environmental Engineering, Suleyman Demirel University, Isparta, Turkey
e-mail: aylauysal@sdu.edu.tr

N. Balkaya, S. Guneysu (eds.), *Recycling and Reuse Approaches for Better Sustainability*, Environmental Science and Engineering,
https://doi.org/10.1007/978-3-319-95888-0_7

and nitrogen loss to the environment. Therefore, phosphorus and nitrogen recovery from sewage sludge could provide a key solution to nutrients shortage [2].

Generally, the first step of phosphorus and nitrogen recovery involves the release of phosphorus and nitrogen from the solid phase into the solution [3]. Many studies of the chemical processes using acid or alkali are under way in order to release phosphorus and nitrogen from sewage sludge [4, 5]. Acid hydrolysis can dissolve Ca and Al combined P (Ca-P, Al-P) but also dissolve heavy metals, which are undesirable when the recovered P is used as fertilizer [6–8]. However, alkali hydrolysis cannot extract Ca-P and heavy metals but can dissolve Al-P and Fe-bound P (Fe-P) [7, 8]. Alkali hydrolysis methods can recover phosphorus with low heavy metal content and are regarded useful compared to acid hydrolysis. However, in alkali hydrolysis, phosphorus release rate is low compared with acid hydrolysis [3, 5, 9]. To solve this problem, a combination of alkali hydrolysis and thermal treatment is considered useful compared with conventional alkali hydrolysis [5, 10]. Takiguchi et al. [11] reported the release of phosphorus from phosphorus accumulating organisms grown in the laboratory at low temperature (50–90 °C). The combination of chemicals with heat seems to be more beneficial than heat alone [12].

Phosphorus and nitrogen are often released from sludge with hydrolysis and recovered with crystallization as struvite ($MgNH_4PO_4$) [13, 14]. Recently, Bi et al. [1] and Xu et al. [8] recovered phosphorus and nitrogen as struvite from the alkali hydrolysis of sludge. However, there are no studies yet on the recovery of nutrients released from sludges, as struvite, by alkali and low-temperature thermal treatment.

The present study was conducted to achieve three goals. The first aim was to investigate the characteristics of digested sludge by alkali hydrolysis at different NaOH concentrations. These characteristics were further evaluated in terms of the release of P, Ca, Mg, Na, Al, Fe, Zn, Cr, Pb, Cu, and Ni. The second aim was to investigate the optimization of conditions for the release of high nutrients and low Ca, Mg, Na, K, Al, Fe, and Zn from the digested sludge using NaOH hydrolysis and low-temperature thermal treatment using the Box-Behnken design. The third aim was to investigate the recovery of phosphorus and nitrogen by struvite crystallization from hydrolyzed liquids resulted from NaOH hydrolysis and low-temperature thermal treatment.

Materials and Methods

Sludge Sample and Preparation

The digested sewage sludge used in this study was obtained from anaerobic digester effluent of a wastewater treatment plant with an anaerobic/anoxic/aerobic (A/A/O) process located in Antalya, Turkey. The sludge samples were oven-dried at 103 ± 2 °C for 42 h and then finely grounded and sieved to 1 mm mesh size. These samples were used for metals, nutrients measurements, and hydrolysis experiments.

Characteristics of Anaerobically Digested Sewage Sludge

The main characteristics of the anaerobically digested sewage sludge were as follows [6]: total solid (TS), 3%; total volatile solid (TVS), 2.90%; total chemical oxygen demand (TCOD), 32,800 ± 1555.63 mg/L; soluble chemical oxygen demand (SCOD), 823 ± 26.87 mg/L; total nitrogen (TN), 4.19%; total phosphorus (TP), 2.00%; and pH 8.03. Table 1 shows the mean concentrations of the inorganic components in digested sludge as measured during the duplicate analysis [6]. The digested sludge contains a high concentration of Ca, Mg, K, Na, Al, Fe, Zn, Cu, Pb, and Ni. However, the digested sludge had low concentrations of Cr. Cd and Hg concentrations were below the detection limit.

Alkali Hydrolysis and Thermal Treatment

The preferred alkali, in most cases, was sodium hydroxide (NaOH) which was reported to yield greater solubilization efficiency than calcium hydroxide ($Ca(OH)_2$), KOH, and $Mg(OH)_2$ [15, 16]. Moreover, NaOH is the preferred alkali leachate in most cases because it releases more dissolved phosphorus [15]. Hence, alkali hydrolysis using NaOH was performed. Hydrolysis was performed using a Biosan MSH-300i magnetic stirrer with a hot plate at a constant mixing rate (250 rpm). The hydrolysis was conducted in two stages: First, NaOH was performed using solutions of different concentrations (0.1, 0.5, and 1.0M) at a liquid (mL)/solid (g) ratio of 10/1 for 1 h at constant room temperature. A suitable NaOH concentration was

Table 1 Elemental composition of the digested sludge

Element	Concentration (mg/kg)
Ca	83,380 ± 112.1
Mg	10,620 ± 18.0
K	11,560 ± 19.7
Na	11,280 ± 21.2
Al	6831 ± 10.2
Fe	6140 ± 11.2
Zn	10,960 ± 30.5
Cu	917 ± 9.35
Cr	70.42 ± 0.16
Pb	487.4 ± 6.28
Ni	363.3 ± 3.48

determined regarding the release of phosphorus and metals. Second, the Box-Behnken design was employed to optimize high nutrients and low metals released from the digested sludge using a suitable NaOH concentration determined in first-step hydrolysis. The effects of parameters, such as liquid/solid ratio (mL/g), temperature (°C), and reaction time (min) on nutrients and metals released from the digested sludge using alkali hydrolysis and thermal treatment, were determined using the Box-Behnken design.

The samples were centrifuged (at 8000 rpm for 10 min) after hydrolysis to precipitate the sludge samples. The hydrolyzed sludge liquid or supernatant was filtered through a 0.45 μm membrane filter, and the concentration of PO_4^{3-} and metals in the solution was analyzed.

Box-Behnken Experimental Design and Statistical Analysis

Box-Behnken, a spherical and revolving design, was used during the optimization of chemical and physical processes, due to its reasoning design and excellent outcomes [17]. The Box-Behnken design was specifically selected because it requires fewer runs than a central composite design in cases with three or four variables. Therefore, selecting the Box-Behnken experimental design in this study allowed the elimination of some additional experiments, as well as time-consuming and exhausting laboratory studies. The optimal conditions for maximizing the nutrients release and minimizing the metals release using NaOH and thermal treatment were determined by a three-factor, three-level Box-Behnken experimental design. The total number of tests required for the three independent variables studied in hydrolysis process (liquid/solid ratio (X_1), temperature (X_2), and reaction time (X_3)) was 15. Experimental levels used for the three factors were coded as −1, 0, and +1. The experimental range and levels of independent variables considered in this study are presented in Table 2.

Analysis of variance (ANOVA) was used to analyze the results with Minitab 16 software (Minitab, Inc., State College, PA, USA). Minitab 16 software was used for regression and graphical analyses of the data. The significance of independent variables and their interactions were tested by ANOVA. It was determined that results with less than 95% confidence interval ($P > 0.05$) were not statistically significant for the model. According to the Minitab 16 software, each of the response could be fitted by a predictive polynomial quadratic model, as shown in Eq. (1):

Table 2 Factors and levels investigated in alkali hydrolysis and thermal treatment tests using 0.5M NaOH by Box-Behnken design

Coded levels	Liquid/solid ratio (mL/g), X_1	Temperature (°C), X_2	Reaction time (min), X_3
Low level (−1)	5/1	40	40
Center level (0)	10/1	55	70
High level (+1)	15/1	70	100

$$Y = a_0 + a_1x_1 + a_2x_2 + a_3x_3 + a_{12}x_1x_2 + a_{13}x_1x_3 + a_{23}x_2x_3 + a_{11}x^2_1 + a_{22}x^2_2 + a_{33}x^2_3 \quad (1)$$

where Y is release of nutrients or metals (mg/L); x_1, x_2, and x_3 are coded non-dimensional factors for liquid/solid ratio, temperature, and reaction time, respectively; a_0 is the constant term; a_1, a_2, and a_3 are linear effect terms; a_{12}, a_{13}, a_{23}, and a_{24} are interaction effect terms; and a_{11}, a_{22}, and a_{33} are quadric terms.

Struvite Formation

Sludge liquid hydrolyzed with NaOH and thermal treatment was used for struvite formation. The batch chemical precipitation experiments for struvite formation were conducted in 70 mL volume with continuous stirring with a magnetic stirrer at room temperature (stirring speed 250 rpm). The formation of struvite requires Mg^{2+}, NH_4^+, and PO_4^{3-} with an ideal molar ratio of 1:1:1. Thus, $MgCl_2.6H_2O$ and H_3PO_4 were used as additional sources of Mg^{2+} and PO_4^{-3} to simultaneously recover PO_4^{3-} and NH_4^+.

The pH value of the sample was adjusted to the desired level with NaOH (20%) and HCl (1M) solutions, and the pH was maintained at the desired level. Samples were stirred for 30 min and then held for 1 h to allow the precipitate of solid matter to form. Then, the contents of the beaker were filtered through a coarse filter. Solid matter left on the coarse filter was dried by holding at room temperature for 48 h.

Analytical Procedure

TS, TVS, and TN were conducted by the procedure described in the standard methods [18]. TCOD and SCOD were determined by the Hach reactor digestion method. PO_4^{3-} and NH_4^+ were determined using the Hach ascorbic acid method and the Hach Nessler method, respectively. To analyze total metal and TP contents, 1.0 g of sludge sample was weighed, and microwave digestion using HNO_3, HCl, and HF was applied. Ca, Mg, K, Na, Al, Fe, Zn, Cu, Cr, Pb, Ni, Cd, and Hg were measured using inductively coupled plasma (ICP-OES) (Perkin Elmer, DV2100) after acidic microwave digestion. SCOD, PO_4^{3-}, and metal contents were determined by filtering the sample through 0.45 μm membrane filters. The pH values of digested sludge were measured using 1:5 water extraction mechanically stirred for 3 h and subsequent determination using a Hanna HI 221 pH meter.

The crystal structure of the dry precipitate was determined using an X-ray diffractometer (XRD, Philips, X'Pert Pro). To determine metal and total phosphorus contents of the dry precipitate, microwave digestion was applied by the addition of HNO_3 and HCl to 0.1 g of solid sample in pressure-resistant Teflon tubes [19].

Results and Discussion

Effects of Different NaOH Concentrations on the Release of Phosphorus and Metals from Digested Sludge by Alkali Hydrolysis

Table 3 shows the concentrations of released P, Ca, Mg, Na, K, Al, Fe, Zn, Cr, Pb, Cu, and Ni from the digested sludge at different NaOH concentrations. A significantly higher amount of phosphorus was hydrolyzed when the NaOH concentration was raised to over 0.1M. However, the concentrations of released Na, K, and Al increased as the NaOH concentration increased. There were no considerable differences in the Ca, Mg, and Fe. Heavy metals such as Zn, Cr, Pb, Cu, and Ni were also released in the hydrolyzed liquid, but their concentration is very low. For phosphorus recovery by struvite formation, metal release such as K, Na, Ca, and Al was unfavorable because metal phosphate can be crystallized under certain conditions, which affects the purity of the struvite [2, 20]. A previous study showed that H_2SO_4, HCl, HNO_3, and citric and acetic acids applications were efficient for releasing Ca in the range of 590–9900 mg/L [6]. Alkali hydrolysis using NaOH seemed to be more beneficial in terms of low Ca release (Table 3). Specifically, Na, K, and Al releases were significantly high at 1M NaOH concentration, and there was no significant change in P release compared to 0.5M NaOH concentration. Therefore, optimization conditions for the release of nutrients and metals with low-temperature thermal treatment were obtained by using 0.5M NaOH.

Optimization of Conditions for the Release of High PO_4^{3-} and NH_4^+ and Low Metals Using 0.5M NaOH and Low-Temperature Thermal Treatment by Box-Behnken Design

The Box-Behnken design was employed to optimize high PO_4^{3-} and NH_4^+ and low metals release from the digested sludge using 0.5M NaOH and low-temperature thermal treatment. Table 4 shows the experimental conditions and the release of PO_4^{3-}, NH_4^+, and metals from the digested sludge of three factors (liquid/solid ratio (X_1), temperature (X_2), and reaction time(X_3)) in the Box-Behnken design. The release of Cr, Pb, Cu, and Ni was not determined, as they were already present in trace concentrations in the hydrolyzed sludge liquid from different NaOH concentrations (Table 3).

PO_4^{3-} release decreased significantly with temperature increase for liquid/solid ratios at and above the central point (Table 4). However, a decrease in the liquid/solid ratio with an increase in temperature enhanced the release of PO_4^{3-}. Kim et al. [10] applied thermal and alkali treatment for the release of phosphorus and nitrogen from waste-activated sludge and studied the effects of NaOH concentration (0.001–

Table 3 The release of phosphorus and metals using NaOH at different concentrations

NaOH (M)	P (mg/L)	Ca (mg/L)	Mg (mg/L)	Na (mg/L)	K (mg/L)	Al (mg/L)	Fe (mg/L)	Zn (mg/L)	Cr (mg/L)	Pb (mg/L)	Cu (mg/L)	Ni (mg/L)
0.1	55.20	6.91	10.57	1689	287.90	4.65	10.30	2.85	0.12	0.14	3.92	0.47
0.5	250.40	24.68	9.34	5838	381.30	70.05	16.00	4.25	0.22	0.28	5.06	0.67
1	292.40	18.44	8.83	20280	435.60	83.25	15.10	5.35	0.15	0.09	0.68	0.44

Table 4 The release of nutrients and metals from digested sludge using 0.5M NaOH and thermal treatment by Box-Behnken design

Test No.	X_1 (mL/g)	X_2 (°C)	X_3 (min)	PO_4^{3-} (mgL)	NH_4^+ (mg/L)	Ca (mg/L)	Mg (mg/L)	Na (mg/L)	K (mg/L)	Al (mg/L)	Fe (mg/L)	Zn (mg/L)
1	5	40	70	872.10	719.82	10.54	4.54	4925	242.40	52.44	52.74	13.65
2	15	40	70	613.50	531.48	18.96	11.70	7686	159.74	135.80	35.75	12.22
3	5	70	70	1161.00	642.42	26.32	9.28	6084	350.80	37.57	112.40	34.00
4	15	70	70	103.80	503.10	25.55	23.00	7784	157.80	130.60	53.00	24.18
5	5	55	40	912.60	820.44	8.80	10.35	4483	262.40	41.38	66.16	18.96
6	15	55	40	376.20	526.32	2.30	6.85	7815	164.45	161.60	42.84	15.94
7	5	55	100	931.50	603.72	16.95	6.75	4856	239.40	34.35	81.04	20.99
8	15	55	100	182.67	510.84	14.40	12.55	7526	156.05	156.70	47.86	21.92
9	10	40	40	921.00	819.15	0.05	4.75	6681	209.90	153.40	44.98	14.44
10	10	70	40	221.40	780.45	24.75	22.35	6689	238.15	147.00	70.96	35.57
11	10	40	100	816.00	483.75	21.55	13.85	6653	213.50	146.80	54.75	19.12
12	10	70	100	134.93	626.65	44.00	27.30	6605	273.00	145.20	75.19	38.62
13	10	55	70	262.00	657.90	44.85	17.60	5781	244.25	118.20	52.63	20.16
14	10	55	70	280.00	677.25	27.30	16.60	7641	295.15	147.80	62.65	26.48
15	10	55	70	288.50	812.70	25.60	23.10	6612	178.10	147.30	63.30	23.92

1.0N NaOH) and temperature (50, 60, 70, and 80 °C). They reported an increase in the release of phosphorus, nitrogen, and organic compounds using combined thermal and alkali treatments. However, the effect of temperature was significant only at low NaOH concentrations but was insignificant at high NaOH concentrations. In our study, however, the effect of temperature on PO_4^{3-} release was found to be related to the liquid/solid ratio.

The effect of temperature on PO_4^{3-} release was positive when liquid/solid ratio was below 10/1; however, the effect was negative at and above 10/1 liquid/solid ratio (Fig. 1a). Nevertheless, the temperature did not appear to have a significant effect on release of NH_4^+ under any test condition (Fig. 1b). On the other hand, NaOH and thermal treatment dissolved not only PO_4^{3-} and NH_4^+ but also K, Al, Fe, and Zn (Table 4). The release of Fe and Zn increased with temperature. The highest PO_4^{3-} release (1161.00 mg/L) was found at a liquid/solid ratio (mL/g) of 5/1, temperature of 70 °C, and reaction time of 70 min (test no. 3) (Table 4). However, Fe and K release was maximum (112.40 and 350.80 mg/L) under these conditions, and Zn release was also high (Table 4). Alkali hydrolysis and temperature were applied together to determine optimum conditions, considering the operating conditions with the high nutrient release and low metal release as the basis. Therefore, the optimum conditions for nutrient and metal release were determined to be 10/1 liquid/solid ratio, 40 °C temperature, and 40-min reaction time (test no. 9) (Table 4).

The model adequacy was checked by the determination coefficient (R^2). The second-order polynomial model for release of PO_4^{3-} and NH_4^+ (Y) was regressed by considering only the significant terms, shown in Eqs. (2) and (3):

$$Y_P = 276.83 - 352.13x_1 - 200.18x_2 + 244.09x_1^2 - 199.65x_1x_2 \left(R^2 = 0.9080\right) \quad (2)$$

$$Y_N = 715.95 - 89.33x_1 - 90.30x_3 - 89.33x_1^2 \left(R^2 = 0.8629\right) \quad (3)$$

As seen in Eq. (2), the first-degree main effects (liquid/solid ratio (x_1) and temperature (x_2)) and the second-degree main effect (x_1^2) on release of PO_4^{3-} were the significant model terms. As seen in Eq. (3), the first-degree main effects (liquid/solid ratio (x_1) and reaction time (x_3)) and the second-degree main effect (x_1^2) on release of NH_4^+ were the significant model terms. The values of the R^2 were greater than 0.86 for both Y_P and Y_N, suggesting that the regression models were appropriate for simulating the experimental data.

Properties of Struvite Produced from the Hydrolyzed Liquid Using 0.5M NaOH and Low-Temperature Thermal Treatment

Nutrient and metal release was optimum at 10/1 liquid/solid ratio, 40 °C temperature, and 40-min reaction time (test no. 9) (section "Optimization of Conditions for the Release of High PO_4^{3-} and NH_4^+ and Low Metals Using 0.5M NaOH and

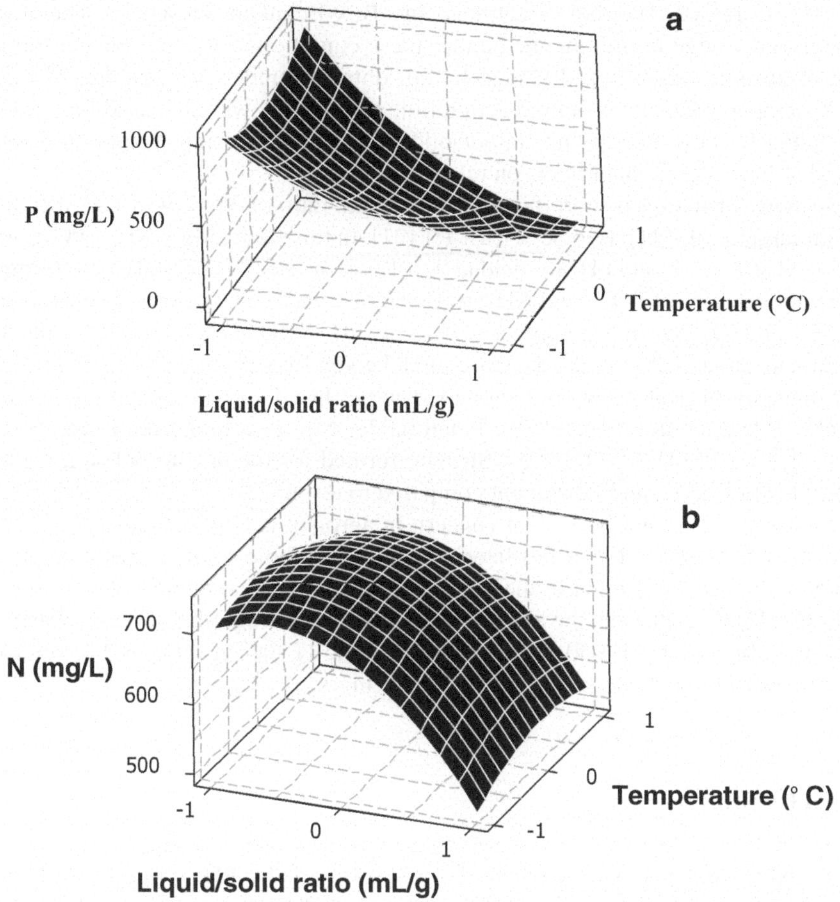

Fig. 1 Effects of interaction of liquid/solid ratio × temperature on release P (**a**) and N (**b**)

Table 5 Characteristics of the struvite precipitate produced from alkali hydrolysis and low-temperature treatment

Produced struvite	P (%)	Mg (%)	K (%)	Ca (%)	Na (%)	Al (%)	Fe (%)
Test number 5	7.7	5.13	0.95	0.20	1.19	0.51	2.54
Test number 9	8.3	6.07	1.28	0.12	2.31	0.12	0.04

Table 6 Heavy metals results obtained for produced struvite in this study

Produced struvite	Cd (mg/kg)	Cu (mg/kg)	Ni (mg/kg)	Pb (mg/kg)	Zn (mg/kg)	Hg (mg/kg)	Cr (mg/kg)
Test number 5	b.d.	7.19 ± 0.02	b.d.	b.d.	505.10 ± 7.1	b.d.	8.49 ± 0.02
Test number 9	b.d.	b.d.	b.d.	b.d.	650.6 ± 7.6	b.d.	b.d.

b.d. below detection limit

Low-Temperature Thermal Treatment by Box-Behnken Design"). Therefore, hydrolyzed sludge liquid obtained under these conditions was used for the formation of struvite. Hydrolyzed sludge liquid, obtained under 5/1 liquid/solid ratio, 55 °C temperature, and 40-min reaction time (test no. 5), was also used for struvite formation because decreasing liquid/solid ratio and increasing temperature were found to have significant effects on release (Table 4).

Struvite formation is more favorable when the ratio of Mg:P was 1:2.5 [2], and the molar ratio of P:N has to be at least 1:1 [21]. In this study, Mg^{2+}:NH_4^+: PO_4^{3-} was arranged at 1.5:1:1, and pH was held at 9.0. Then, the struvite crystals were formed. The removal rates of PO_4^{3-} and NH_4^+ were 97.93% and 62.34% in test number 5 and 95.27% and 77.95% in test number 9, respectively. XRD analysis results of the dry precipitates matched with the database model given for struvite in terms of position and intensity of peaks (data not shown). The produced struvite precipitate for test number 9 had a composition 8.3% P and 6.07% Mg for a molar ratio Mg^{2+}:NH_4^+: PO_4^{3-} 1.5:1:1 at pH 9.0 (Table 5). Struvite formed for the test no. 9 had a significantly lower Ca, Al, and Fe content compared to test no. 5.

Heavy metals could be a great concern in agronomic applications if the recovered struvite is applied as a fertilizer. As shown in Table 6, the concentrations of heavy metals except for Zn in the struvite produced for test number 9 were below detection limits. Although the anaerobically digested sewage sludge had a considerably high Zn content (10960 ± 30.5 mg/kg in Table 1), the Zn value of the produced struvite was below the legal limit value (1100 mg/kg).

Conclusions

PO_4^{3-}, NH_4^+, and metals were released from the digested sludge during 0.5M NaOH hydrolysis and thermal treatment. The quantity of PO_4^{3-} released depended on the temperature and liquid/solid ratio. This study demonstrated that increasing temperature had a remarkable effect on PO_4^{3-} release when the liquid/solid ratio was under the central point. However, the determination of optimum conditions was based upon metal release. When the temperature was greater than 40 °C, some metals such as Ca, Fe, and Zn were hydrolyzed. Liquid/solid ratio at the central point (10/1) and low temperature (<50 °C) is recommended for digested sludge hydrolysis to achieve the release of high PO_4^{3-} and low metal content. PO_4^{3-} and NH_4^+ recovery as struvite with high-quality could be achieved from the hydrolyzed liquid of NaOH and thermal treated digested sludge.

Acknowledgments This study was supported by the Scientific and Technological Research Council of Turkey (TUBITAK 2209-A program, project no. 1919 B011502482).

References

1. Bi W, Li Y, Hu Y (2014) Recovery of phosphorus and nitrogen from alkaline hydrolysis supernatant of excess sludge by magnesium ammonium phosphate. Bioresour Technol 166:1–8
2. Wang Y, Xiao Q, Zhong H, Zheng X, Wei Y (2016) Effect of organic matter on phosphorus recovery from sewage sludge subjected to microwave hybrid pretreatment. J Environ Sci 39:29–36
3. Xu Y, Hu H, Liu J, Luo J, Qian G, Wang A (2015) pH dependent phosphorus release from waste activated sludge: contributions of phosphorus speciation. Chem Eng J 267:260–265
4. Chen Y, Jiang S, Yuan H, Zhou Q, Gu G (2007) Hydrolysis and acidification of waste activated sludge at different pHs. Water Res 41:683–689
5. Takahaski M, Takemoto Y, Onsihi K (2015) Phosphorus recovery from carbonized sewage sludge by hydrothermal processes. J Mater Sci Eng 5(1-2):58–62
6. Uysal A, Tuncer D, Kir E, Sardohan Koseoglu T (2016) Phosphorus recovery from hydrolysed sewage sludge liquid containing metals using Donnan dialysis. 6th International Conference on Environmental Pollution and Remediation (ICEPR'16), Proceedings of the 2nd World Congress on New Technologies, August 2016, Budapest, Hungary, ICEPR 125, p 1–7
7. Ali TU, Kim DJ (2016) Phosphorus extraction and sludge dissolution by acid and alkali treatments of polyaluminum chloride (PAC) treated wastewater sludge. Bioresour Technol 217:233–238
8. Xu DC, Zhong CQ, Yin KH, Peng SH, Zhu TT, Cheng G (2018) Alkaline solubilization of excess mixed sludge and the recovery of released phosphorus as magnesium ammonium phosphate. Bioresour Technol 249:783–790
9. Sano A, Kanomata M, Inoue H, Sugiura N, Xu KQ, Inamori Y (2012) Extraction of raw sewage sludge containing iron phosphate for phosphorus recovery. Chemosphere 89:1243–1247
10. Kim M, Han DW, Kim DJ (2015) Selective release of phosphorus and nitrogen from waste activated sludge with combined thermal and alkali treatment. Bioresour Technol 190:522–528
11. Takiguchi N, Kishino M, Kuroda A, Kato J, Ohtake H (2007) Effect of mineral elements on phosphorus release from heated sewage sludge. Bioresour Technol 98:2533–2537
12. Suarez-Iglesias O, Urrea JL, Oulego P, Collado S, Diaz M (2017) Valuable compounds from sewage sludge by thermal hydrolysis and wet oxidation. A review. Sci Total Environ 584–585:921–934
13. Doyle JD, Parsons SA (2002) Struvite formation, control and recovery. Water Res 36:3925–3940
14. Uysal A, Tuncer D, Kir E, Sardohan Koseoglu T (2017) Recovery of nutrients from digested sludge as struvite with a combination process of acid hydrolysis and Donnan dialysis. Water Sci Technol 76(10):2733–2741
15. Torres ML, Liorens MDE (2008) Effect of alkaline pretreatment on anaerobic digestion of solid wastes. Waste Manage 28(11):2229–2234
16. Kim J, Park C, Kim TH, Lee M, Kim S, Kim SW, Lee J (2003) Effects of various pretreatments for enhanced anaerobic digestion with waste activated sludge. J Biosci Bioeng 95(3):271–275
17. Dong CH, Xie XQ, Wang XL, Zhani Y, Yao YJ (2009) Application of Box-Behnken design in optimisation for polysaccharides extraction from cultured mycelium of Cordyceps sinensis. Food Bioprod Process 87(2):139–144
18. Standard Methods for the Examination of Water and Wastewater (2005) 21st edn, American Public Health Association/American Water Works Association/Water Environment F, Washington DC, USA
19. Uysal A, Yılmazel YD, Demirer GN (2010) The determination of fertilizer quality of the formed struvite from effluent of a sewage sludge anaerobic digester. J Hazard Mater 181:248–254
20. Jaffer Y, Clark TA, Pearce P, Parsons SA (2002) Potential phosphorus recovery by struvite formation. Water Res 36:1834–1842
21. Cieslik B, Konieczka P (2017) A review of phosphorus recovery methods at various steps of wastewater treatment and sewage sludge management. The concept of "no solid waste generation" and analytical methods. J Clean Prod 142:1728–1740

Evaluation of Contaminant Transport Through Alternative Liner Systems from Leachate to Groundwater Using One-Dimensional Mass Transport Model

Gamze Varank, Ahmet Adiller, Senem Yazici Güvenç, Elanur Adar, and Ahmet Demir

Abstract One-dimensional (1D) advection–dispersion transport modeling was conducted as a conceptual approach for evaluation of organic (phenolic compounds) contaminant transport through alternative liner systems from leachate to groundwater. In this study, ten identical pilot-scale landfill reactors with different alternative composite liners were simultaneously operated for a period of about 290 days. The results of 1D transport model showed that the highest molecular diffusion coefficients for 2,3,4-TCP and 2,3,4,5-TeCP and PCP were determined to be with the average values of 54.25×10^{-9}, 44.17×10^{-9}, and 15.19×10^{-9} m^2/sn and the lowest molecular diffusion coefficients for 2,4-DCP and 2,3,5-TCP were obtained to be with the average values of 1.107×10^{-9} and 1.115×10^{-9} m^2/sn approximately in all reactor systems. The results indicate that liner systems have no significant effect on organic contaminant migration from leachate to groundwater and the dominant mechanism in transportation of organic contaminant is molecular diffusion and geomembrane layer is ineffective in organic contaminant transport through composite liners.

Keywords Landfill leachate · Liner systems · 1D mass transport model · Contaminant transport · Phenolic compounds

---

G. Varank (✉) · S. Y. Güvenç · E. Adar · A. Demir
Department of Environmental Engineering, Yildiz Technical University, Istanbul, Turkey
e-mail: gvarank@yildiz.edu.tr

A. Adiller
Department of Environmental Health, Uskudar University, Istanbul, Turkey

N. Balkaya, S. Guneysu (eds.), *Recycling and Reuse Approaches for Better Sustainability*, Environmental Science and Engineering,
https://doi.org/10.1007/978-3-319-95888-0_8

Introduction

Landfill technology has gone through significant development. Although municipal solid waste (MSW) landfills have evolved from uncontrolled city dumps to highly engineered structures in the last several decades, they still pose an environmental threat to groundwater resources because of leachate consisting of a wide range of toxic inorganic and organic compounds. Contaminants in MSW landfill leachate can be classified in groups of dissolved organic matter, inorganic macro components, heavy metals, and xenobiotic organic compounds [1]. Researchers have concluded that heavy metal concentrations in landfill leachate are not a major concern but the issue of xenobiotic organic compounds (XOCs) is of a great concern although XOC concentrations are reported to be very low [1–7]. Phenol and substituted phenols categorized in xenobiotic organic compounds are common transformation products of several pesticides. Phenol and many substituted phenols have been designated as priority pollutants by the US Environmental Protection Agency [8, 9].

Liner system is one of the most important elements of a modern engineered landfill [10]. Composite liners consisting of a geomembrane (usually high-density polyethylene (HDPE) 1–2.5 mm thick) overlying a compacted soil liner (0.30–1.5 m thick) are the standard design for liner systems in municipal solid waste landfills [11]. The predominant pathways for contaminant transport through composite liners depend on the type of contaminant: (1) inorganic and organic contaminant transport through defects in the geomembrane and through the clay liner by advection and dispersion and (2) transport of organic contaminants through the intact geomembrane and through the clay liner by diffusion [11–14]. Geomembranes are essentially impervious to diffusion of inorganic solutes [15, 16] and liquid flow [17, 18] but organic solutes can rapidly diffuse through geomembranes [10, 19–23]. Moreover, the mass flow rate of organic solutes from defects can be ignored because the mass flow rate through an intact composite liner can be several orders of magnitude greater than that through defects [11, 14, 24].

Though the geomembrane is believed to be the primary barrier to contaminant transport, the clay component usually controls the rate of toxic organic contaminant transport since XOCs are shown to diffuse through geomembrane at appreciable rates. Additionally, xenobiotic organic compound (XOC) transport is generally more critical than transport of inorganic compounds (e.g., toxic heavy metals), even though XOCs are often found at lower concentrations in leachates. Since diffusive transport is often the dominant mode of contaminant transport through engineered barriers including compacted clay liners [25–28] and composite liners [10, 23, 29], comparing composite liner systems based on leakage rate may not be sufficient unlike the current approach of assuming that an alternative liner is equivalent to the prescriptive liner if it discharges less liquid [10, 30].

Equivalency of alternative composite liners can be evaluated by using numerical models that represented diffusion through intact composite liners and advection and diffusion through defects in composite liners [14, 31]. Diffusion of contaminants through porous media is often modeled by numerical methods [32]. Rowe and Booker

[33] have developed the finite layer methods, and Leo and Booker [34] developed the boundary element method. Foose et al. [31] and Foose [29] used a multiple-layer finite-difference model to simulate contaminants' transport diffusion through composite liners. Although numerical solutions must be used for many transport problems, people can better understand the mechanism of contaminant diffusion, predict the movement of contaminant plumes, measure the field parameters related to contaminant diffusion, and verify the results of numerical modeling using analytical solutions [35]. Foose et al. [21], Booker and Rowe [36], and Shackelford and Lee [37] have provided the analytical solutions of one-dimensional (1D) contaminant diffusion for homogeneous media. Foose [29] has given the analytical solution of 1D diffusion for composite liner with semi-infinite bottom boundary, constant concentration at top boundary, and zero concentration in initial conditions.

This paper presents a systematic and comprehensive approach to determine mass transport parameters for phenolic compounds through alternative landfill liners. An analytical solution of 1D contaminant diffusion through composite liner systems is presented under conditions of steady-state flow. Transport of phenolic compounds representing organic contaminants was investigated to determine the transport parameters of organic compounds migrating through alternative composite liners and assess the applicability of simple diffusion models for the preliminary comparison of the performance of different liner systems.

Materials and Methods

Reactor Setup and Operation

Ten pilot-scale landfill reactors (R1–R10) were operated for about 290 days to investigate diffusive transport of phenolic compounds representing organic contaminants in leachate. Reactors were made of HDPE pressurized pipes with a wall thickness of 0.005 m, diameter (D_R) of 0.40 m, and height (H_R) of 0.80 m, respectively. The diameter and the height of the reactors made of HDPE pressurized pipes were 40 cm and 80 cm, respectively. The reactors were comprised of two parts: consisting leachate in the upper part and distilled water simulating groundwater in the lower part. The bottom of the both parts of the reactors consisted of perforated pipes inserted to collect samples from leachate and groundwater.

Ten alternatives with different composite liners were employed for the reactor liners. The clay material and geomembrane used in this study were obtained from Odayeri Sanitary Landfill in in Istanbul, Turkey. Liner systems were constructed at the top of the lower part, and groundwater was separated from the liner system by screenlike plastic material. The thicknesses of the liners were 20 cm. The total area of geomembrane defects was about 1.0% of the surface area of the reactor.

Equivalent hydraulic conductivity (k_e) values for the liners consisting of different materials with different hydraulic conductivity values were determined by using the arithmetic average method proposed by Golan and Whitson [38] and given in Table 1.

Table 1 Liner systems of the pilot-scale reactors and hydraulic conductivity values (m/s)

Reactor	Liner	k_e
R1	%100 clay	6.3×10^{-8}
R2	%100 bentonite	2.7×10^{-10}
R3	%100 kaolinite	3.2×10^{-7}
R4	%100 zeolite	8.8×10^{-8}
R5	%50 clay + %50 bentonite	3.16×10^{-8}
R6	%50 clay + %50 kaolinite	1.92×10^{-7}
R7	%80 clay + %20 zeolite	6.8×10^{-8}
R8	%80 bentonite + %20 zeolite	1.78×10^{-8}
R9	%90 clay + %10 lime	4.8×10^{-8}
R10	%90 bentonite + %10 lime	2.2×10^{-10}

One-Dimensional Mass Transport Model

Contaminant transport and flow through composite liners are three-dimensional (3D) processes [11]. The conceptual approach used herein is to analyze a 1D system approximating the 3D system characteristics, as conducted by Foose [11]. In this study, the following assumptions are made for the estimation of transport parameters of phenolic compounds [11]:

1. Porous medium is isotropic and saturated.
2. Composite liner and underlying geologic layers are horizontal and homogeneous.
3. Solute transport in the underlying aquifer occurs only via advection; flow in the aquifer occurs in one-dimensional motion to the layers and is uniform and steady state.
4. Solute transport in the underlying aquifer occurs only via advection, and Darcy's law is valid.
5. No contaminant decay occurs in the composite liner and underlying geologic layers or aquifer.

Selected parametric values used in 1D advection–dispersion model for the estimation of the transport parameters are given in Table 2.

Analytical Procedure

All analyses were conducted according to standard methods of APHA [39]. Solid-phase microextraction (SPME) method proposed by Ribeiro et al. [40] was used to determine concentrations of phenolic compounds in leachate and groundwater samples.

Table 2 Parametric values used in 1D advection–dispersion model for the estimation of the transport parameters

Parameter	Value
Head of liquid on the top of the geomembrane (h_w)	0.30 m
Thickness of the HDPE geomembrane (L_g)	0.002 m
Thickness of the composite liner (Le)	0.20 m
The quality of the intimate contact between the geomembrane and its underlying clay liner (C_f)	0.01
Retardation factor (R)	
Organics	1.0
Inorganics	2.0
Frequency of geomembrane defects (N)	1
Area of single defect in geomembrane (a)	5×10^{-5} m^2
Surface area of the reactor (A_R)	0.126 m^2

Results and Discussion

A simplified one-dimensional (1D) advection–dispersion transport model was readily conducted to simulate transport of organic contaminants existing in leachate through alternative landfill liners and the solute as well as to estimate the transport parameters of the studied organic contaminants. Molecular diffusion coefficients of phenolic compounds are given in Fig. 1.

As can be seen in Fig. 1, the highest molecular diffusion coefficients were estimated for 2,3,4-TCP and 2,3,4,5-TeCP and PCP approximately in all reactor systems. Average molecular diffusion coefficients of 2,3,4-TCP, 2,3,4,5-TeCP, and PCP were determined to be 54.25×10^{-9}, 44.17×10^{-9}, and 15.19×10^{-9} m^2/sn, respectively. The lowest molecular diffusion coefficients were estimated for 2,4-DCP and 2,3,5-TCP with the values of 1.107×10^{-9} and 1.115×10^{-9} m^2/sn. As average molecular diffusion coefficients were evaluated based on reactor liner systems, it can be concluded that no significant differences were obtained. The highest average molecular diffusion coefficients were estimated for R2 and R10 with the values of 21.28×10^{-9} m^2/sn and 15.46×10^{-9} m^2/sn. The lowest average molecular diffusion coefficient was estimated for R4. For the rest of the reactors, average molecular diffusion coefficients were found to be in the range of 5.34×10^{-9} to 9.64×10^{-9} m^2/sn. The results indicate that liner systems have no significant effect on organic contaminant migration from leachate to groundwater. Results of the study showed that molecular diffusion is the dominant mechanism in organic contaminant transport and also geomembrane layer is ineffective in preventing contaminant migration.

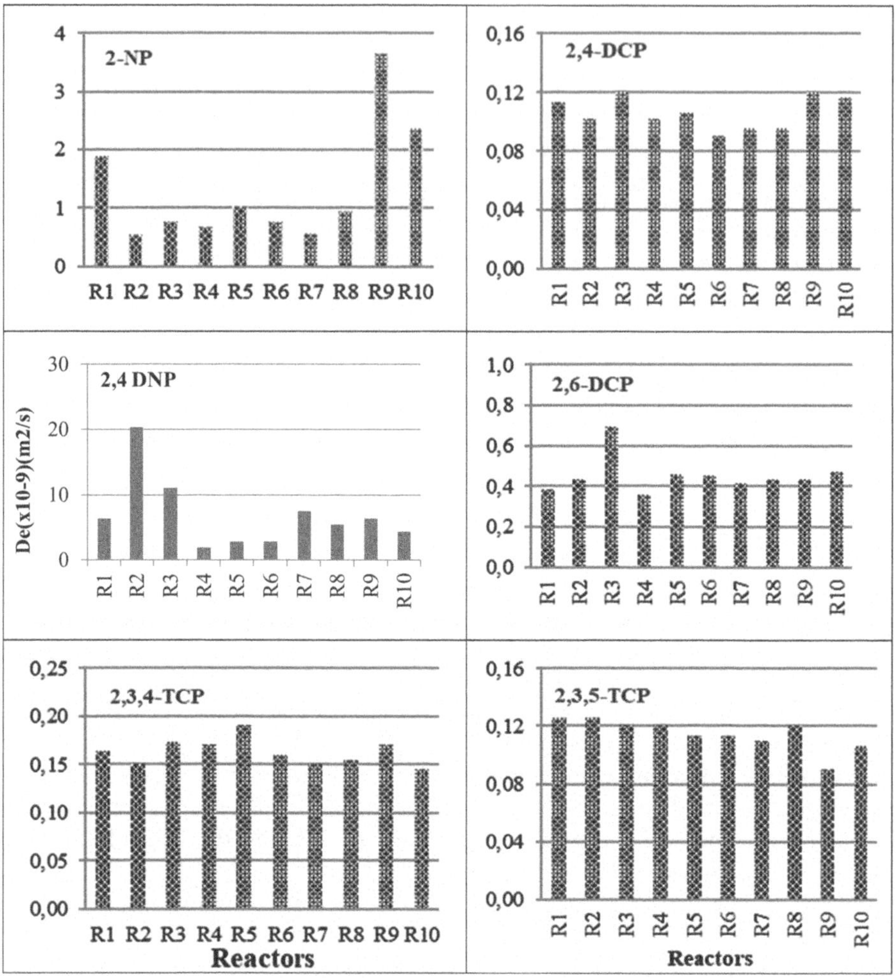

Fig. 1 Schematic presentation of diffusion coefficients of phenolic compounds

Conclusions

In this study, simplified 1D advection–dispersion transport model was performed to evaluate organic contaminant transport from leachate to groundwater through composite liners. Ten pilot-scale landfill reactors were operated for a period of about 290 days to examine diffusive transport of phenolic compounds. Results of the study indicated that the highest molecular diffusion coefficients were estimated for 2,3,4-TCP and 2,3,4,5-TeCP and PCP and the lowest molecular diffusion coefficients were estimated for 2,4-DCP and 2,3,5-TCP approximately in all reactor systems. Average molecular diffusion coefficients were found to be in the range of 54.25×10^{-9} to 1.107×10^{-9} m^2/sn. As average molecular diffusion coefficients

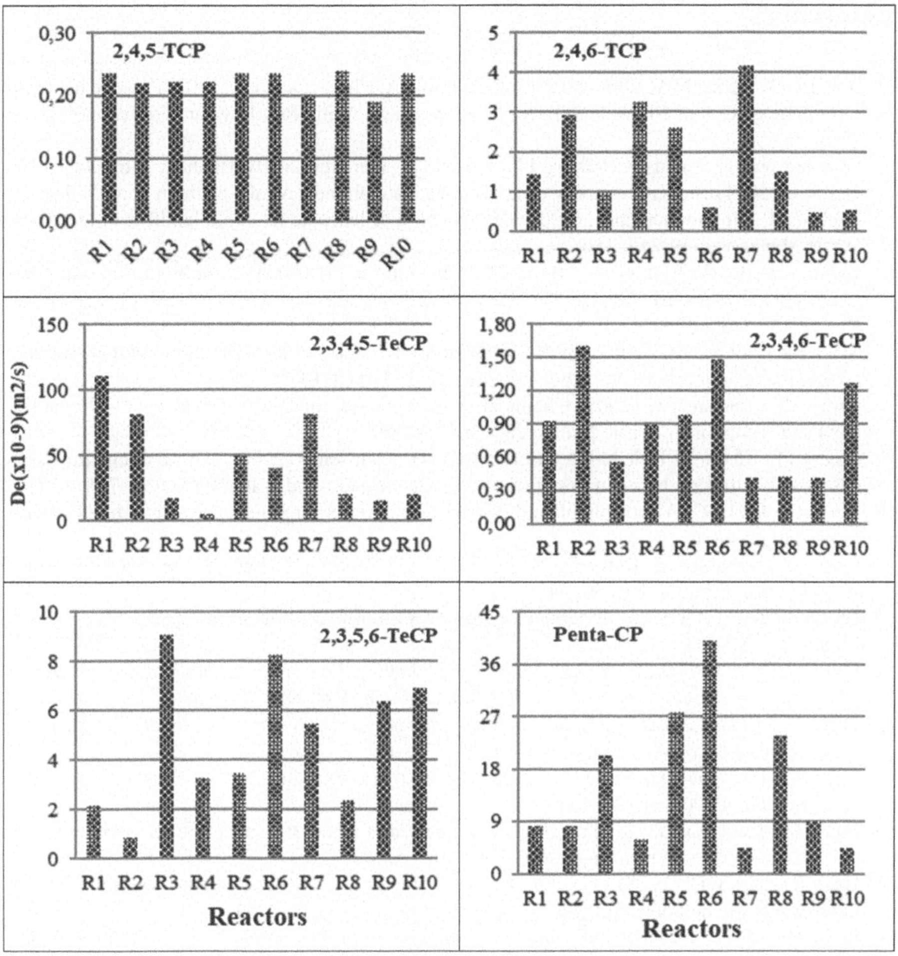

Fig. 1 (continued)

were evaluated based on reactor liner systems, results of the study showed that molecular diffusion is the dominant mechanism in organic contaminant transport and also geomembrane layer is ineffective in preventing contaminant migration. Further investigations related with different operating conditions affecting organic contaminant transport through composite liners should be conducted.

Acknowledgments This research has been supported by the Scientific Research Projects Coordinatorship of Yildiz Technical University (Project Number: 2011-05-02-KAP01), Istanbul, Turkey. The authors wish to acknowledge ISTAC (Istanbul Greater Metropolitan City Municipality Environment Protection and Waste Material Recycling Industry and Trade J.S.Co.).

References

1. Kjeldsen P, Barlaz MA, Rooker AP, Baun A, Ledin A, Christensen TH (2002) Present and long-term composition of MSW landfill leachate: a review. Critic Rev Environ Sci Technol 32(4): 297–336
2. Christensen TH, Kjeldsen P, Bjerg PL, Jensen DL, Christensen JB, Baun A, Albrechtsen HJ, Heron G (2001) Biogeochemistry of landfill leachate plumes. Appl Geochem 16:659–718
3. Kjeldsen P, Christophersen M (2001) Composition of leachate from old landfills in Denmark. Waste Manag Res 19:249–256
4. Baun A, Ledin A, Reitzel LA, Bjerg PL, Christensen TH (2004) Xenobiotic organic compounds in leachates from ten Danish MSW landfills-chemical analysis and toxicity tests. Water Res 38(18):3845–3858
5. Slack RJ, Gronow JR, Voulvoulis N (2005) Household hazardous waste in municipal landfills: contaminants in leachate. Sci Total Environ 337(1–3):119–137
6. Oman CB, Junestedt C (2008) Chemical characterization of landfill leachates – 400 parameters and compounds. Waste Manag 28(10):876–1891
7. Bejerg PL, Tuxen N, Reitzel LA, Albrechtsen HJ, Kjeldsen P (2009) Natural attenuation processes in landfill leachate plumes at three Danish sites. Ground Water 49(5):688–705
8. Boopathy R (1997) Anaerobic phenol degradation by microorganisms of swine manure. Curr Microbiol 35:64–67
9. Boyd SA, Shelton DR, Berry D, Tiedje JM (1983) Phenolic compounds in digested sludge. Appl Environ Microbiol 46:50–54
10. Edil TB (2003) A review of aqueous-phase VOC transport in modern landfill liners. Waste Manag 23:561–571
11. Foose GJ (2010) A steady-state approach for evaluating the impact of solute transport through composite liners on groundwater quality. Waste Manag 30(8–9):1577–1586
12. Foose GJ (1997) Leakage rates and chemical transport through composite landfill liners. Ph.D. thesis, University of Wisconsin-Madison
13. Katsumi T, Benson CH, Foose GJ, Kamon M (2001) Performance-based design of landfill liners. J Eng Geol 60(1–4):139–148
14. Foose GJ, Benson CH, Edil TB (2002) Comparison of solute transport in three composite liners. J Geotech Geoenviron Eng 128(5):391–403
15. Haxo JH, Lahey TP (1988) Transport dissolved organics from dilute aqueous solutions through flexible membrane liners. Hazard Waste Hazard Mater 5(4):275–294
16. Rowe RK (1998) Geosynthetics and the minimization of contaminant migration through barrier systems beneath solid waste. In: Proceedings of the sixth international conference on geosynthetics, Atlanta, USA, vol 1. IFAI, Roseville, MN, pp 27–103
17. Giroud JP, Bonaparte R (1989) Leakage through liners constructed with geomembranes. Part II – composite liners. Geotext Geomembr 8(2):71–111
18. Lambert S, Touze-Foltz N (2000) A test for measuring permeability of geomembranes. In: Proceedings Eurogeo 2, second European conference on geosynthetics, Bologna, Italia. Pàtron, Bologna, pp 15–18
19. Park JK, Nibras M (1993) Mass flux of organic chemicals through polyethylene geomembranes. Water Environ Res 65(3):227–237
20. Müller W, Jakob I, Tatzky-Gerth R, August H (1998) Solubilities, diffusion and partition coefficients of organic pollutants in HDPE geomembranes: experimental results and calculations. Paper presented at the sixth international conference on geosynthetics, Atlanta, GA, Industrial Fabrics Association International, St. Paul, MN, p 239–248
21. Foose GJ, Benson CH, Edil TB (2001) Predicting leakage through composite landfill liners. J Geotech Geoenviron 127(6):510–520
22. Sangam HP, Rowe RK (2001) Migration of dilute aqueous organic pollutants through HDPE geomembranes. Geotext Geomembr 19(6):329–357

23. Kalbe U, Muller W, Berger W, Eckardt J (2002) Transport of organic contaminants within composite liner systems. Appl Clay Sci 21:67–76
24. Park JK, Sakti JP, Hoopes JA (1996) Transport of organic compounds in thermoplastic geomembranes I: mathematical model. J Environ Eng 122(9):800–806
25. Shackelford CD (1990) Transit-time design of earthen barriers. Eng Geol 29:79–94
26. Toupiol C, Willingham TW, Valocchi AJ, Werth CJ (2002) Long-term tritium transport through fieldscale compacted soil liner. J Geotech Geoenviron Eng 128(8):640–650
27. Willingham TW, Werth CJ, Valocchi AJ, Krapac IG (2004) Evaluation of multidimensional transport through a fieldscale compacted soil liner. J Geotech Geoenviron Eng 130(9):887–895
28. Bezza A, Ghomari F (2008) Nondestructive test to track pollutant transport into landfill liners. Environ Geol 57(2):285–290
29. Foose GJ (2002) Transit-time design for diffusion through composite liners. J Geotech Geoenviron 128(7):590–601
30. Richardson GN (1997) GCLs: alternative subtitle D liner systems. Geotechnical Fabrics Report, Industrial Fabrics Association, St. Paul, MN. 15(5): 36–42
31. Foose GJ, Benson CH, Edil TB (1999) Equivalency of composite geosynthetic clay liners as a barrier to volatile organic compounds. In: Proceedings of Geosynthetics 99, Boston, MA. Industrial Fabrics Association International, St. Paul, MN, pp 321–334
32. Paseka AL, Lqbal MZ, Walters JC (2000) Comparisons of numerical simulation of solute transport with observed experimental data in a silt loam subsoil. Environ Geol 39(9):977–989
33. Rowe RK, Booker JR (1984) The analysis of pollutant migration in a non-homogeneous soil. Geotechnique 34:601–612
34. Leo CJ, Booker JR (1999) A boundary element method for analysis of contaminant transport in porous media-II: non-homogeneous porous media. Int J Numer Anal Meth Geomech 23:1701–1715
35. Park E, Zhan H (2001) Analytical solutions of contaminant transport from finite one-, two-, and three-dimensional sources in a finitethickness aquifer. J Contam Hydrol 53:41–61
36. Booker JR, Rowe RK (1987) One dimensional advective–dispersive transport into a deep layer having a variable surface concentration. Int J Numer Anal Meth Geomech 11:131–142
37. Shackelford CD, Lee JM (2005) Analyzing diffusion by analogy with consolidation. J Geotech Geoenviron Eng 131(11):1345–1359
38. Golan M, Whitson CH (1986) Well performance. Springer, Berlin, p 617
39. APHA (American Public Health Association), AWWA (American Water Works Association), WEF (Water Environment Federation) (2005) Standard methods for the examination of water and wastewater, 21st edn. APHA, AWWA, WEF, Washington DC
40. Ribeiro A, Neves MH, Almeida MF, Alves A, Santos L (2002) Direct determination of chlorophenols in landfill leachates by solid-phase micro-extraction–gas chromatography–mass spectrometry. J Chromatogr A 975(2):267–274

An Assessment of Pyrolysis Process for the Treatment of Agricultural and Forest Wastes

Atakan Öngen, Nazlıcan Karabağ, Selda Yiğit Hunce, Hüseyin Kurtuluş Özcan, Emine Elmaslar Özbaş, and Serdar Aydın

Abstract As a result of social and economic activities of people, materials that completed useful life that should be removed from the environment we live in are classified as waste. Nowadays, except for local uses (burning, warming, etc.), there are few uses for general purposes (alternative fuel, etc.). Energy usage and waste amount are increasing rapidly with the increase of the population in the world and the development of industry and technology. Control of energy consumption and waste is crucial for a sustainable future. In this study, the use of pyrolysis for agriculture and forest wastes was investigated based on the literature. For this purpose, many studies on pyrolysis of organic wastes such as agriculture and forest wastes have been compiled and presented on a common basis. The methods followed in the selected studies are examined and given in the form of tables under the common headings in order that different researchers can reach easily. It has been generally agreed that the pyrolysis process, which is considered as an alternative waste management method, is an energy and environmentally friendly application for waste and that this application can be used to reduce waste and recycle waste.

Keywords Pyrolysis · Waste-to-energy · Agricultural waste · Forest waste · Waste management

Introduction

The substances, formed by domestic, commercial, and industrial activities and considered as useless by the consumer but which are regularly removed or landfilled to protect environment and human health, are defined as waste [1].

The need for energy is increasing with the increase in world population. The need of energy due to the growing population and industrialization can't be met by

A. Öngen (✉) · N. Karabağ · S. Y. Hunce · H. K. Özcan · E. E. Özbaş · S. Aydın
Department of Environmental Engineering, Istanbul University-Cerrahpaşa, Istanbul, Turkey
e-mail: aongen@istanbul.edu.tr

N. Balkaya, S. Guneysu (eds.), *Recycling and Reuse Approaches for Better Sustainability*, Environmental Science and Engineering,
https://doi.org/10.1007/978-3-319-95888-0_9

the limited resources of the world, and the gap between production and consumption of energy is growing rapidly. On the other hand, conventional energy production methods based on fossil fuels are an important factor in increasing environmental pollution. It is a fact that the need for clean energy is significant, that fossil fuels are rapidly consumed, and that global warming is an undeniable danger. Decreasing the effects of global warming especially caused by greenhouse gas emissions and the conversion of these emissions into energy by using smart solutions are significant for the welfare of the community [2, 3].

Recently, the energy demand also increases continuously with the increase of the population and technological developments. The decline in energy resources, the demand for fossil fuel reserves which are limited and reaching the ending point, the increment in production costs, and the global warming increase the tendency toward alternative energy sources. This trend has also increased the availability of new, renewable, and clean energy sources [4, 5].

New and renewable energy sources are mainly categorized as biomass, solar, hydraulic, wind, geothermal, tidal, and wave energy, and most of these energy sources have been widely used in the world. The most common new and renewable energy source is biomass energy [4, 5].

Given the immense potential of biomass, it has an important place among the renewable energy sources for the future due to various social and economic benefits [6].

Biomass is defined as all organic matter containing various products, wastes, and plants, which can be renewed in less than 100 years, and converting solar energy into chemical energy through photosynthesis. Biomass, with an industrial view, is concerned with the acquisition and use of fuel, gas, and chemicals from alive or recently dead entities. Biomass includes plants grown to produce biofuels and also organic matter and wastes that are used to produce fiber, heat, and chemicals [6–8].

Biomass energy and biomass technology are seen as an appropriate and important source of energy and technology due to its advantages because biomass is an inexhaustible source, can be grown everywhere, enables energy production at every scale, has storage capacity, causes less greenhouse effect than other energy sources, does not cause acid rain, provides CO_2 equilibrium in the atmosphere, helps socio-economic development especially for rural areas, needs low light intensity and temperatures between 5 and 35 °C, and also can be used in chemical production, fuel production, and electricity generation. By means of these advantages, energy production from biomass sources and their share in total energy consumption are increased worldwide [7–9].

Energy demand of Turkey is growing rapidly due to increasing population and economic growth. Turkey is an energy-dependent country having limited energy resources, and for this reason, the needed energy is imported in an expensive manner. According to data provided by the Energy and Natural Resources Ministry, Turkey produces only 1/3 of the consumed energy. The external dependency in primary energy sources is also 72.6% [2].

Due to these reasons, the simplest way of energy conversion from solid organic wastes, especially forest and agricultural wastes, is possible by directly burning them. Pyrolysis and/or gasification processes are rapidly developing processes in terms of technology and application areas as an alternative to burning, which is under criticism due to its known effects on air pollution. By these processes, liquid, solid, and gaseous products with economic value can be produced from organically rich fuels. For centuries charcoal was produced by pyrolysis; it is also produced by gasification, heating the biomass under a certain pressure with air so as not to leave anything other than ash and sludge behind.

Biogas (depending on the nature of the organic material and the applied process) is a colorless, flammable gas containing 60–70% CH_4, 30–40% CO_2, and a low amount of H_2S, N, H, and CO. Biogas as a precautionary move against global climate change is a significant waste control method; its use in heat and electricity production reduces external dependence on energy and contributes to economic development [10].

The aim of this study is to investigate the feasibility of pyrolysis of agriculture and forestry wastes in order to ensure energy consumption and waste management which are very important for a sustainable future. For this purpose, many studies on the evaluation of pyrolysis on agriculture and forest wastes have been examined and collected under the common titles, thus providing a holistic view for researchers working on the subject.

General Information

Pyrolysis

Pyrolysis is the phenomenon of decomposition of organic materials in an inert atmosphere or vacuum. Pyrolysis is the process of thermal decomposition of organic materials by heating them in an oxygen-free environment at temperatures between 300 and 900 °C and the formation of gas, solid, and liquid products as a result of this decomposition [11, 12]. While at low temperatures, there is a tendency to produce wood charcoal, the tendency to form gaseous products increases as temperature increases. Pyrolysis is a kind of hydrocarbon recycling process in which solid, liquid, and gaseous products can be obtained [13]. It is important to separate pyrolysis from gasification. Gasification breaks down biomass with synthesis by carefully controlling the amount of oxygen present [14]. Pyrolysis has advantages compared to direct burning such as low transportation and storage costs and production of high energy-density products, and these advantages increase the usage area [15].

The main products of pyrolysis are pyrolysis gas, semicoke, and oil. Gas and semicoke can be removed or burned as fuel. Oil can be used as fuel or as a raw material for the chemical industry.

Pyrolysis of wastes can also be considered highly environmentally friendly in terms of heat and power generation in turbines, obtaining gas and fuel for turbines, green energy, electricity production, and raw material production.

Potential of Agricultural Wastes

The agricultural sector has a great importance for Turkey in terms of providing input to industry, improving export, and creating new job opportunities. Although the sector is economically and environmentally so important, agricultural waste cannot be fully utilized in Turkey due to disposal, collection, transportation, and labor costs. Agricultural wastes remaining in the field after harvest in Turkey are being left to decay or are burned by farmers. Energy demand is met by imports in Turkey instead of recycling waste and waste resources and developing new technologies for this purpose. This energy import has a negative contribution to the country's economy [3–16].

With the use of biomass energy, low-cost and environmentally less harmful production which is the main basic of renewable energy production is provided with the best utility and higher quality. Advantages of biomass and energy production from agricultural wastes are known as the increase in crop production, the increase in the economic and environmental impact of rural areas, and the increase in social welfare in rural areas. Biomass energy has also importance in terms of being one of the best ways to obtain energy in our agriculturally rich country [8].

Our country has agriculturally very fertile regions such as Central Anatolia, Southeastern Anatolia, and Çukurova. After the production of cereals and seeds obtained from these areas, the remaining parts of the plants are burnt or are remained under the soil. Our country is located in a region which has a significant agricultural waste potential such as wheat, cereal stalks, barley, hemp stalks, and other agricultural wastes. Figure 1 displays the main amounts of agricultural wastes in our country [17].

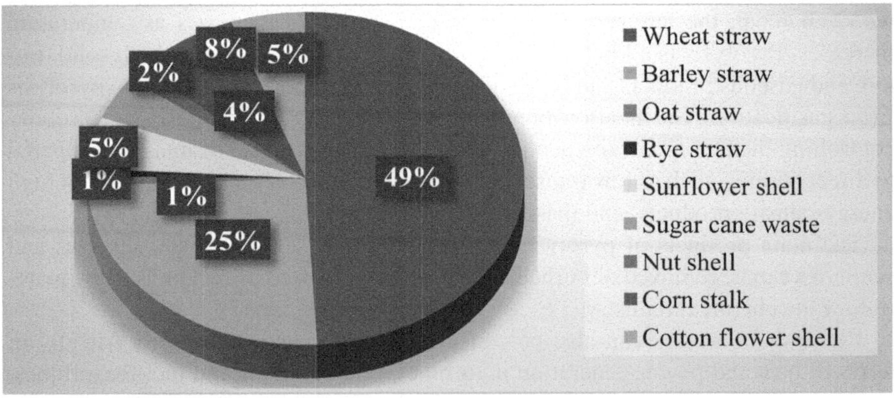

Fig. 1 The main agricultural waste amounts in Turkey [17]

Potential of Forest Wastes

Forest waste is an important biomass resource to produce energy due to its chemical structure (Figs. 2 and 3). Turkey has a great chance to produce bioenergy as an alternative energy resource to improve national economy with its approximately 7×10^6 m^3 biomass potential [18].

A significant portion of the total annual energy available from biomass in the world is produced from forest wastes. The potential of the forest and agricultural

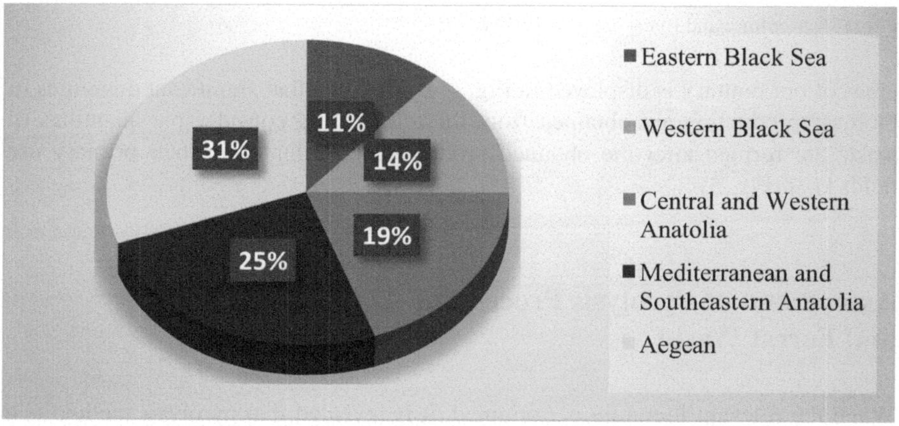

Fig. 2 Distribution of biogas energy potential of agricultural wastes according to geographical regions of Turkey [19]

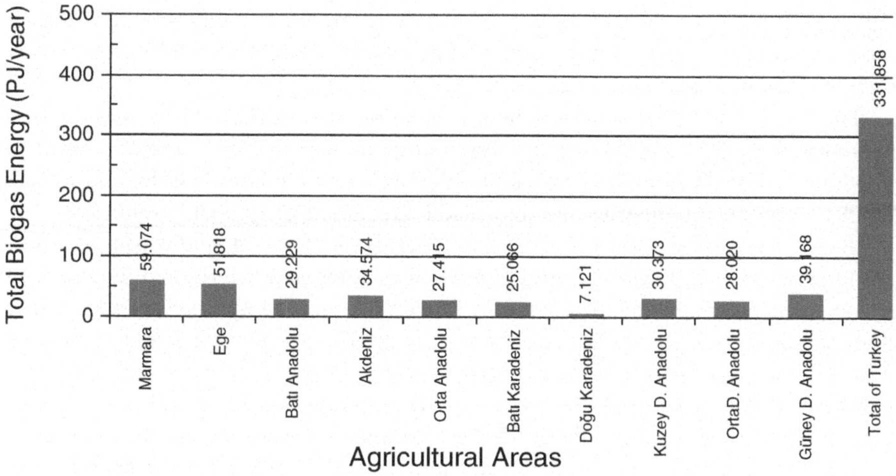

Fig. 3 Annual total biogas energy values that can be obtained from animal manure and grain stalk wastes depending on Turkey's agricultural regions [18]

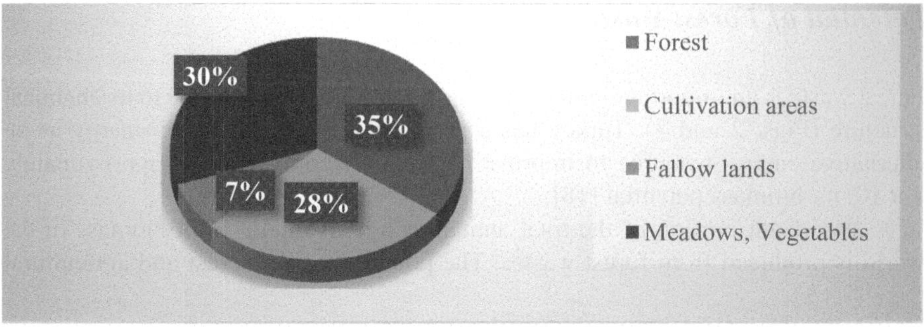

Fig. 4 Agriculture and forest areas of Turkey [17]

areas of our country is displayed in Fig. 4. It is known that significant quantities of biomass production are obtained from these areas and considerable quantities of waste are formed after the obtained products are evaluated in their primary use fields [17] (Fig. 5).

Assessment of Pyrolysis Process for Agricultural Waste and Forest Waste

When the relevant literature is examined, it is revealed that pyrolysis method is a frequently used method in the management of agricultural and forest wastes. There are many studies related to these management alternatives.

In the management of agricultural wastes by pyrolysis method, Önal and Pütün [20] employed rapid pyrolysis on tobacco field wastes and aimed to produce artificial fuel. Pyrolysis process was carried out at different temperatures and different nitrogen gas flow rates, and the effect of pyrolysis temperature and nitrogen flow rate was determined on the yield of pyrolysis products.

The highest liquid product efficiency was obtained as 32.63% by mass with a heating rate of 500 °C/min at a pyrolysis temperature of 550 °C and at a nitrogen flow rate of 200 mL/min. It has been revealed that the final liquid product may be an input for chemical industries and may also be regarded as synthetic liquid fuel.

In another study, Poyraz [21] performed pyrolysis of cotton field wastes at different temperatures and different nitrogen gas flow rates and determined the effect of pyrolysis temperature and nitrogen flow rate on the yield of pyrolysis products. The highest liquid product yield was achieved as 26.0% at a pyrolysis temperature of 550 °C and a nitrogen flow rate of 100 mL/min. It was concluded that this slow pyrolysis (5 °C/min) provides optimum pyrolysis conditions for the average 1.33 mm particle size of the cotton stalks at 550 °C, 100 mL/min nitrogen gas flow rate.

Aboagye et al. [22] determined the rapid pyrolysis of the orange shell wastes from the fruit processing industries in Ghana was sufficient to meet the expected fuel demand from 2020 to 2030 and calculated the potential yield of biofuels. The

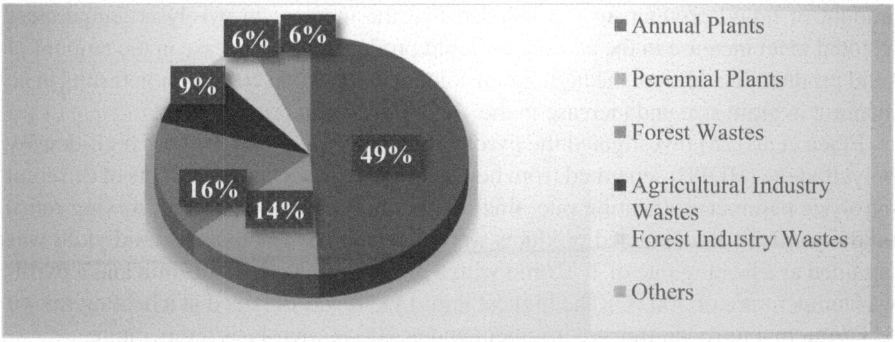

Fig. 5 Annual biomass potential of Turkey [17]

study found that these wastes would cover a large part of the country's fuel needs for 2020 and 2030.

Soysa et al. [23] pyrolyzed the Ceylon waste tea in order to produce biological crude oil. Pyrolysis was carried out at various pyrolysis reaction temperatures, and optimum mass and energy yields were calculated for biological crude oil production. While the reaction temperature was 500 °C, the biological crude oil has had the highest energy efficiency with the highest thermal value of 3725 kcal/m^3 and 36.8% water content.

Lee et al. [24] aimed at energy recovery and biofuel production by pyrolyzing agricultural wastes that use CO_2. It was focused on N_2 and CO_2 characterization of produced biochar. In the pyrolysis of the red pepper stalk, CO_2 has caused the production of synthesis to be increased, thus reducing tar production.

Hawash et al. [25] conducted pyrolysis of some different agricultural wastes in their study and examined the formation quantities, the content, and the quality of each produced liquid and solid products. Some agricultural wastes, jatropha seed, cake, olive oil, peach kernel, and Spirulina microalga species, were used as raw materials. Experimental results show that the jatropha seeds have the highest biofuel weight percentage and highest calorific value and that the history stone produces the highest solid product percentage with the highest calorific value.

Kung et al. [26] examined the feasibility and application of pyrolysis and biochar applications in China to understand the issues of bioenergy production, agricultural cost savings, and atmospheric quality enhancement. A great number of agricultural and animal wastes have been analyzed and quantitative measures have been provided for economic and environmental benefits. The results show that pyrolysis yields of both slow and rapid pyrolyzed agricultural wastes provide the greatest economic benefit, but animal wastes reduce greenhouse gas emissions.

When the studies about forest waste management by pyrolysis method were examined, it is seen that Yılgın et al. [27] investigated the flash pyrolysis of wood species such as oak, pine, and poplar at different temperatures. Also the effects of temperature on grain size and wood type were analyzed on the yield of obtained liquid and other products. As a result, the liquid product can be obtained from the pyrolysis of the wood, and it is possible to select a sudden heating and short residence time to increase the amount of liquid product. It is concluded that the increase in pyrolysis temperature resulted in an increase in the amount of liquid product and a decrease in the amount of solid product. In addition, the high lignin content in the wood composition results in an increase in grain size and increase in the yield of the solid product.

Erşen et al. [28] investigated the pyrolysis of pine wood powder with high-density polyethylene (HDPE) separated from household plastic waste. The effects of different pyrolysis parameters (heating rate, final temperature, sweeping speed, mixing ratio) on the yields of the obtained products were determined. The highest solid yield was obtained at a heating rate of 1 °C/min with a nitrogen flow of 10 mL/min and a pyrolysis temperature of 500 °C. The highest liquid yield was obtained at a heating rate of 5 °C/min in a nitrogen free environment and at 700 °C pyrolysis temperature.

Özçifçi et al. [29] evaluated the yellow pine wood flour as a biomass source and investigated the effects of parameters such as temperature and catalyst type on the yield of pyrolysis products in the experiments. As a result, it has been determined that temperature and catalyst type are effective parameters in solid, liquid, and gas product yields. It has been ascertained that the efficiency of liquid and gaseous products increases with the increase in temperature and the yield of solid product decreases and all the catalysts used decrease the yield of liquid product.

In another study, Duranay and Yılgın [30] examined the effect of different biomass samples (orange shell, apple bark, poplar wood, poplar wood shavings, and walnut shells) on pyrolysis product yields. Biomass type has been found to be effective on pyrolysis product yield. It has been found that the yield of the liquid product is high in the wood pellet and product of the poplar tree leaf having high ash content.

Demiral et al. [31] have performed Cr(VI) removal by adsorption from aqueous solutions using activated carbon of the olive residue pyrolysis solid product obtained by water vapor activation method at 850 °C. As a result of the experimental studies, Cr(VI) adsorption was maximum when the initial pH value was 1.5. When the results obtained were evaluated, it was found that the activated carbon obtained from solid product of olive residue pyrolysis was an effective adsorbent for Cr(VI) removal from aqueous solutions.

Solar et al. [32] investigated the effect of temperature and residence time on the quality of wood obtained in the pyrolysis of biomass waste. The peak temperature had an effect on the yield and composition of the pyrolysis fractions but greater variations in yield and composition in pyrolysis products were observed at 900 °C. The higher peak temperature resulted in the higher quality of wood charcoal, but the pyrolysis solid product yield decreased.

On the other hand, Carrasco et al. [33] have demonstrated that as a result of technological economic analysis of liquid fuels produced from forest wastes, this waste can be used as feedstock for pyrolysis to obtain composition of hog fuel (sawdust, bark, and secondary woody residue produced from milled by-products such as sawdust) and product yields.

The biomass is converted to bio-oil, solid residue, and gas with 17% energy requirement in pyrolysis reactor with 61%, 24%, and 15% efficiency, respectively. The final mass yield of gasoline/diesel hydrocarbons was 16% with 40% energy yield based on dry biomass, and this yield represented a fuel production of 51.9 gallons per dry metric ton of raw material.

	Operational conditions			Results					Products (%)			
	Study	Aim	Method	Reactor type	Temperature (°C)	Inert gas flow rate (mL/min)	Particle size (mm)	Heating rate (°C/min)	Solid	Liquid	Gas	Cal. value (kcal/m)
Agricultural waste	Duranay and Yılgın [30]	Fuel	Pyrolysis	Fixed bed	500	100	13	–	25–39	15–60	15–65	–
	Demiral et al. [31]	Cr(VI) removal	Pyrolysis	Fixed bed	500	–	0.224–0.6	50	–	–	–	–
	Poyraz [21]	Waste recycling	Slow pyrolysis	Fixed bed	400, 500, 550, 700	50, 100,200	1.33	5	–	26.0	–	5089
	Önal and Pütün [20]	Waste recycling Fuel	Fast pyrolysis	Fixed bed	400, 500, 550, 700	50, 100, 200, 400	0.425–0.850	500	33.5	32.6	–	7485
	Hawash et al. [25]	Waste recycling	Pyrolysis	Fixed bed	500	–	1–2	–	43	24.2	–	9076
	Lee et al. [24]	Waste recycling	Pyrolysis	Fixed bed	220–400	–	3	500	21.5	31.8	46.8	–
	Soysa et al. [23]	Waste recycling Fuel	Pyrolysis	Fluidized bed	450, 500, 550, 600	–	0.85–2	–	38.5	33.3	–	3726
	Aboagye et al. [22]	Waste recycling Fuel	Fast pyrolysis	Fixed bed	–	–	–	–	–	–	–	4382
	Kung et al. [26]	Waste recycling	Fast and slow pyrolysis	Fixed bed	–	–	–	–	15–35	30–70	13–35	–

	Operational conditions			Results					Products (%)			
	Study	Aim	Method	Reactor type	Temperature (°C)	Inert gas flow rate (mL/min)	Particle size (mm)	Heating rate (°C/min)	Solid	Liquid	Gas	Cal. value (kcal/m)
Forest waste	Yılgın et al. [27]	Fuel Raw material	Flash pyrolysis	Fixed bed	400, 500, and 600	100	8, 10, 12	–	26–32	62–66	24–29	–
	Erşen and Pehlivan [28]	Fuel Raw material	Flash pyrolysis	Fixed bed	500, 600, and 700	0, 10, 20	13	1, 5, 10, and flash	10.3	74	30	–
	Özçiftçi and Özbay [29]	Waste recycling	Pyrolysis	Fixed bed	400, 500, and 600	30	0.850–1.60	5	39.8	56.4	29.6	–
	Carrasco et al. [33]	Technoeco analysis	Pyrolysis	Fluidized bed	500	105,000	5	–	24	61	15	4370
	Solar et al. [32]	Waste recycling	Pyrolysis	Pressurized—Continuous	300–900	–	0.6	–	19–31	5–22	47	7714

Conclusions

In this study, the applicability of pyrolysis of organic wastes and product efficiency have been researched and compiled in the light of relevant literature. The compiled results are given below.

- In the studies on the evaluation of agricultural wastes, pyrolysis process has been carried out in the temperature range of 220–700 °C, and it has been reported that the temperature significantly affects the pyrolysis efficiency. It has been also noted that the increase in temperature increases yield of liquid and gaseous products and decreases the solid residue amount.
- The flow rate of the inert N_2 gas is ranged from 50 to 400 mL/min in literature data, and the increase in flow rate increases the yield of liquid and gaseous products and decreases the amount of solid product.
- Particle sizes have been reported between 0.225 and 13 mm in relevant studies, and it has been assessed that the higher efficiency is observed when the particle size is smaller.
- Calorific values of agricultural wastes vary between 3580 and 9080 kcal/kg. The utilization of agricultural wastes with high calorific value by pyrolysis method can be considered as a quite efficient waste management alternative.
- Pyrolysis process has been carried out in the temperature range of 300–900 °C for forest wastes, and it has been reported that the pyrolysis around 900 °C yields the highest efficiency. The yield of liquid and gaseous products is increased with the increase in temperature, and the amount of solid residue is very small.
- Particle size affects the efficiency of pyrolysis process. Thermal degradation of organic materials with particle size between 0.850 and 13 mm may achieve higher calorific values up to 7714 kcal/m^3.
- There are also studies where the flow rate of the inert N_2 gas is chosen to be 105,000 mL/min. As the flow rate increases, the yield of liquid and gaseous products increases and the amount of solid product decreases.
- Products with economical value can be obtained by the pyrolysis process from the organic matter and/or wastes, and it is an effective method in reaching the targets of "minimization and reduction" in waste management.
- The operational parameters such as temperature, heating rate, agent gas flow, particle size, and reactor type have significant effects on pyrolysis process. Therefore, the need for future studies on identification of system performances depending on changing conditions of each parameter is inevitable in order to ensure widespread use of pyrolysis as a waste management alternative.

Discussions and Recommendations

The energy demand has also increased with the growth of population and technology in the world. Turkey is one of the countries experiencing the energy demand problem and importing a large part of its energy. A large part of this need will be met in economically and environmentally friendly way by the use and development of new and renewable energy sources. These renewable energy sources will also benefit globally by reducing global warming, desertification, greenhouse effect, and the disappearance of endemic plant species.

In this study, recycling of agriculture and forest wastes by pyrolysis was investigated based on the literature, and a number of studies about the use of pyrolysis process for organic wastes such as agriculture and forest wastes have been collected and presented in a common basis. The methods used in the selected studies are examined and given in the form of tables under the common headings in order to be easily accessible by different researchers. It has been generally agreed that the pyrolysis process, which is considered as an alternative waste management method, is an energy and environmentally friendly application for waste and that this application can be used to reduce waste and recycle waste.

The methods followed by selected studies were reviewed under common headings and given in tabular forms in order to be clear and comprehensible to the researchers. A general point of view has been reached that the pyrolysis process, which has been accepted as an alternative waste management method, is an environmentally friendly process for wastes and that this application can be used for waste reduction and waste utilization. Use of energy carriers such as biomass as a renewable energy source to produce energy and valuable by-products will help in the management of organic wastes and will be an environmentally benign processing and sustainability approach for the future.

References

1. Tepecik M, Ongun AR (2017) Lead, cobalt and cadmium content of corn plant and effect on garbage compost mass. J Tekirdag Agric Fac 14(2):1–8
2. Ersoy AD (2017) Turkey's animal manure greenhouse gas emissions status and potential of biogas energy. Master Thesis, Department of Environmental Engineering, Institute of Science and Technology, Hacettepe University
3. Diken B (2017) A research on the determination of the gasification performance of grass pellets. Master Thesis, Biosystems Engineering, Institute of Science and Technology, Namik Kemal University
4. Kılıç F (2011) Biogas, importance, location and general condition in Turkey. Eng Mach 52(617):94–106
5. Deviren H, İlkılıc C, Aydın S (2017) Materials used in biogas production and usage areas of biogas. Batman Univ J Life Sci 7(2/2)
6. Saraçoğlu S (2017) As the world of renewable energy sources biomass production and availability in Turkey. Fiscaoeconomia 1(3):126–155
7. Demir M (2017) Potential and usability of biomass energy in Kars. Turk J Geogr 68:31–41

8. Kurt G, Koçer N (2010) Biomass potential and energy production in Malatya. Erciyes Univ Sci Inst J 26(3):240–247
9. Topal M, Arslan I (2008) Biomass energy and Turkey. VII. National clean energy symposium, UTES'2008, İstanbul, 17–19 Dec 2008
10. Tınmaz Köse E (2017) Determination of biogas energy potential of animal fertilizers in Trakya region and creating digital maps. Pamukkale Univ Eng Sci J 23(6):762–772
11. http://www.csb.gov.tr/db/ced/editordosya/OTL%20NAYLON%20PLASTIK%20VE%20POLIMER%20ATIKLARIN%20GERI%20KAZANIMI.pdf. Accessed 21 Jan 2018
12. Öztürk K (2012) Biogas technology, production - use - project, 1st edn. Umuttepe Publications, Kocaeli, p 268
13. Önal E (2007) Pyrolysis of different biomass and their synthesis with synthetic polymers, determination of product properties obtained. Ph.D. thesis, Institute of Science, Anadolu University, Eskişehir, p 153
14. Mohan D, Pittman CU, Steele PH (2006) Pyrolysis of wood/biomass for bio-oil: a critical review. Energy Fuel 20(3):848–889
15. Gürleyik E (2006) Investigation of Pyrolysis and Products Produced by Oil Seeds of Fossil Resources. Yüksek Lisans Tezi, Anadolu University Fen Bilimleri Enstitüsü, Eskişehir, p 144
16. Acaroğlu M, Hacıseferoğulları H (2018) New test procedures for briquetting and pelletizing in biomass energy. Winter School Biomass Energy Technologies, İzmir, 5–16 Feb 2018
17. Yılmaz M (2012) The energy potential of Turkey and its importance of renewable energy sources in terms of electricity production. Ankara Univ J Environ Sci 4(2):33–54
18. Aybek A, Üçok S, İspir MA, Bilgili ME (2015) Digital mapping and determination of biogas energy potential of usable animal manure and cereal straw wastes in Turkey. J Tekirdag Agric Fac 12(3):109–120. ISSN: 1302-7050
19. Türkoğlu T, Gökoğlu C (2017) Determination of fuel properties of wood pellets made from red pine Forest harvest residues. Süleyman Demirel Univ J Nat Appl Sci 21(1):58–63
20. Önal E, Pütün AE (2009) Rapid pyrolysis of tobacco field wastes. Anadolu Univ J Sci Technol 10(2):557–564
21. Poyraz Z (2012) Pyrolysis of cotton field wastes. DPÜ Journal of Institute of Science, Thesis, p 28
22. Aboagye D, Banadda N, Kiggundu N, Kabenge I (2017) Assessment of orange peel waste availability in Ghana and potential bio-oil yield using fast pyrolysis. Renew Sust Energ Rev 70:814–821
23. Soysa R, Yeon Seok C, Seock Joon K, Sang Kyu C (2016) Fast pyrolysis characteristics and kinetic study of Ceylon tea waste. Int J Hydrog Energy 41:16436–16443
24. Lee J, Xiao Y, Seong-Heon C, Jae-Kon K, Sang Soo L, Daniel CW, Tsang Yong S, Eilhann EK (2017) Pyrolysis process of agricultural waste using CO_2 for waste management, energy recovery, and biochar fabrication. Appl Energy 185:214–222
25. Hawash SI, Farah JY, El-Diwani G (2016) Pyrolysis of agriculture wastes for bio-oil and char production. J Anal Appl Pyrolysis 124:367–372
26. Kung C, Fanbin K, Yongrok C (2015) Pyrolysis and biochar potential using crop residues and agricultural wastes in China. Ecol Indic 51:139–145
27. Yılgın M, Deveci Duranay N, Pehlivan D (2005) Flash pyrolysis of wood. Paper presented at III. Yenilenebilir Enerji Kaynakları Sempozyumu ve Sergisi, Mersin, 19 Oct 2005
28. Erşen T, Pehlivan D (2011) Combined pyrolysis of high density polyethylene-wood mixtures. J Gazi Univ Eng Arch Fac 26(3):607–612
29. Özçifçi A, Özbay G (2013) Production of biofuels by catalytic pyrolysis method in furniture industry wastes. Gazi Univ J Eng Arch 28(3):473–479
30. Duranay Deveci N, Yılgın M (2006) Pyrolysis of different biomass species in fixed bed. Seventh national chemical engineering congress, Anadolu University, Eskişehir, 5–8 Sept 2006

31. Demiral H, Demiral İ, Tümsek F (2006) Removal of Cr (VI) from activated carbon from olive crack pyrolysis solid product. Seventh national chemical engineering congress, Anadolu University, Eskişehir, 5–8 Sept 2006
32. Solar J, Marco I, Caballero BM, Lopez-Urionabarrenechea A, Rodriguez N, Agirre I, Adrados A (2016) Influence of temperature and residence time in the pyrolysis of woody biomass waste in a continuous screw reactor. Biomass Bioenergy 95:416–423
33. Carrasco JL, Sampath G, Akwasi AB, Charles AM, William JD (2017) Pyrolysis of forest residues: an approach to techno-economics for bio-fuel production. Fuel 193:477–484

Part III
Renewable Energy

Excess Pressure in Municipal Water Supply Systems as a Renewable Energy Source: Antalya Case Study

Ayse Muhammetoğlu and Habib Muhammetoğlu

Abstract Pressure management is one of the most effective methods for water loss reduction. Pump used in turbine mode (pump as turbine, PAT) is a viable option for reducing excess water pressure, water losses, and pipe failures in water distribution networks in addition to energy production at microscale. In this study, a PAT system recently installed in Antalya City of Turkey is presented, and its performance is evaluated for its initial operational period of approximately 5 months. This full-scale PAT system was implemented in a parallel pipeline with a pressure-reducing valve (PRV). The operation of the PAT system was continuously monitored online for flow rate, power, and pressure levels. The PAT system proved to work efficiently in a wide range of inflows (130–300 m^3/h) where the produced energy varied between 0.7 and 8.4 kWh for a reduction of approximately 1 bar pressure head with an average efficiency of 60%. Environmental benefits of green energy production, reduction in physical water losses, and carbon dioxide emissions were evaluated. The payback period of the PAT application was computed as 53 days based on the costs and revenues of the PAT application.

Keywords Energy recovery · Water supply · Urban network · Pump

Introduction

Safe water supply is among the priority tasks of municipalities and water utilities. However, sustainability of water supply systems (WSSs) is an important issue besides continuous and hygienic supply of water. Within the context of sustainable WSS, improvement of energy efficiency, reduction of energy, and water losses in addition to reduction of greenhouse gas (GHG) emissions are of great concern. High amounts of energy are consumed and GHG emissions are encountered during

A. Muhammetoğlu (✉) · H. Muhammetoğlu
Department of Environmental Engineering, Akdeniz University, Antalya, Turkey
e-mail: aysemuh@akdeniz.edu.tr; muhammetoglu@akdeniz.edu.tr

N. Balkaya, S. Guneysu (eds.), *Recycling and Reuse Approaches for Better Sustainability*, Environmental Science and Engineering,
https://doi.org/10.1007/978-3-319-95888-0_10

water abstraction, treatment, transmission, and distribution stages. Pressure in water pipes is one of the main design parameters for WSSs where it is controlled by upper and lower limits. Excess pressure usually occurs during water transmission from high elevations of water resources to low elevations, and this high water pressure has to be reduced. Excess pressure at transmission lines is commonly reduced using traditional techniques including break pressure tanks, water storage tanks, or pressure-reducing valves (PRVs). However, a more sustainable technique of energy recovery from excess pressure is possible by using turbines. Applications of turbines for energy production at water transmission lines are available at many countries such as the United States, Canada, Honduras, Switzerland, Austria, and Italy.

One of the main factors causing leakage and pipe failure in water distribution networks (WDNs) is the excess pressure. An increase in the water pressure always implies an increase of leakage from the holes and cracks of the network. Consequently, pressure management is the most important and commonly applied action for physical water loss reduction [1–4]. Commonly, the excess pressure in WDNs is controlled and reduced to optimum operational levels by the use of PRVs at selected locations without any power generation. However, both pressure reduction and energy production from the available excess water pressure could be achieved by the use of turbines or pumps as turbines (PATs). In the literature, mainly two alternative schemes were proposed for hydropower generation in WDNs: only PAT [5] or combined PAT and PRV [6–8]. The PAT systems mainly include a centrifugal water pump, operated in reverse mode, and a synchronous electrical generator with constant rotational speed to produce energy. There are several advantages for using PATs instead of turbines: (1) PATs are usually cheaper than turbines, (2) turbines are usually designed specifically for application sites, (3) PATs are usually maintained easily from the market for different power and flow rate requirements, and (4) spare parts of PATs are easily available. The main reported disadvantage of PATs is the lower efficiencies compared with turbines. PATs contribute to water saving in WDNs by reducing leakage due to pressure reduction, besides energy production and reduction of GHG emissions [6]. In this respect, energy production in WDNs is widely discussed in the literature [9–12]. Energy production potential in WDNs is usually at microscale (<100 kW). Due to their low operational and capital costs, use of PATs is commonly advised [13, 14].

Due to the increasing global awareness on the adverse impacts of high energy consumption and GHG emissions on climate change, all sectors are usually forced to reduce their energy consumptions and GHG emissions. Metropolitan cities have priorities to accomplish sustainable WSSs and to take precautions to limit the GHG emissions [15]. Consequently, optimization of energy in water supply sector has gained an increasing interest [16], and several research studies were conducted to evaluate the technical and economic feasibilities for energy recovery in WSSs. Some of the presented cases for leakage reduction and energy production indicate good opportunities for actual implementations with relatively short payback periods and environmental benefits [17, 18]. As an example, the PAT systems at pico scale (<4 kW) were recommended as viable options with payback periods of 4–22 months [19]. Turbines generally have very high efficiency levels, while efficiency of the

PAT systems could be raised up to 85% [13]. One of the main challenges related to the PAT systems is the variations in the flow rates where the systems are claimed to be sensitive to medium and high flow rate variations as discussed by several researchers [5, 6, 19, 20]. Theoretical, experimental, and numerical investigations are presented for possible applications of micro-hydropower and PAT systems. Moreover, several research studies are focused on the design of the PATs; analysis of cavitation, force, and efficiency; comparison of the conventional turbines with the PATs; and cost analysis for evaluating economic feasibility [14, 21]. To the best of authors' knowledge, there is no published full-scale application of the PAT systems in WDNs for water loss reduction and energy recovery [7]. The following sections present design details, installation scheme, and initial operational results of a full-scale PAT system installed at Antalya City of Turkey. This innovative system was installed on a bypass line and in parallel with a PRV. The performance of the installed PAT system was continuously monitored by online measurement equipment for flow rate, power, and pressure levels at upstream and downstream of the PAT. The PAT system proved to work effectively in a wide range of inflows (130–300 m^3/h), while the produced energy varied between 0.7 and 8.4 kWh for the reduction of approximately 1 bar pressure head.

The Case Study

Site Selection

The full-scale PAT system was implemented at a pilot study area (PSA), namely, *Aksu* district WDN of Antalya Metropolitan City, located at the south of Turkey. The PSA was selected by close cooperation with Antalya Water and Wastewater Authority (ASAT) and was evaluated to have excess water pressure and frequent pipe bursts. The WDN data sets such as elevations, coordinates, lengths, diameters, types, and ages of the pipes besides locations of service connections and valves were obtained from ASAT and verified on site. The main components of the PSA with digital elevation model (DEM) are depicted in Fig. 1.

The population of the study area is approximately 25,000 people, while the total length of the WDN is approximately 80 km. A supervisory control and data acquisition (SCADA) station is located at the entrance to the PSA where continuous online measurements of flow rate and pressure are realized and the measured data sets are recorded at the headquarters of ASAT. The existing water pressures at the entrance to the PSA were always in the range of 3–5 bars. The measured flow rates in 2015 demonstrated a wide variation where the minimum, the mean, and the maximum values were 130 m^3/h, 252 m^3/h, and 552 m^3/h, respectively. The occurrence of the maximum flow rate was a very rare condition that lasted shortly due to pipe breaks. Additionally, the minimum flow rate occurred for very short time periods, which was mainly due to some interventions in the WDN for maintenance

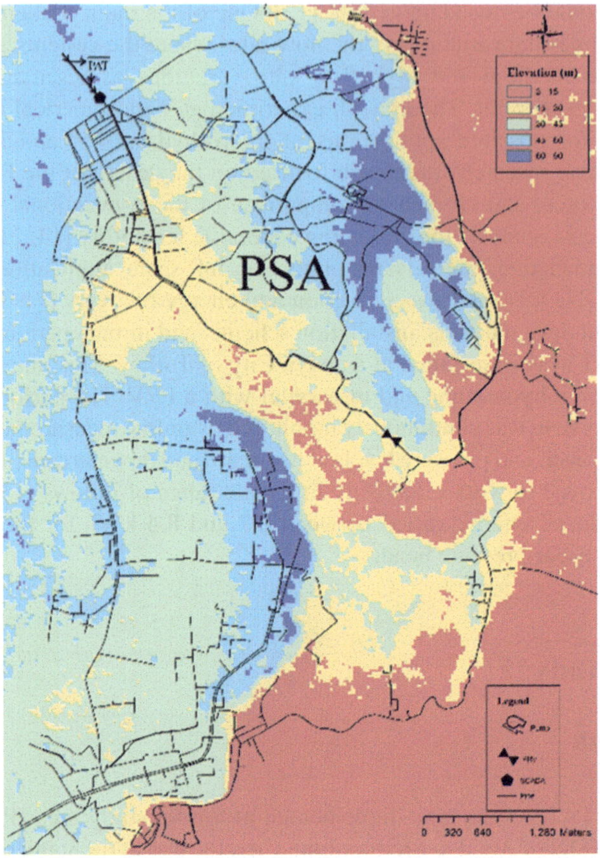

Fig. 1 Main components of the pilot study area (PSA) with digital elevation model [8]

and repair. The topography of the PSA exhibits wide spatial variations, between 3 and 90 m above the mean sea level, and this situation causes high variations in the pressure levels in the WDN. Due to the high variation of topography in the PSA, a booster pump is operated mostly in summer months to supply water to high elevations whereas a PRV is continuously operated to reduce excess water pressure at lower elevations. The excess water pressure level to be reduced by the PAT system was determined approximately as 1 bar as obtained by the results of the hydraulic modeling study carried out for the PSA besides the technical advice of ASAT [7].

Initial Setup and Test Operations at Lab

The initial setup and test operations were accomplished at the research and development laboratory (R&D lab) of the manufacturer factory of the PAT (Fig. 2). The PAT system was designed in parallel with a PRV to maintain the continuous and constant reduction of approximately 1 bar water pressure for this specific application.

Fig. 2 The initial setup and test operations at the research and development laboratory

Various flow rates and inlet pressure levels were tested for about 3 months. Consequently, the head, power, and efficiency curves of the PAT were prepared (Fig. 3). Based on the flow rate data sets of the PSA in Antalya, the PAT system was designed to operate between 130 and 330 m^3/h. According to the R&D lab tests, the PAT system could generate 0.2 kWh energy at 128 m^3/h flow rate and 6.2 m net head, while it could generate 9.6 kWh energy at 308 m^3/h flow rate and 17.9 m net head. The PAT system was designed and tested to operate at highly varying inflows and pressure levels.

Field Installation and Operation of PAT

The full-scale installation of the PAT system was designed on a separate bypass line of the main pipe where the total flow rate could be directed to the bypass line using the existing control valve on the main pipe (Fig. 4).

Two additional isolation valves were installed at the inlet and the outlet of the bypass line in order to stop operation of the PAT system when needed. The cross-sectional view of the bypass line for the installed PAT system and the PRV is presented in Fig. 5. The pipe diameters are 650 mm for the main pipe and 250 mm for the bypass line. The installed PAT system, shown in Fig. 6, consists of the following equipment: a PAT (10 kW capacity), a synchronous alternator (11 kVA, 750 rpm), a flow modulated actuated valve, an electronic load controller, a PRV, online mea-

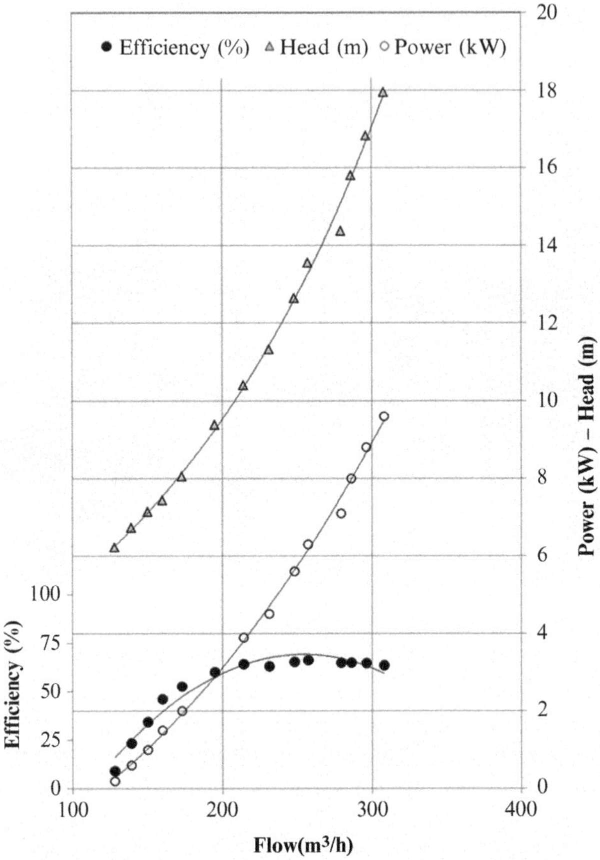

Fig. 3 Head, power, and efficiency curves of the PAT designed for the PSA [7]

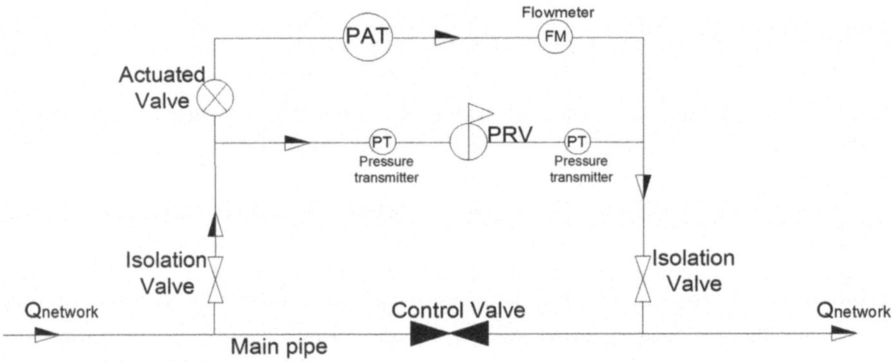

Fig. 4 Installation scheme of bypass line for the PAT and PRV [7]

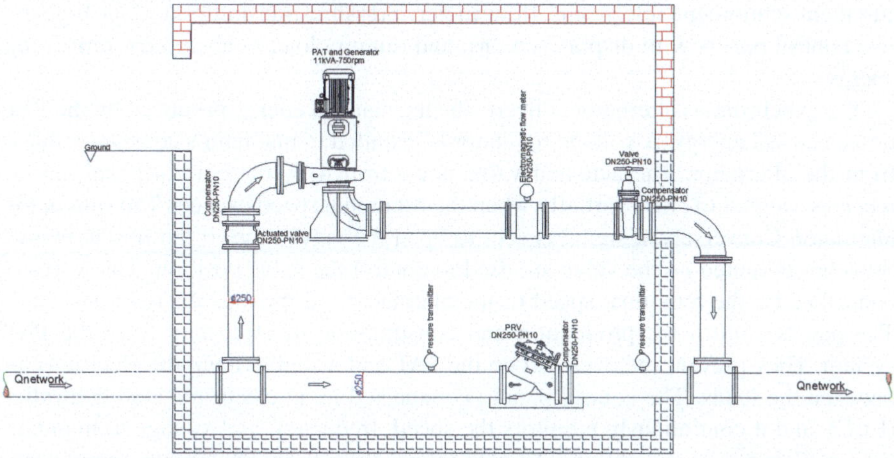

Fig. 5 Cross-sectional view of the bypass line for the installed PAT system and PRV [7]

Fig. 6 Photographs of the installed PAT system [7]

surement equipment (electronic flow meter, pressure transmitters), data loggers, two control panels with display screens, and dummy load to absorb the remaining energy.

The synchronous alternator converts the mechanical energy produced by the PAT into electrical energy. A voltage regulator is installed to maintain a constant voltage from the alternator. The actuated valve is controlled by the control panel, and it receives a signal to close partially when the rotational speed exceeds 750 rpm in the alternator. Conversely, a signal is sent to open the valve more if the inflow is low. Switches mounted on the valve are used to control the valve position. This valve is controlled by the rotational speed of the alternator and the total energy consumed. The panels control the production and consumption of electricity from the PAT system. They prevent the overload on the PAT and also distribute the electricity in between the loads. The control panel is connected to an electronic load controller (ELC), and it continuously monitors the speed, frequency, and voltage to maintain high-quality power generation. Detailed information about the system components were presented elsewhere [7]. The full-scale PAT system was in operation starting from January 26, 2016.

Energy Recovery Analysis

Generated power from hydroelectric turbines or PATs could be calculated using Eq. 1 where the notations are as follows: P, power (watt); Q, flow rate (m^3/s); ρ, density of water (kg/m^3); g, acceleration due to gravity (m/s^2); H, available head (excess pressure) (m); and e_0, total efficiency of power generation system. Total efficiency includes all losses from conversion of kinetic energy into mechanical energy such as turbine/PAT losses, energy conversion, and distribution [22].

$$P = Q \rho g H e_0 \tag{1}$$

The performance of the installed PAT system was continuously monitored by the online measurement of the flow rate, power production, and pressure levels at upstream and downstream of PAT. The installed system operated efficiently under the highly varying flow rates and pressure levels, and the automation system of PAT was able to manage with all the operational conditions. The continuous online flow rate and the pressure measurements from the SCADA station, located at the downstream of the PAT system, are presented in Fig. 7 for the period of December 17, 2015, till June 14, 2016 (1 month before the operation of the PAT in addition to about 5 months in operation). Within this period, there were even a few occasions when water was cut for some construction and maintenance works and the installed PAT system was automatically off-line and online without any personal intervention. There were no complaints from the water subscribers in the PSA regarding the reduction of the water pressure. The pressure values, measured continuously at the upstream and downstream of the PAT system, are presented in Fig. 8. Continuously measured hydropower production from the installed PAT system is presented in Fig. 9.

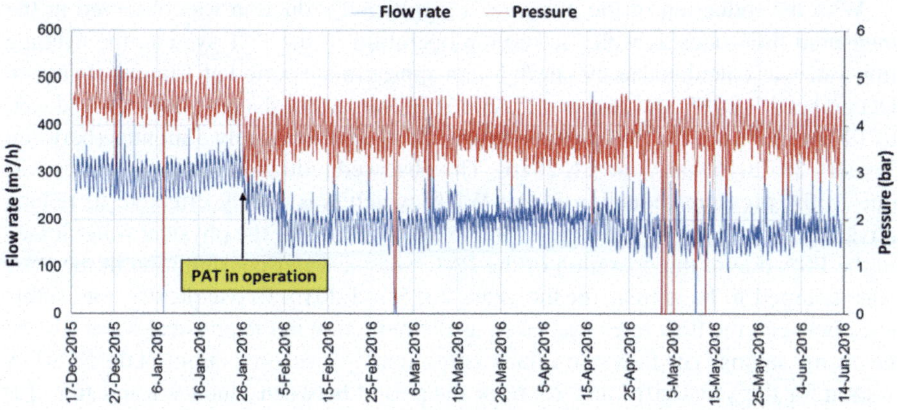

Fig. 7 Flow rate and pressure values at the PSA before and after the operation of PAT [8]

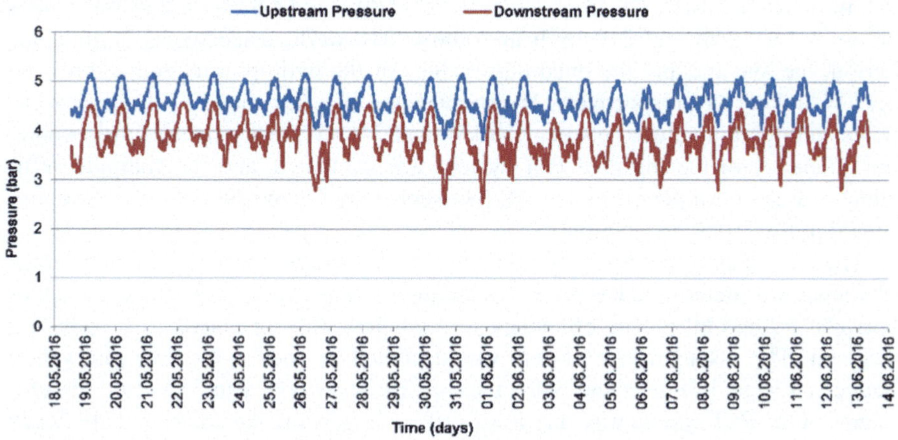

Fig. 8 Pressure values measured at upstream and downstream of PAT system [8]

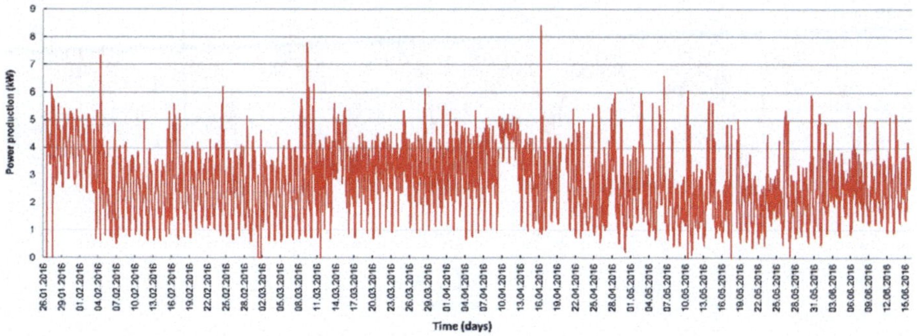

Fig. 9 Continuously measured hydropower production from the installed PAT system [8]

With the reduction of the pressure, a significant reduction was observed in the measured flow rates as well. Before the operation of the PAT system, the average flow rate was calculated as 293 m^3/h for an approximate period of 1 month (between December 17, 2015, and January 25, 2016). However, the average flow rate reduced to 190 m^3/h during the first operational period of approximately 5 months (between January 26, 2016, and June 14, 2016). The observed reduction in the flow rates was due to the pressure reduction in the WDN which was highly effective to reduce physical water losses. In order to estimate the reduction of the physical water losses in the PSA due to the operation of the PAT system, water consumption in the PSA was assumed to be similar for the years 2015 and 2016. Consequently, the difference between the flow rates was taken as the reduction in the physical water losses, i.e., water saving. The flow rate values, continuously measured online at the SCADA station for the years 2015 and 2016 for the period between January 1 and June 14, were compared. During this period, reduction in the physical water losses in the PSA was estimated at about 59 m^3/h. Within the same period, the average and maximum flow rates that passed through the installed PAT system were 169 m^3/h and 289 m^3/h, respectively. The frequencies were 90.89% and 4.83% for flow rate values between 130 m^3/h and 210 m^3/h and above 210 m^3/h, respectively. Within this period, the average and the maximum values of the hydropower production were 2.7 kW and 8.4 kW, respectively. The frequencies were 92% and 58% for the hydropower production values between 1-5 kW and 2-4 kW, respectively. Furthermore, the efficiency of the installed PAT system was evaluated, and the computed efficiency values for a period of 1 week (between May 25 and June 2, 2016) are presented in Fig. 10 as an example.

The efficiency of the PAT system demonstrated a high variation and the average efficiency was determined as 60%. The frequency was 77.14% for efficiency values between 50 and 80%. The efficiency levels below 40% or above 80% were less frequent. The installed PAT system was designed to operate for flow rate values between 130 to 330 m^3/h and when the flow rate was less than 130 m^3/h, the efficiency of the PAT system was very low (<10%). In general, the power and efficiency values obtained from the full-scale application of the PAT system were consistent

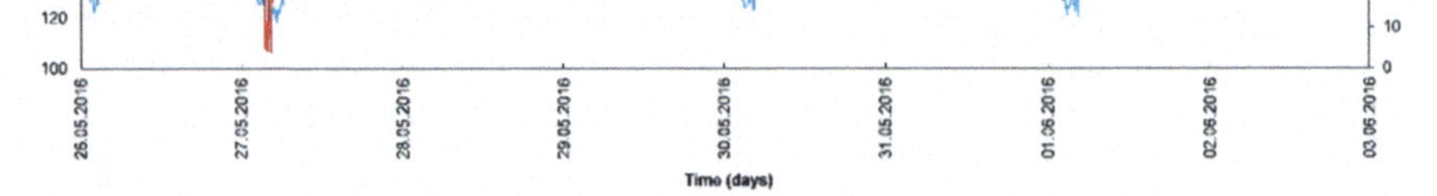

Fig. 10 Efficiency of the installed PAT system and its variation with flow rate [8]

with the characteristic curve of the PAT. The installed full-scale PAT system was observed to work efficiently and without any failure for highly varying operational conditions, with respect to flow rate, pressure, and even interruptions due to the maintenance and repair of the WDN.

Cost-Benefit Analysis

The implementation cost of the PAT system, with 10 kW capacity, includes electrical and mechanical equipment, construction costs for the underground bypass room and the aboveground protection house, and assembly and installation of all electrical and mechanical parts. The costs were presented in Table 1 for the main components, based on the real local prices in Turkey. Environmental benefits of the PAT system were evaluated in four major groups: (1) green energy production by the PAT system by reducing the excess pressure, (2) reduction in carbon dioxide (CO_2) emissions due to the green energy production, (3) water saving in the WDN due to the reduced pressure and the physical water losses, and (4) energy saving and CO_2 emission reductions from the transmission and distribution of the saved water. The reference value of CO_2 emission for energy production in Turkey was estimated at 0.53426 kg CO_2/kWh as an average value [23]. The revenues of the direct economic benefits were estimated/computed based on the local and national economic values. The electricity tariff for municipal use in Turkey was taken as 0.1269 Euro/kWh including all taxes. The energy consumption for transmission and distribution of the drinking water in Antalya City WSS was estimated at 0.67 kWh/m^3. The average water tariff of 0.50 Euro/m^3, applied by ASAT for the water subscribers in Antalya City, was considered. The payback period was calculated by dividing the total implementation cost by the total amount of revenues, as presented in Table 2. All the direct economic benefits and their revenues were evaluated for the period between February 1 and June 13, 2016, which covered 133 days of operation. The total revenue for this period was approximately 95,391 € or 717 €/day, and the payback period was computed as 53 days or 1.8 months. The considerably short payback period in this specific application was mainly due to the high revenues obtained from the water savings.

Table 1 Implementation cost of the installed full-scale PAT system [8]

Items	Cost (€)
Construction costs	8,450
Mechanical installation of PAT system	5,050
Cost of PAT system (generator, ELC panel, hydraulic unit, display panel, three data loggers, two pressure sensors)	15,680
Other mechanical equipment (PRV, valves, electromagnetic flow meter, all pipe fittings)	9,030
Total cost	38,210

Table 2 Environmental benefits and revenues of the installed PAT system [8]

Main groups	Environmental benefit	Revenue (€)
Energy production by the installed PAT system	8,599 kWh	1,091
Reduction of CO_2 emissions due to energy production of 8599 kWh	4,594 kg CO_2 reduction	–
Water saving due to pressure reduction by the installed PAT system	188,600 m^3	94,300
Energy saving from transmission and distribution of water (due to water saving of 188,600 m^3)	126,362 kWh	[a]
Reduction of CO_2 emissions from energy saving of 126,362 kWh	67,510 kg CO_2 reduction	–
Total revenues of environmental benefits for 133 days of operation	95,391 €	
Revenues from environmental benefits for each day of operation	717 €/day	
Payback period (days)	53	

[a]The revenue for energy saving from transmission and distribution of saved water (126,362 kWh) was already accounted in the water saving revenue

The installed PAT system proved to be efficient in reducing water pressure and leakage and producing energy. However, there are still some challenges related to the operation of the full-scale PAT system: (1) the efficiency of the PAT system reduces considerably with the reduced flow rate, (2) the optimum design of the PAT system depends on the levels of the flow rates and the pressure levels, and (3) Produced electricity from the PAT system needs to be consumed immediately on-site for operational control.

Conclusion

One of the challenging operational problems of WDNs is water loss which is mainly due to excess water pressure. Consequently, different types of PRVs are commonly installed at WDNs for pressure management to reduce excess pressure and water losses. However, PAT is an alternative system that improves efficiency of WDNs by reducing pressure and water losses besides producing energy at the same time. Theoretical PAT systems are widely discussed in the literature for their advantages, disadvantages, operational difficulties, and technical and economic feasibility compared with PRVs. In this study, implementation and operational performance of a full-scale PAT system in Antalya City WDN in Turkey were presented. To the best of authors' knowledge, there is no full-scale application of PAT systems in WDNs for water loss reduction and energy recovery [7]. The performance of the installed PAT system was continuously monitored by online measurement equipment for flow rate, power, and pressure levels. The PAT system, being in operation since January 2016, works efficiently in a wide range of inflows (130–300 m^3/h), while the produced energy varies between 0.7 and 8.4 kWh for a reduction of approximately 1 bar pressure head with an average efficiency of 60% and a payback period of 1.8 months. Implementation of PAT system leads to many further environmental benefits and operational advantages such as reducing CO_2 emissions and improvement of energy efficiency for sustainable management of WDNs; reducing number and frequency of pipe failures and bursts; reducing time, money, and efforts spent for maintenance and repair of pipes; improving water supply services by reducing the frequency of interventions; improving satisfaction of water subscribers; and also delaying the need for new water supply projects. However, there is a need for further studies to improve efficiency of PAT systems at fluctuating flow rates and to design PAT systems with a frequency regulator for operating at variable rotational speeds.

Acknowledgments This study was supported by the Scientific and Technological Research Council of Turkey (TUBITAK) (Project No.114Y203), Antalya Water and Wastewater Administration (ASAT), ALDAS Infrastructure Construction & Contracting Co., Standart Pompa ve Makina Sanayi Tic. A.S., and Akdeniz University. The authors would also like to thank members of the research team, Prof. Afsin Gungor and Dr. I. Ethem Karadirek from Akdeniz University and Cezmi Nursen from Standart Pompa ve Makina Sanayi Tic. A.S. Thanks to Asst. Prof. Ziyad Abunada for his valuable editing review and comments.

References

1. Araujo LS, Ramos H, Coelho ST (2006) Pressure control for leakage minimization in water distribution systems management. Water Resour Manag 20:133–149
2. Thornton J, Lambert A (2006) Managing pressures to reduce new break. Water 21 Magazine IWA Publishing 8(6):24–26
3. Karadirek IE, Kara S, Yılmaz G, Muhammetoglu A, Muhammetoglu H (2012) Implementation of hydraulic modelling for water-loss reduction through pressure management. Water Resour Manag 26:2555–2568
4. Vicente DJ, Garrote L, Sanchez R, Santilla D (2016) Pressure management in water distribution systems: current status, proposals, and future trends. ASCE J Water Resour Plann Manage 142(2). https://doi.org/10.1061/(ASCE)WR.1943-5452.0000589
5. Carravetta A, Del Giudice G, Fecarotta O, Ramos HM (2012) Energy production in water distribution networks: a PAT design strategy. Water Resour Manag 26:3947–3959
6. Fontana N, Giugni M, Portolano D (2012) Losses reduction and energy production in water-distribution networks. ASCE J Water Resour Plann Manage 138(3):237–244
7. Muhammetoglu A, Karadirek IE, Ozen O, Muhammetoglu H (2017) Full-scale PAT application for energy production & pressure reduction in a water distribution network. ASCE J Water Resour Plann Manage 143(8):04017040
8. Muhammetoglu A, Nursen C, Karadirek IE, Muhammetoglu H (2018) Evaluation of performance and environmental benefits of a full scale pump as turbine system in Antalya water distribution network. Water Sci Technol-Water Supply 18(1):130–141
9. Ramos H, Borga A (1999) Pumps as turbines: an unconventional solution to energy production. Urban Water 1(3):261–263
10. Giugni M, Fontana N, Portolano D (2009) Energy savings policy in water distribution networks. In: Proceedings of the International Conference on Renewable Energies and Power Quality (ICREPQ'09), Valencia, Spain

11. Samora I, Manso P, Franca MJ, Schleiss AJ, Ramos HM (2016) Energy recovery using micro-hydropower technology in water supply systems: the case study of the City of Fribourg. Water 8(8):344. https://doi.org/10.3390/w8080344
12. Samora I, Manso P, Franca MJ, Schleiss AJ, Ramos HM (2016) Opportunity and economic feasibility of inline microhydropower units in water supply networks. ASCE J Water Resour Plan Manage 142(11):04016052. https://doi.org/10.1061/(ASCE)WR.1943-5452.0000700
13. Ramos HM, Vieria F, Covas DIC (2010) Energy efficiency in a water supply system: energy consumption and CO_2 emission. Water Sci Eng 3(3):331–340
14. Jain SV, Patel RN (2014) Investigations on pump running in turbine mode: a review of the state-of-the-art. Renew Sust Energ Rev 30:841–868
15. Jenerette GD, Larsen L (2006) A global perspective on changing sustainable urban water supplies. Glob Planet Chang 50(3–4):202–211
16. Brown LE, Mitchell G, Holden J, Folkard A, Wright N (2010) Priority water research questions as determined by UK practitioners and policy makers. Sci Total Environ 409(2):256–266
17. Colombo A, Kleiner Y (2011) Energy recovery in water distribution systems using microturbines. NRC Construction, Canada, pp 1–9
18. Vilanova MRN, Balesteri JAP (2014) Hydropower recovery in water supply systems: models and case study. Energy Convers Manag 84:414–426
19. Lopes RE, Martinez CB (2006) Use of pumps as turbines for systems with water discharge. In: Proceedings of VISEREA- Seminário Iberoamericano sobre Sistemas de Abastecimento Urbano de Água, João Pessoa, Brazil
20. Caxaria GA, Sousa DM, Ramos HM (2011) Small-scale hydropower: generator analysis and optimization for water supply systems. In: Proceedings of World Renewable Energy Congress, Linköping, Sweden
21. Binama M, Su W, Li X, Li F, Wei X, An S (2017) Investigation on pump as turbine (PAT) technical aspects for micro hydropower schemes: a state-of-the-art review. Renew Sust Energ Rev 79:148–179
22. McNabola A, Coughlan P, Corcoran L, Power C, Williams AP, Harris I, Gallagher J, Styles D (2014) Energy recovery in the water industry using micro-hydropower: an opportunity to improve sustainability. Water Policy 16:168–183
23. Muhammetoğlu H, Soyupak S, Karadirek İE (2011) İçme Suyu Dağıtım Şebekelerinde Optimum Klorlama Uygulamalarının Matematiksel Modeller Kullanılarak Gerçekleştirilmesi Ve Dezenfeksiyon Sistemlerinin Yönetimi: Tübitak Kamag 107g088 Handbook, Technical Report, p 49

Biofuel Production from Carbon Dioxide Gas in Polluted Areas

Delia Teresa Sponza, Cansu Vural, and Gokce Güney

Abstract Although carbon dioxide (CO_2) in the air is at a low level (between 0 and 0.03%), the concentration of it is significantly higher in industrial regions. The CO_2 concentration in the atmosphere increases 2–3 ppm every year because of the burning of fossil fuels. Global studies have focused on reducing the carbon dioxide level to the minimum limit (450 ppm) by reducing CO_2 emissions 50–80% by the year 2050. In this study, in order to minimize the CO_2 levels in the Aliağa and Atatürk industrial districts in Izmir, Turkey, *S. elongatus* from cyanobacteria were isolated from the Gölcük Lake in Ödemiş, Izmir, and were used to produce 1-butanol from CO_2 via photosynthesis as a fuel source, instead of gasoline, for cars. The maximum 1-butanol concentration produced was 79 mg/L, and the 1-butanol$_{produced}$/$CO_{2utilized}$ efficiency was 87.6% in the *S. elongatus* species isolated from the Gölcük Lake at a temperature of 30 °C, at 60 W light intensity, at pH = 7.1, at 120 mV redox potential, at a flow rate of 0.083 m^3/min using CO_2 from the Aliağa industrial region, and at 0.5 mg/L dissolved O_2 concentration. The maximum 1-butanol concentration produced was 59 mg/L, and the 1-butanol$_{produced}$/$CO_{2utilized}$ efficiency was 67.9% in the Atatürk industrial district due to low levels of polluted air in this region. In order to produce 10.000 m^3 1-butanol from 1000 g/L CO_2, the cost was calculated as 0.13 euro, while the addition of plasmid increased the cost to 0.66 euro to produce 10.000 m^3 1-butanol.

Keywords 1-Butanol · CO_2 · *S. elongatus* · Cyanobacteria · Biofuel

D. T. Sponza (✉) · G. Güney
Department of Environmental Engineering, Dokuz Eylul University, Izmir, Turkey
e-mail: delya.sponza@deu.edu.tr; gokce.pehlivaner@deu.edu.tr

C. Vural
Fundamental and Industrial Microbiology Department, Ege University, Izmir, Turkey

N. Balkaya, S. Guneysu (eds.), *Recycling and Reuse Approaches for Better Sustainability*, Environmental Science and Engineering,
https://doi.org/10.1007/978-3-319-95888-0_11

Introduction

Global climate change and an increase in greenhouse gas emissions, as well as the depletion of conventional fuel reserves, have become a growing concern over the years. The combustion of coal, oil, and natural gas emits more than six billion tonnes of CO_2 annually in the atmosphere [1]. The United Nations promoted the Kyoto Protocol (1997) with the objective of reducing greenhouse gases such as CO_2, CH_4, SO_2, and NO_2 by 5.2% on the basis of the emission in 1990, and more than 170 countries have ratified the protocol [1]. According to the Intergovernmental Panel on Climate Change, the average concentration of CO_2 increased from 315 ppm in 1960 to 380 ppm in the year 2007, and there has been a 35% increase in CO_2 emission worldwide since 1990 [2–5]. Although carbon dioxide in the air is at a low level (between 0 and 0.03%), the concentration is significantly higher in municipal polluted areas, and the carbon dioxide levels are increased in industrial regions. Sometimes the carbon dioxide levels in industrial regions can reach 600–700 ppm [6]. The carbon dioxide concentration in the atmosphere increases 2–3 ppm every 7 years because of the burning of fossil fuels. 80–85% of atmospheric carbon dioxide is produced by fossil fuels (petroleum, coal, and natural gas), and 15–20% of the atmospheric carbon dioxide originates from the respiration of humans and other organisms and from microorganisms transforming organic compounds. The pollution level continuously increases as energy sources are utilized. Global studies have focused on reducing the carbon dioxide level to the minimum limit (450 ppm) by reducing CO_2 emissions to 50–80% by the year 2050. Photosynthetic microorganisms such as cyanobacteria, which directly fix carbon dioxide as their primary carbon source, have been the primary organisms of interest for this fuel production strategy. They can grow much faster than plants and do not need to be grown on arable land. Furthermore, these organisms are grown submerged in water, which allows for the use of CO_2 at higher concentrations than that of ambient air [6, 7] and could potentially allow for the use of concentrated CO_2 emissions from industrial waste sources. Research on eukaryotic algae has primarily centered on their ability to produce large amounts of lipids for the production of biodiesel [6–8]. Cyanobacteria, prokaryotic organisms, combine the advantages of both eukaryotic algae, as a photosynthetic microorganism and as a tractable and naturally transformable host. *Synechococcus elongatus* sp. strain PCC 7942 was successfully engineered to produce ethanol through the addition of a pyruvate decarboxylase and an alcohol dehydrogenase, redirecting carbon from pyruvate [9, 10]. Cyanobacterial production of ethanol has since been significantly improved [11, 12]. Though compatible with the current fuel infrastructure as a supplement to gasoline, ethanol serves as a rather poor gasoline substitute due to its hygroscopicity and low energy density. For these reasons, efforts have shifted toward 1-butanol, an important chemical feedstock for hydrocarbons normally derived from petroleum, which was successfully produced in *S. elongatus* by diverting carbon flux from the valine biosynthesis pathway through the addition of a keto acid decarboxylase, reaching 1.1 g/L in 8 days [13–16]. Isobutanol, a promising candidate for a gasoline substitute, can easily be obtained from isobutyraldehyde via chemical conversion. The most recent studies which included reducing CO_2 sources from the air by the utilization of cyanobacteria were as follows: Cyanobacteria have been demonstrated to produce isobutyraldehyde (1100 mg/L), isobutanol (450 mg/L) [15, 17–20], ethanol (0.021–550 mg/L) [16, 17] (~37 mg/L) [18], isoprene (0.05 mg/g dry cell/day) [15–17], sugars (~36 mg/L), and lactic acid (~56 mg/L) [21–23]. The relatively high-flux production of isobutanol and isobutyraldehyde [17–19] demonstrates the feasibility for commercial scale synthesis of 1-butanol from CO_2 [15, 21, 24, 25].

In this study, it was aimed to remove the high CO_2 concentrations in the Aliağa and Atatürk industrial districts atmospheres, using *S. elongatus* cyanobacteria isolated from the Gölcük Lake in Ödemiş, İzmir, via photosynthesis. The effects of CO_2 concentrations on *S. elongatus* numbers, on 1-butanol concentrations, on 1-$\text{butanol}_{\text{produced}}/CO_{2\text{utilized}}$ efficiencies, and on enzyme concentrations were studied in two districts, one polluted (Aliağa) and one non-polluted (Atatürk industrial). The effects of plasmid addition to *S. elongatus* species based on 1-butanol productions and enzyme activities were investigated. A cost analysis was performed for this study. The 1-butanol produced was cheap, could be used as a biofuel, and could be an alternative to gasoline in cars.

Material and Methods

Microorganisms Isolated and Used in this Study

In this study, photosynthetic cyanobacteria *S. elongatus* were isolated from Gölcük Lake in Ödemiş, İzmir, Turkey.

Culture Medium and Conditions

All *S. elongatus* cultures were grown on BG-11 (1.5 g $NaNO_3$, 0.036 g $CaCl_2$, 0.006 g ferric ammonium citrate, 0.001 g EDTA, 0.04 g K_2HPO_4, 0.075 g $MgSO_4$, 0.02 g Na_2CO_3), (1000 × trace mineral (2.86 g H_3BO_3, 1.81 g $MnCl_2{\cdot}4H_2O$, 0.22 g $ZnSO_4{\cdot}7H_2O$, 0.390 g $Na_2MoO_4{\cdot}2H_2O$, 0.0790 g $CuSO_4{\cdot}5H_2O$, 0.0494 g $Co(NO_3)_2{\cdot}6H_2O$) medium) [22–24]. The pH of the liquid was adjusted to 7.1 by adding 1 mM $NaHCO_3$ and 100 mM KOH to the BG-11 medium. All *S. elongatus* strains were cultured in BG-11 medium in a shaking incubator at 120 rpm. Cultures were grown under 60 W light condition at 30 °C. Under laboratory conditions, 100 mL of isolated *S. elongatus* culture grown on BG-11 liquid medium was added to the 250 mL sterile glass bottles. To each culture, purified CO_2 gas trapped from the Aliağa and Atatürk organized regions was given between 10.00 and 18.00 five times a day at a flow rate of 0.083 m^3/s during a period of 30 days. In order to identify the *S. elongatus*, cyanobacteria isolated from the atmospheres were analyzed under a microscope, and their dimensions were measured. Then, their growths in urea, ammonia, nitrite, and nitrate media were monitored. A Bradford test was performed to determine the protein content.

Purification of CO_2 from the Trapped Air

The atmospheric gas trapped by vacuum pumps from the Atatürk industrial and Aliağa regions was stored in polyethylene air bags. The atmospheric gas was passed through an absorbent pad containing *Azotobacter* bacteria. In this step, the N_2 gas present in the air samples was eliminated since these bacteria used N_2 gas for growth. The *Azotobacter* sp. medium consisted of 0.8% K_2HPO_4, 0.02% KH_2PO_4, 0.02% $MgSO_4$, 0.01% $CaSO_4$, 0.0015% $FeSO_4{\cdot}7H_2O$, 0.00025% g $NaMoO_3$, and 0.5% sucrose. Then the remaining gas was passed through an absorbant pad containing *Nitrosomonas* sp. bacteria [25, 26]. These bacteria used the O_2 in the air. As *Nitrosomonas* medium, Medium A, Medium B, Medium A–Z, and A-2 Trace medium were used. A medium consisting of $(NH_4)_2SO_4$ 0.1 g, K_2HPO_4 0.1 g, NaCl 0.2 g, $MgSO_4{\cdot}7H_2O$ 0.05 g, $FeCl_3$ trace, $CaCO_3$ 1.0 g, and tap water of 100 mL was used. Medium B was made up as follows: NaCl 0.3 g, $MgSO_4{\cdot}7H_2O$ 0.14 g, $FeSO_4{\cdot}7H_2O$ 0.03, and $(NH_4)_2SO_4$ 0.66 g. They were dissolved in 90 mL distilled water. The trace element mixture contained $LiSO_4$ 0.01 g, $CuSO_4$ 0.02 g, $ZnSO_4$ 0.02 g, H_3BO_4 0.22 g, $Al_2(SO_4)$ 0.02 g, $SnCl_2$ 0.01 g, $MnCl_2$ 0.14 g, $NiCl_2$ 0.02 g, $CoSO_4$ 0.02 g, $TiCl_4$ (15% solution) 0.13 mL, Kl 0.01 g, KBr 0.01 g, and distilled water 360 mL [27–29]. As a result, the air was purified of N_2 and O_2 and contained 99.9% CO_2 with the exception of some NO_x gases at trace levels.

Measurement of 1-Butanol

1-Butanol was measured in an Agilent GC-MS 7890-A system with a flame ionization detector and DB-FFAP capillary column. 1-Butanol standard of 0.001% v/v was used and dilutions were performed. Helium gas was used as the carrier gas. The injector and detector temperatures were maintained at 250 °C. Injection volume was 1 μL. The GC oven temperature was initially held at 40 °C for 4 min and then raised to 250 °C. Column flow rate was 3 mL/min. 1-Butanol typically had a retention time of 2.483 min [30–32].

CO_2 Measurement in GC-MS

For CO_2 quantification, 1 mL of gas was drawn by a 1MR-VLL-GT flod syringe. Helium gas was used as the carrier gas. The injector and detector temperatures were maintained at 40 °C. Injection volume was 1 μL. The GC oven temperature was initially held at 40 °C for 4 min then increased to 250 °C [33].

Enzyme Activity Measurements

3-Hydroxybutyryl-CoA dehydrogenase (Hbd) activity measurement: Hbd activity was measured in a spectrophotometer AquaMate by monitoring the decrease in absorbance at 340 nm, which corresponded to consumption of NADH [6]. 3-Hydroxybutyryl-CoA dehydrogenase (Hbd) measurement: Hbd activity was measured in a spectrophotometer AquaMate by monitoring the decrease in absorbance at 340 nm, which corresponded to consumption of NADH [6]. Bradford protein measurement: Take into consideration the blue color production with interaction of Coomassie brilliant blue and proteins at an adsorbance of 595 nm as measured in an AquaMate (Helios Aquamate) spectrophotometer.

Dissolved oxygen was measured in a WTW dissolved oxygen meter. The light power was measured by a luminometer (OASIS). Additional pure CO_2 was added with a CO_2 tube with a purity of 99.9%.

Calculation of 1-Butanol Production Yields Versus CO_2 Utilized

The calculation of the yield for 1-butanol concentration produced from the CO_2 utilized was done via the formula given in (1), and the yield percentages were calculated via the formula given in (2). In the calculations, the theoretical value of 1-butanol$_{\text{produced}}$/$CO_{2\text{used}}$ is 0.25.

$$\text{Yield}\left(\text{Y}\right) = \frac{1 \;\; \text{butanol produced}}{\text{CO}_2 \text{ utilized}} \tag{1}$$

$$\text{Yield\%}\ \left(\text{Y\%}\right) = \frac{1 \;\; \text{butanol produced} / \text{CO}_2 \text{ utilized}}{\text{Theoric 1} \;\; \text{butanol produced} / \text{CO}_2 \text{ utilized}} \times 100 \tag{2}$$

Table 1 Identification of *S. elongatus* cyanobacteria

Studies	Cyanobacteria sizes lengths (μm)	Number of cyanobacteria (count/mL) widths (μm)	CO_2 concentration (mg/L)	Ammonia test	Nitrate test	Nitrite test	Urea test	Bradford test	Protein concentration (μg/mL)
Aliağa district	3–18	0.6–5	10,000	203	Blurred-Positive (+)	Blurred-Positive (+)	Blurred-Positive (+)	Blurred-Positive (+)	94
Aliağa district	2–17	0.5–5	9500	180	Blurred-Positive (+)	Blurred-Positive (+)	Blurred-Positive (+)	Blurred-Positive (+)	70
Aliağa district	1–16	0.4–4	9000	127	Blurred-Positive (+)	Blurred-Positive (+)	Blurred-Positive (+)	Blurred-Positive (+)	62
Atatürk organized area	1–15	0.4–3.8	8300	113	Blurred-Positive (+)	Blurred-Positive (+)	Blurred-Positive (+)	Blurred-Positive (+)	49
Atatürk organized area	1–14	0.4–3	6500	99	Blurred-Positive (+)	Blurred-Positive (+)	Blurred-Positive (+)	Blurred-Positive (+)	32
Atatürk organized area	1–12	0,3–3	8500	97	Blurred-Positive (+)	Blurred-Positive (+)	Blurred-Positive (+)	Blurred-Positive (+)	38

Results

Identification and Enumeration of S. elongatus

The identification of *S. elongatus* cyanobacteria was performed by taking into account the ammonia, nitrate, urea, and Bradford tests results (Table 1). The identification of *S. elongatus* cyanobacteria was done at the end of 30 days (t = 30 °C, pH = 7.1, Eh = 120 mV, light conditions = 60 W, CO_2 flow rate = 0.083 m^3/s). The dimensions of the *S. elongatus* cells differed from 1.5 up to 2.0 μm of length with a width of 0.4–6 μm as is reported in the literature studies. In the ammonia test, the reproduction of blurry showed that ammonia was used by *S. elongatus* at 30 °C, which is a positive result, as given in the literature (Table 1). In the nitrate test, *S. elongatus* produced blur by using nitrate reductase enzymes in nitrate medium. For nitrite test, *S. elongatus* used the reductase enzyme and utilized the nitrite and converted it to nitrate. On the other hand, *S. elongatus* used urea and turned it into ammonia and caused blur in tubes (Table 1). The Bradford test was done at the end of 30 days in order to determine the amount of protein in *S. elongatus* (Table 1).

According to the results obtained from the biochemical tests, it can be concluded that all the isolated bacteria from Gölcük Lake in Ödemiş, İzmir, were *S. elongatus* from cyanobacteria, and they were used in Aliağa and Atatürk industrial district atmospheres. The maximum *S. elongatus* cells were reached in the Aliağa region which is this region with a higher level of measured CO_2 concentration. The numbers of *S. elongatus* were reduced in the Atatürk industrial district, and they were lower than those of the Aliağa region since in this region the CO_2 concentrations were at the lowest level. Therefore, the 1-butanol concentrations produced would be lower from the district with low CO_2 concentrations. This correlated with the maximum protein concentration measured in the *S. elongatus* cells isolated from Gölcük Lake.

$Butanol_{produced}/CO_{2utilized}$ Yields

The maximum 1-butanol concentration produced from the CO_2 was recorded as 79 mg/L with a 1-$butanol_{produced}/CO_{2utilized}$ yield of 87.6% at 30 °C temperature, at 60 W light intensity, at pH = 7.1, at 120 mV redox potential, and at a 0.083 m^3/s flow rate together with 0.5 mg/L CO_2 and 0.1 mg/L dissolved O_2 concentration as the optimum operational conditions (Table 2). Water, N, P, and CO_2 play an important role in the growing of photosynthetic cyanobacteria [22, 23]. It was determined that cyanobacteria grow faster in a high CO_2 concentration, and more CO_2 increases the photosynthesis in cyanobacteria. In this study, it was determined that *S. elongatus* removed more CO_2 in the Aliağa industrial district where the CO_2 concentrations are high due to pollution (Table 2). The CO_2 concentrations are low in the Atatürk industrial region since this area is near the coast. Therefore, the CO_2 levels are low

Table 2 1-Butanol productions from CO_2

	Utilized concentration of CO_2 (mg/L)	Concentration of 1-butanol (mg/L)	Yield[a]	Yield%
Aliağa district	203	79	0.198	87.6
Aliağa district	180	69	0.182	79.2
Aliağa district	127	64	0.172	69
Atatürk organized area	113	12	0.162	56
Atatürk organized area	99	10	0.152	49
Atatürk organized area	97	10	0.150	40

[a]($1\text{-Butanol}_{productions}/CO_{2utilized}$)

and the CO_2 utilized is low because this is not a polluted area. To produce 1 mol 1-butanol with Calvin-Benson-Bassham cycle, 6 mol CO_2 is required. This amount of CO_2 is calculated with the numbers of glyceraldehyde-3-phosphate (G3P) and NAD(P)H which are necessary for 1-butanol production. For 6 mol NADPH, 1 mol G3P availability from 3 mol CO_2 is required. Consequently, 2 mol G3P is required to produce 1 mol 1-butanol. Thus, 6 mol CO_2 is required to produce 1 mol of 1-butanol [15]. Cyanobacteria fix carbon to provide the skeletons needed to assimilate N into amino acids and build protein and cellular biomass; fixed carbon can also be used to accumulate carbohydrate storage products (carbohydrate ballasting) in order to make the cell heavier during buoyancy regulation.

The maximum yield obtained in this study was 87.6 and 79.2% in the Aliağa region (Table 2). Lan and Liao [15] found 68–70% 1-butanol yields by producing 64–79 mg/L 1-butanol from CO_2 using *S. elongatus* cyanobacteria [15]. In this study, the 1-butanol concentrations produced were around 10–12 mg/L in the Atatürk industrial region. Among the studies carried out, the lowest 1-butanol yields were obtained from the Atatürk industrial district, around 0.150–0.162 $1\text{-butanol}_{produced}/CO_{2used}$.

Enzyme Activities

The enzyme activities were high at high $1\text{-butanol}_{produced}/CO_{2utilized}$ yields (Table 3). The enzymes involved in the photosynthesis of CO_2 to produce 1-butanol are HdB, Ter, AdHE2, AtoB, and Crt. Among these enzymes, Ter and Crt were catalyzing crotonyl-CoA and 3-hydroxybutyryl-CoA, respectively, and they were measured at low concentrations (Table 3).

Table 3 Enzyme activity measurements

Studies	Concentration of HdB (μg/mL)	Concentration of Ter (μg/mL)	Concentration of AdHE2 (μg/mL)	Concentration of AtoB (μg/mL)
Aliağa district	0.016	0.0022	0.0026	0.0048
Aliağa district	0.013	0.0018	0.0023	0.0046
Aliağa district	0.011	0.0015	0.0019	0.0042
Atatürk organized area	0.008	0.0012	0.0016	0.0039
Atatürk organized area	0.005	0.0008	0.0011	0.0027
Atatürk organized area	0.009	0.0013	0.0015	0.0034

Effect of Plasmid Addition on 1-Butanol Production

Plasmid transformation: this was done by carrying the atoB and AdHE2 genes in pEL5 plasmid to the genes of *S. elongatus* PCC 7942 by using recombinant DNA techniques. The pEL5 plasmid DNA was done with the recombination of *S. elongatus* PCC 7942's neutral zone. Plasmid pEL5 consists of the genes atoB and AdHE2. These plasmids were used because they include atoB and AdHE2, which are necessary to produce 1-butanol from CO_2 via *S. elongatus* PCC 7942. The genes atoB and AdHE2 in plasmid pEL5 were transferred into *S. elongatus* PCC 7942 DNA via BamH1 and Notl. areas in pAM2991. pEL5 genotype is an AdHE2 that is resistant to spectinomycin ($Spec^R$), intended for NSI, PtrcatoB [15]. After pAM2991 plasmid was transferred to *S. elongatus*, a recombination was recorded. PCR photos for AdHE2 (a) and atoB (b) genes primers were given in Fig. 1a, which exhibited the acetyl-CoA acetyltransferase (atoB) while Fig. 1b showed the aldehyde/alcohol dehydrogenase (AdHE2).

As shown in Table 4, the addition of plasmid to the *S. elongatus* significantly increased 1-butanol production, the number of *S. elongatus*, and 1-butanol yields, and the atob and AdHE2 enzymes' activities were high in *S. elongatus* containing plasmid compared to *S. elongatus* without plasmid.

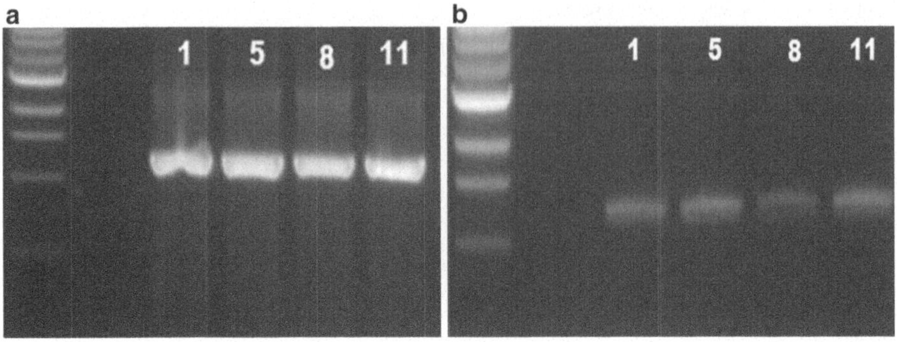

Fig. 1 PCR photos AdHE2 (a) and atoB (b) genes primers: (**a**) acetyl-CoA acetyltransferase (atoB), (**b**) aldehyde/alcohol dehydrogenase (AdHE2)

Table 4 Effect of plasmid addition on 1-butanol production, 1-butanol yields, and enzyme activities

	Without plasmid (mg/L)	With plasmid (mg/L)	*S. elongatus* number/mL with plasmid	1-Butanol yield (%) with plasmid	atoB (μg/L) with plasmid	AdHE2 (μg/L) with plasmid
Aliağa district 1-butanol (mg/L)	153	189	15,000	93	12×10^{-1}	9×10^{-1}
Aliağa district 1-butanol (mg/L)	140	162	14,000	90	12×10^{-1}	8×10^{-1}
Atatürk organized area 1-butanol (mg/L)	124.38	134	12,000	84	9×10^{-1}	6×10^{-1}

Cost Analysis

The isolation of *S. elongatus*, labor, chemical, purification, and electricity costs were recorded as 1.06 euro to produce 1 L of 1-butanol from 180 mg/L CO_2 trapping from Table 5. In order to produce 10,000 m^3 1-butanol from 1000 g/L CO_2, the cost was calculated as 0.13 euro. With the addition of plasmid, the total cost to produce 10,000 m^3 1-butanol was 0.66 euro. The cost of 1 L of the 1-butanol produced was compared with gasoline. The cost of the 1-butanol produced was approximately 3 times lower than that of 1 L of conventional gasoline (Fig. 2).

Table 5 Cost Analysis for 1-butanol production from CO_2

Isolation cost for *S. elongatus*	0.10 euro
Labor cost	0.06 euro
Electricity	60 W light power: 0.2 euro illumination for 30 days
CO_2 purification	0.3 euro
Total cost without plasmid	Total cost: 0.66 euro/L
Addition of plasmid (0.4 euro)	Total cost: 1.06 euro/L
Total cost without plasmid	0.13 euro/10000 m^3 1-butanol
Addition of plasmid (0.4 euro)	Total cost: 0.66 euro/10000 m^3 1-butanol

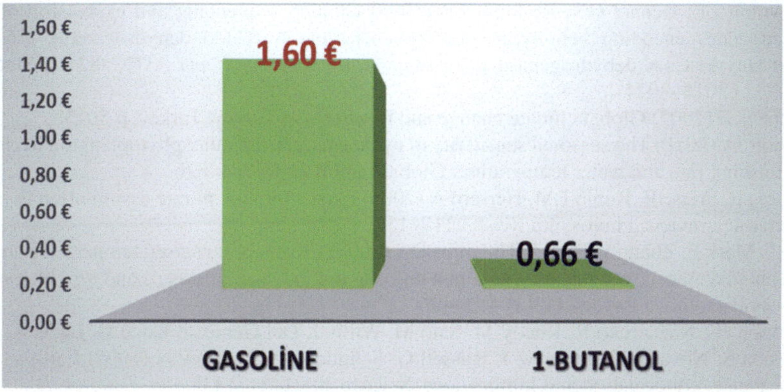

Fig. 2 Comparison of 1-butanol and gasoline costs

Conclusions

The eco-friendly 1-butanol is an economic alternative fuel. It can be produced from the waste CO_2 in the atmosphere which causes pollution; it is suitable to be used as fuel for heating, power, and as an alternative motor fuel. From a pollutant, a renewable energy (biofuel) was produced. The maximum 1-butanol concentration produced was 79 mg/L, and the 1-butanol$_{produced}$/$CO_{2utilized}$ efficiency was 87.6%. When compared with the prices of gasoline, the cost of 1-butanol is 3 times cheaper. At the present time, 1 L of gasoline is approximately 0.93 euro/L while the maximum cost of 1 L of 1-butanol is 0.28 euro/L. This biofuel could be a viable alternative to gasoline.

Acknowledgments This study was prepared in the scope of master of science studies in biotechnology and at the same time it was supported by the DEU Scientific Research Foundation (2014. KB.FEN.035) in Dokuz Eylul University Graduate School of Natural and Applied Sciences. In addition, the author acknowledges The Scientific and Technological Research Council of Turkey (TUBİTAK) for the financial support to the project numbered 114Y72.

References

1. Microbe Wiki (2014) Title of *Synechococcus*. https://microbewiki.kenyon.edu/index.php/Synechococcus. Accessed 15 Apr 2014
2. Armbrust EV, Bowen JD, Olson RJ, Chisholm SW (1989) Effect of light on the cell cycle of a marine *synechococcus* strain. Appl Environ Microbiol 55(2):425–432
3. Atsumi S, Hanai T, Liao JC (2008) Non-fermentative pathways for synthesis of branched-chain higher alcohols as biofuels. Nature 451:86–89
4. Devaki B, Watanabe N, Ogawa T (1999) National center for biotechnology information. U.S. National Library of Medicine 96(6):3188–3319
5. Binder BJ, Chisholm SW (1995) Cell cycle regulation in marine *synechococcus sp.* strains. Appl Environ Microbiol 61(2):708–717
6. Boynton ZL, Bennet GN, Rudolph FB (1996) Cloning sequencing and expression of clustered genes encoding beta-hydroxybutyryl-coenzyme A (CoA) dehydrogenase, crotonase and butyryl-CoA dehydrogenase from *clostridium acetobutylicum* ATCC 824. J Bacteriol 178(11):3015–3024
7. Denhez F (2007) Global climate change and its effects in Turkey. Turkey, p 56–58
8. Elliott JA (2010) The seasonal sensitivity of cyanobacteria and other phytoplankton to changes in flushing rate and water temperature. Glob Chang Biol 16:864–876
9. Flores E, Frías JE, Rubio LM, Herrero A (2005) Photosynthetic nitrate assimilation in cyanobacteria: a review. Photosynth Res 83:117–133
10. Fu F, Mark E, Zhang Y, Feng Y, Hulchins DA (2007) Effects of increased temperature and CO_2 on photosynthesis, growth and elemental ratios in marine *synechococcus* and *prochlorococcus* (cyanobacteria). J Phycol 443(3):485–496
11. Hansen JL, Nazaernko R, Ruedy M, Sato M, Willis J, Del Genio A, Koch D, Lacis A, Lo K, Menon S, Novakov T, Perlwitz J, Russell G, Schmidt GA, Tausnev N (2005) Earth's energy imbalance: confirmation and implications. Science 308:1431–1435
12. Kuan D, Duff S, Posarac D, Bi X (2015) Growth optimization of *synechococcus elongatus* PCC 7942 in lab flasks and 2-D photobioreactor. Can J Chem Eng 93:640–647
13. Kajiwara S, Yamada H, Ohkuni N (1997) Design of the bioreactor for carbon dioxide fixation by *synechococcus* PCC 7942. Energy Convers Manag 38:529–532
14. Kamennaya NA, Anton FP (2011) Characterization of cyanate metabolism in marine synechococcus and prochlorococcus *sp.* Appl Environ Microbiol 77(1):291–301
15. Lan EI, Liao JC (2011) Metabolic engineering of cyanobacteria for 1-butanol production from carbon dioxide. Metab Eng 13:353–363
16. Lan EI, Liao JC (2012) ATP drives direct photosynthetic production of 1-butanol in cyanobacteria. Proc Natl Acad Sci U S A 109(16):6018–6023
17. Liu H, Robert R, Laws E, Landry M, Camphell L (1999) Cell cycle and physiological characteristics of *synechococcus* (WH7803) in chemostat culture. Mar Ecol Prog Ser 189:17–25
18. Luque I, Flores E, Herrero A (1993) Nitrite reductase gene from *synechococcus sp.* PCC 7942: homology between cyanobacterial and higher-plant nitrite reductases. Plant Mol Biol 21:1201–1205
19. Maeda S, Kawaguchi Y, Ohe TA, Omata T (1998) Cis-acting sequences required for NtcB-dependent, nitrite-responsive positive regulation of the nitrate assimilation operon in the cyanobacterium *synechococcus sp.* strain PCC 7942. J Bacteriol 180:4080–4088
20. Standard methods for water and wastewater engineering (2012) APHA – AWWA, USA
21. Murray R, Badger T, Andrews J (1982) Photosynthesis and inorganic carbon usage by the marine cyanobacterium *synechococcus sp.* Plant Physiol 70(2):517–523
22. NCBI (2014) U.S. National Library of Medicine. https://www.ncbi.nlm.nih.gov/Taxonomy/Browser/wwwtax.cgi?mode=Tree&id=1129&lvl=3&lin=f&keep=1&srchmode=1&unlock. Accessed 15 Apr 2014
23. Palenik B, Brahamsha B, Larimer FW (2003) The genome of a motile marine *synechococcus*. Nature 424:1037–1042

24. Parmar A, Singh NK, Pandey A, Gnansounou E, Madamwar D (2011) Cyanobacteria and microalgae: a positive prospect for biofuels. Bioresour Technol 102:10163–10172
25. Prince RC, Khsheegi HS (2005) The photobiological production of hydrogen: potential efficiency and effectiveness as a renewable fuel. Crit Rev Microbiol 31(1):19–31
26. Yan R, Zhang Z, Zhu D (2009) Carbon and energetic metabolism of *synechococcus sp.* PCC7942 under photoautotrophic conditions. Springer Science and Business Media 25(9):1352–1359
27. Rippka R, Deruelles J, Waterbury JB, Herdman M, Stainer RY (1979) Generic assignments, strain histories and properties of pure cultures of cyanobacteria. J Gen Microbiol 111:1–61
28. Romo S, Soria J, Fernandez F, Ouahid Y, Sola A (2013) Water residence time and the dynamics of toxic cyanobacteria. Freshw Biol 58:513–522
29. Pope D (1975) Effects of light intensity, oxygen concentration, and carbon dioxide concentration on photosynthesis in algae. Microb Ecol 2(1):1–16
30. Aravinthan DT, Jennifer DP, Thomas SR (2001) Envelope structure of *synechococcus sp.* WH8113, a nonflagellated swimming cyanobacterium. BMC Microbiol 1:4. https://doi.org/10.1186/1471-2180-1-4
31. Shen CR, Lan EI, Dekishima Y, Baez A, Cho KM, Liao JC (2011) Driving forces enable high-titer anaerobic 1-butanol synthesis in *escherichia coli*. Appl Environ Microbiol 77:2905–2915
32. Snyder DS, Brahamsha B, Azadi P, Palenik B (2009) Structure of compositionally simple lipopolysaccharide from marine *synechococcus*. J Bacteriol 191(17):5499–5509
33. Riming Y, Zhu D, Zhang Z (2012) Carbon metabolism and energy conversion of *synechococcus sp.* PCC 7942 under mixotrophic conditions: comparison with photoautotrophic condition. J Appl Phycol 24(4):657–688

Artificial Intelligence-Based Prediction Models for Energy Performance of Residential Buildings

Ersin Namlı, Hamit Erdal, and Halil Ibrahim Erdal

Abstract Although energy sources on the environment are limited, in all parts of life, energy requirement increases rapidly which depends on the increasing technology and population. This problem enforces researchers to study on energy efficiency, performance, and optimization. This paper presents artificial intelligence-based (AI) prediction models to estimate energy loads for residential buildings. The model was developed by using eight input parameters (relative compactness, surface area, wall area, roof area, overall height, orientation, glazing area, glazing area distribution) related to two output parameters (heating and cooling loads).

The dataset contains 768 residential building information. In the proposed model, advanced machine learning (ML) regression algorithms (tree-based and lazy learning algorithms) compared with multilayer perceptron (MLP) and support vector regression (SVR) algorithms. According to coefficient of determination (R^2), mean absolute error (MAE), and root mean squared error (RMSE) values best prediction model obtained for energy-efficient building design.

Keywords Energy performance · Machine learning · Building energy loads · Artificial intelligence

E. Namlı (✉)
Department of Industrial Engineering, Istanbul University-Cerrahpaşa, Istanbul, Turkey
e-mail: enamli@istanbul.edu.tr

H. Erdal
Institute of Social Sciences, Ataturk University, Erzurum, Turkey

H. I. Erdal
Turkish Cooperation and Coordination Agency, Republic of Turkey Prime Ministry, Ankara, Turkey

N. Balkaya, S. Guneysu (eds.), *Recycling and Reuse Approaches for Better Sustainability*, Environmental Science and Engineering,
https://doi.org/10.1007/978-3-319-95888-0_12

Introduction

The energy issue can be described as the most competitive area between countries. All government policies in the short, medium, and long term are planned to be energy-oriented. In this context, states support the private sector with high budgets in energy investments or in R&D activities related to energy.

Heating and cooling loads can be calculated in detail using advanced software, with the following heat gain factors: solar, lighting, electrical devices, and human and heat loss parameters due to building characteristics. Prediction of heating and cooling loads in buildings is of high importance for engineers, and these estimates also enable building energy efficiency to be improved.

Numerous developments are experienced in energy-efficient areas, which was the aim of the use of efficient energy and its sources, in many countries around the world. On this subject, the construction sector has conducted various important studies for the purpose of determining the precise energy consumption of buildings and to provide optimizations in the usage rates of energy types. Yıldız and Arsan in 2011 studied about the most important factors that affect the heating and cooling energy loads for buildings, in hot and humid climates, by considering an existing apartment's design [1]. In their study, Cheng and Cao in 2014 mentioned a method called EMARS, the combination of two different methods which are the artificial bee colony and multivariate adaptive regression splines (MARS), which predicts heating and cooling loads of buildings [2]. Khayatian et al. in 2016 used ANNs, by considering those that were easy and not time-consuming, to predict heat energy demand of buildings according to the energy consumption of existing buildings [3]. Yücel and Namlı in 2018 developed a machine learning-based model for predicting energy performances of buildings [4].

Artificial Intelligence-Based Prediction Models

Artificial intelligence (AI) algorithms estimate unrecognized interactions between the inputs and output from a dataset [5]. Table 1 shows AI regression techniques, used in this paper. These regression algorithms have been commonly utilized for modelling numerous real-world regression problems.

Table 1 AI techniques utilized in this study

Categories	Techniques	Abbreviation
Functions	Multilayer perceptron	MLP
	Support vector regression	SVR
Lazy learning techniques	IBk linear NN search	IBk
	Locally weighted learning	LWL
Tree-based learning techniques	Model trees regression	M5P
	Reduced error pruning tree	REPTree

Multilayer Perceptron (MLP)

MLP is the most common artificial neural network (ANN) technique and contains neurons with practically weighted interconnections, where signals all the time flow to the output layer [6]. The architecture of the MLP comprises of input layer, hidden layers, and output layer. Input signals flow from the input layer to the hidden layer without implementing any functions [7]. A classical MLP structure may be described in Eqs. (1)–(5) [8–13]:

$$u_j = \sum_{N_{\text{input}}}^{i=1} X_i a_{ij} + a_{0j} \tag{1}$$

Equation (1) specifies the total yields of the inputs (X_i), a bias term of the hidden layer (a_{0j}), and the weight vectors (a_{ij}). In Eq. (2), the output of the hidden layer (Z_j) is provided by converting the total that is specified in Eq. (1), by utilizing activation function g.

$$Z_j = g\left(u_j\right) \tag{2}$$

One of the most common activation functions is sigmoid function, specified in Eq. (3) for input x.

$$g\left(x\right) = \text{sigmoid}\left(x\right) = \frac{1}{\left(1+e^{-x}\right)} \tag{3}$$

Equation (4) specifies the total yields of the hidden layer's outputs (Z_j), a bias term of the output layer (b_{0k}), and the weight vectors (b_{jk}).

$$v_k = \sum_{N_{\text{hidden}}}^{j=1} Z_j b_{jk} + b_{0k} \tag{4}$$

In Eq. (5), the outputs of the output layer (Y_k) are provided by converting the total, computed in Eq. (4), utilizing the sigmoid function g, specified in Eq. (3).

$$Y_k = g\left(v_k\right) \tag{5}$$

Support Vector Regression (SVR)

SVR is a specific usage of support vector machines. The basics of the support vector regression (SVR) are to map (transfer) the input data x into a higher-dimensional feature space [14]. Mapping could be handled by utilizing a kernel function. The most widely used kernel functions can be sorted as linear, polynomial, Gaussian (RBF), and sigmoid (MLP) kernels [15–18].

IBk Linear NN Search (IBk)

IBk algorithm operates as a k-nearest neighbor (k-NN) classifier, which could be defined as a widely used lazy-/instance-based technique for both regression and classification problems. In this study, we utilized it for a regression problem. The technique can do distance weighting and normalizes input values by default [11].

Locally Weighted Learning (LWL)

The locally weighted learning algorithm assigns instance weights for classification and regression problems [14]. In this study, we utilized it for a regression problem. The most impressive factor behind the LWL method is that any nonlinearity can be estimated by a linear model, when the output surface is smooth. So, instead of a complicated method, it is easy to predict nonlinear functions by using simple local procedures [5].

Model Trees Regression (M5P)

The M5P algorithm constructs a model tree (for the reason that each leaf nodes comprises a linear regression model to provide the estimated output, the tree is named a model tree [19] utilizing the M5 algorithm. An M5 tree is constructed utilizing a divide-and-conquer technique. The major advantages of M5P are (1) the regression functions don't usually include many variables, (2) the decision strength of M5P is net, and (3) the M5P is smaller compared to other trees [20].

Reduced Error Pruning Tree (REPTree)

The REPTree produces a regression tree utilizing the node statistics such as variance decrease or information gain computed in the top-down period and prunes it utilizing reduced error pruning [19]. Optimized for speed, the REPTree just classifies values for numeric attributes once and handles with missing values by splitting input data into small pieces [5].

Application and Discussion of Results

Dataset

Simulations of the energy aspects of residential buildings are extensively used approach [21]. Tsanas and Xifara associated heating load (HL) (y_1) and cooling load (CL) (y_2) with eight input variables: "relative compactness (x_1), surface area (x_2), wall area (x_3), roof area (x_4), overall height (x_5), orientation (x_6), glazing area (x_7), and glazing area distribution (x_8)." These inputs are utilized in an environmental analysis tool ECOTECH, which enables architects to simulate energy performance of residential buildings from the earliest phases of conceptual design: 768 different buildings with the exact volume of 771.75 m^2 but diverse surface areas and dimensions [21]. Dataset contains all numeric variables. There are no missing values. The basic statistics of the dataset are given in Table 2.

Performance Statistics and Evaluation Process

The AI model techniques, suggested in this paper, were evaluated by utilizing the three well-known performance statistics (i.e., coefficient of determination (R^2), mean absolute error (MAE), root mean squared error (RMSE)). The performance statistics were computed to analyze the relation between the original data and the estimated data with maximum predilection for R^2 closest to unity whereas for MAE and RMSE closest to zero [22].

The mathematical representation of R^2 is

$$R^2 = \left(\frac{n\sum y \cdot y' - (\sum y)(\sum y')}{\sqrt{(\sum y^2) - (\sum y)^2}\sqrt{(\sum y'^2) - (\sum y')^2}} \right)^2 \tag{6}$$

Table 2 Variables and their main characteristic

Variables	Min	Max	Mean	Std. dev.	Number of possible values
x_1	0.62	0.98	0.764	0.106	12
x_2	514.5	808.5	671.708	88.086	12
x_3	245	416.5	318.5	43.626	7
x_4	110.25	220.5	176.604	45.166	4
x_5	3.5	7	5.25	1.751	2
x_6	2	5	3.5	1.119	4
x_7	0	0.4	0.234	0.133	4
x_8	0	5	2.813	1.551	6
y_1 (kW)	6.01	43.1	22.307	10.09	586
y_2 (kW)	10.9	48.03	24.588	9.513	636

where y' is the estimated value, y actual value, and n number of instances [23].

The MAE is presented by the following equation [24]:

$$\mathrm{MAE} = \frac{1}{n} \sum_{i=1,n} \left| y - y' \right| \tag{7}$$

The RMSE is the square root of the mean square error. Mathematical representation of the RMSE is [8]

$$\mathrm{RMSE} = \sqrt{\frac{\Sigma \left(y' - y \right)^2}{n}} \tag{8}$$

In this paper, two evaluation processes (ten- and fivefold cross-validation) were used to analyze predictive models. For the proposed model evaluation, k-fold cross-validation reduces bias caused by randomness in selecting testing cases [25].

Empirical Results

The results of the machine learning models are summarized in Table 3. The performance of all eight predictors except IBk is relatively close, with the R^2 values ranging between 0.9019 and 0.9984. The best prediction performance subject to R^2 is provided by the REPTree and M5P, closely followed by MLP and SVR, respectively, for HL. The best performance for CL is obtained by the M5P and REPTree, followed by MLP and SVR, respectively. The obtained results were almost identical to R^2 results. Table 3 demonstrated that REPTree, M5P, MLP, and SVR produce the lowest MAE and RMSE, respectively, for both HL and CL.

Four evaluation processes, tenfold cross-validation (CV10), and fivefold cross-validation (CV5) were utilized to evaluate the predictive models. Table 4 summarized the best evaluation process according to performance statistics. The best results of MLP, M5P, and REPTree are obtained by using CV10 evaluation process for HL, and the best results of SVR, IBk, and LWL are obtained by using CV5 evaluation process for CL, individually. On the other hand, the best results of REPTree is obtained by using CV5 evaluation process for CL, dissimilarly.

Conclusions

In this study, AI techniques have been investigated to propose models that predict residential buildings' energy performance via utilizing experimental data. For this purpose, eight artificial intelligence (AI) models were compared for prediction.

Table 3 Statics of predictive models

Model	Heating load						Cooling load					
	CV10			CV5			CV10			CV5		
	R^2	MAE (kW)	RMSE (kW)	R^2	MAE (kW)	RMSE (kW)	R^2	MAE (kW)	RMSE (kW)	R^2	MAE (kW)	RMSE (kW)
MLP	**0.9941**	**0.8401**	**1.1111**	0.9873	1.2195	1.6372	**0.9771**	**1.8381**	**2.0825**	0.9687	1.9771	2.4416
SVR	0.9546	2.0402	3.0051	**0.9549**	**2.034**	**2.9944**	0.9363	2.2057	3.3605	**0.9388**	**2.1748**	**3.2921**
IBk	0.8464	3.3267	5.5392	**0.8667**	**3.0384**	**5.1683**	0.8259	3.5802	5.5225	**0.8457**	**3.3614**	**5.2003**
LWL	0.9019	3.3033	4.3576	**0.9025**	**3.2923**	**4.3442**	0.9037	3.0091	4.0708	**0.9047**	**2.9977**	**4.0524**
M5P	**0.996**	**0.6497**	**0.9164**	0.9952	0.7119	1.0002	**0.9822**	**1.1868**	**1.7919**	0.9821	1.9995	1.7956
REPTree	**0.9984**	**0.3864**	**0.5725**	0.9965	0.4447	0.8453	0.9805	1.1799	1.8674	**0.9808**	**1.1403**	**1.8544**

Table 4 Best evaluation processes

	Heating load			Cooling load		
Model	R^2	MAE	RMSE	R^2	MAE	RMSE
MLP	CV10	CV10	CV10	CV10	CV10	CV10
SVR	CV5	CV5	CV5	CV5	CV5	CV5
IBk	CV5	CV5	CV5	CV5	CV5	CV5
LWL	CV5	CV5	CV5	CV5	CV5	CV5
M5P	CV10	CV10	CV10	CV10	CV10	CV10
REPTree	CV10	CV10	CV10	CV5	CV5	CV5

Data for 768 cases of HL and CL and two evaluation processes (i.e., tenfold and fivefold cross-validations) were utilized to examine the predictive models. The obtained findings present the applicability of AI models for predicting energy consumption of residential buildings. Clearly, the tree-based (i.e., M5P and REPTree) models were determined to be the best, and the lazy learning algorithm (i.e., IBk and LWL) was determined to be the weakest performing predictive model.

References

1. Yıldız Y, Arsan ZD (2011) Identification of the building parameters that influence heating and cooling energy loads for apartment buildings in hot-humid climates. Energy 36(7):4287–4296
2. Cheng MY, Cao MT (2014) Accurately predicting building energy performance using evolutionary multivariate adaptive regression splines. Appl Soft Comput 22:178–188
3. Khayatian F, Luca S (2016) Application of neural networks for evaluating energy performance certificates of residential buildings. Energy Build 125:45–54
4. Namlı E, Yücel M (2018) Yapay zekâ modelleri ile betonarme yapilara ait enerji performans siniflarinin tahmini. Uludağ Univ J Faculty Eng 22(3):325–346
5. Demirdogen O, Erdal H, Akbaba AI (2017) Comparing various machine learning methods for prediction of patient revisit intention: a case study. Selcuk Univ J Eng Sci Technol 5(4):386–401
6. Yumurtaci-Aydogmus H, Erdal HI, Karakurt O, Namli E, Turkan YS, Erdal H (2015) A comparative assessment of bagging ensemble models for modeling concrete slump flow. Comput Concr 16(5):741–757
7. Erdal H (2015) Makine öğrenmesi yöntemlerinin inşaat sektörüne katkısı: basınç dayanımı tahminlemesi. Pamukkale Üniv Müh Bilim Dergi 21(3):109–114
8. Yumurtaci-Aydogmus H, Ekinci A, Erdal HI, Erdal H (2015) Optimizing the monthly crude oil price forecasting accuracy via bagging ensemble models. J Econ Int Finance 7(5):127–136
9. Erdal HI, Ekinci A (2015) Bank failure prediction using hybrid classifier ensembles of random sub-spaces and bagging. In: The Second Yandex School of Data Analysis, machine learning: prospects and applications, pp 5–8
10. Yaprakli TS, Erdal H (2016) Firma başarısızlığı tahminlemesi: makine öğrenmesine dayalı bir uygulama. Int J Inf Technol 9(1):21–31
11. Namli E, Erdal HI, Erdal H (2016) Dalgacık dönüşümü ile beton basınç dayanım tahmininin iyileştirilmesi. Politeknik Dergisi 19(4):471–480
12. Ozturk H, Namli E, Erdal HI (2016) Modelling sovereign credit ratings: the accuracy of models in a heterogeneous sample. Econ Model 54:469–478

13. Ozturk H, Namli E, Erdal HI (2016) Reducing overreliance on sovereign credit ratings: which model serves better? Comput Econ 48(1):59–81
14. Turkan YS, Yumurtaci-Aydogmus H, Erdal H (2016) The prediction of the wind speed at different heights by machine learning methods. Int J Optim Control Theories Appl 6(2):179–187
15. Erdal HI, Ekinci A (2013) A comparison of various artificial intelligence methods in the prediction of bank failures. Comput Econ 42(2):199–215
16. Ekİncİ A, Erdal Hİ (2011) Türkiye'de banka başarısızlıklarının tahmini üzerine bir uygulama. Iktisat Isletme Finans 26(298):21–44
17. Erdal HI, Karakurt O (2013) Advancing monthly streamflow prediction accuracy of cart models using ensemble learning paradigms. J Hydrol 477:119–128
18. Erdal HI, Baray A, Esnaf S (2014) Estimation of the manufacturing industry sub-sectors' capacity utilization rates using support vector machines. Artif Intell Res 3(1):1–11
19. Wang Y, Witten I (1997) Inducing model trees for continuous classes. In: Ninth European conference on machine learning, Prague, Czech Republic, pp 128–137
20. Ekinci S, Celebi UB, Bala M, Amasyali MF, Boyaci UK (2011) Predictions of oil/chemical tanker main design parameters using computational intelligence techniques. Appl Soft Comput 11:2356–2366
21. Tsanas A, Xifara A (2012) Accurate quantitative estimation of energy performance of residential buildings using statistical machine learning tools. Energy Build 49:560–567
22. Erdal HI (2013) Two-level and hybrid ensembles of decision trees for high performance concrete compressive strength prediction. Eng Appl Artif Intell 26(7):1689–1697
23. Erdal HI, Karakurt O, Namli E (2013) High performance concrete compressive strength forecasting using ensemble models based on discrete wavelet transform. Eng Appl Artif Intell 26(4):1246–1254
24. Erdal HI, Namli E, Yumurtaci-Aydogmus H, Turkkan YS (2013) Comparing ensembles of decision trees and neural networks for one-day-ahead streamflow prediction. Sci Res J I(IV):43–55
25. Christopher MB (2016) Pattern recognition and machine learning (information science and statistics). Springer-Verlag New York Inc., New York, NY

Green Roofs and Urban Life Sustainability

Hatice İnan

Abstract As the population coming to the cities increases, the surface textures of the natural landscapes around the city deteriorate and are quickly transformed into urban areas. At the local level, ecosystem and climate change also occur when materials such as concrete and asphalt with higher thermal capacities and thermal conductivity replace plant cover and water masses. As a result, the continuity of urban life becomes dangerous. Sustainable cities make the city more efficient and provide people with a high-quality living environment, without consuming too much natural resources. These local effects can be limited by city-level harmonization policies such as cool plots in cities, cool and green roofs, and expanding vegetation. In this study, the effects of the presence of green roofs instead of the traditional roof on the sustainability of life in the cities were examined.

Keywords Global warming · Climate effects · Stormwater management · Urban heat island · Energy · Air pollution

Introduction

The population of urban areas worldwide is becoming more crowded, and nowadays 54% of the world's population is in cities. By the year 2050, it is predicted that the world population will be about 9 billion, and two-thirds of this population (about 66%) will live in cities [1–4]. The surplus population in urban areas would dramatically increase the problems such as high pollution and temperatures ("urban heat islands"), degraded water and air quality, and intensified biodiversity loss in local and global natural ecosystems [5–7]. These problems are faced when there is an excessive use of the natural resources; an increase in the request for food, water, and energy; and, therefore, an increase in the pollution of water, atmosphere, and soil

H. İnan (✉)
Department of Environmental Engineering, Gebze Technical University, Kocaeli, Turkey
e-mail: inan@gtu.edu.tr

N. Balkaya, S. Guneysu (eds.), *Recycling and Reuse Approaches for Better Sustainability*, Environmental Science and Engineering,
https://doi.org/10.1007/978-3-319-95888-0_13

[6–8]. In addition to all pressures to the city life, global warming makes all these problems more dramatic [9, 10]. Cities constitute about 80% of the gross world product and contribute to the world CO_2 emissions, which are more than 60% and about 80% of the total greenhouse gas emissions [6–13].

Concerning ecological, environmental, economic, and social experiences, it can be said that it is essential to create a sustainable life or development. It must be understood that "sustainable living" is a situation in which the needs for the current and future generations can be supplied without any risk [10–15]. In this context, sustainability of the urban life issue gains attention.

Today, the global climate change and its effects are becoming challenging threats to civilization. The main reasons for the climate change are the increased burning of fossil fuels because of tremendous energy demand and also changing the land use such as deforestation, etc., all of which increase the greenhouse gas (GHG) emission. As reported by [16], GHG emissions consist of from mainly energy supply sector, transport, and industry, and then changes in land use patterns [16–19]. Climate change also affects lots of urban activities including urban planning, secure water supply, stormwater management, waste management and transportation, buildings, energy, health, and food security. Furthermore, the other risks are also expected: shifts or extinction of species, phenology changes, increased damage from storms and wildfires, and losing of coastal zones [11].

Currently, green roofs are the green infrastructures for urban areas, and they can mitigate the global warming effects and many other environmental problems [20]. It can be a good alternative in sustainable urban planning because of its several benefits including the improved air quality, stormwater management, lowering building energy consumptions, decreasing climate impacts, and mitigating the urban heat island effect, etc. [21–27]. Therefore, they are taken attention; interests of them and their number are increasing in many countries. Green or vegetative roofs can have a considerable function in sustainable urban life buildings [28]. In fact, the roof of the urban buildings is unused spaces. By vegetating the roof area, it can turn into a green area and supply attractive green space [29–33].

Green Roof Construction

The green roof, which is built on the roof of a building, is purposely fitted or cultivated with vegetation (vegetative roof). This roofing style is an old technique seen in the Babel Hanging Gardens built around BC 500. Recently, this roof construction technique has been used for insulation purposes from extreme climates in Nordic countries. In the last few decades, modern green roof technology started and developed firstly in Germany [27]. Contemporary technical knowledge and practice about green roof have emerged from Germany in the 1970s and 1980s. Switzerland, Austria, the UK, and France, Canada and parts of the USA followed to Germany by increasingly installing green roofs. Recently, developed countries like the USA, Canada, Australia, Singapore, and Japan are implementing and widening green roof system by doing either building new green roofs or by adapting to the old buildings [25–27].

Benefits of Green Roof

Mostly the green roofs are mainly categorized as intensive and extensive. They have different habitats, affect storm water runoff to varying degrees, and have impact on energy usage and thermal performance differently [26, 34, 35]. Intensive and extensive green roofs have different substrate thickness and vegetative types. Intensive green roofs have thick substrate layer (20–200 cm), a broad variety of plants (shrubs, and small trees), high maintenance, high capital cost, and greater weight. Because they need fertilizing, weeding, and watering, their maintenance cost will be high. However, extensive green roofs have low weight due to thin substrate layer (less than 15 cm). In addition, they have less capital cost and maintenance. The typical plants are grasses, moss, and few succulents. Because of lighter, so no additional structural support is not desired since it is lighter by weight [25, 27, 28, 32]. Basically, all green roofs are made up of five main key components:

- The planted layer (sedum and grasses up to shrubs and trees).
- The substrate (growing medium (from 20 mm (extensive roof) to 1500 mm deep (intensive roof))).
- The filter layer (prevents the substrate from blocking the drainage layer).
- The drainage layer (affect the amount of stormwater retaining on the roof at any time).
- The root barrier (protect the integrity of the roof from damage).

Stormwater Management and Rainwater Harvesting

As the cities become crowded and permeable places become impermeable, stormwater, the runoff created by the rainfall, is an important problem that can make a city life suffer and dangerous [36]. For this reason, stormwater management is an important need. Stormwater effects could be effectively controlled by reducing surface runoff as possible and maximizing infiltration and subsurface runoff. If the city has a combined sewer system that comprised of a wastewater and stormwater, in heavy rain, local rivers and streams result in the pollution because of an overflow from a combined sewer system [37]. The green infrastructure is a good alternative for stormwater management to protect rivers and natural systems [23, 26, 38].

Stormwater cannot penetrate impervious surfaces such as paving or roofs. When precipitation reaches an impervious surface, it flows directly to downhill. If the surface is a pervious surface in which water can penetrate such as soil or a green roof system, the stormwater infiltrates by the surface until it is saturated and then flows off according to the slope. While commonly used impervious roofs cause instant runoff flow, green roof system holds stormwater and retains it, thereby helping to reduce peak flows [39]. In addition, decreasing the volume of stormwater conveyed to the sewer system after rainfall will buffer the first flush of stormwater to regulate the flow of water into sewage treatment facilities, and it helps to protect

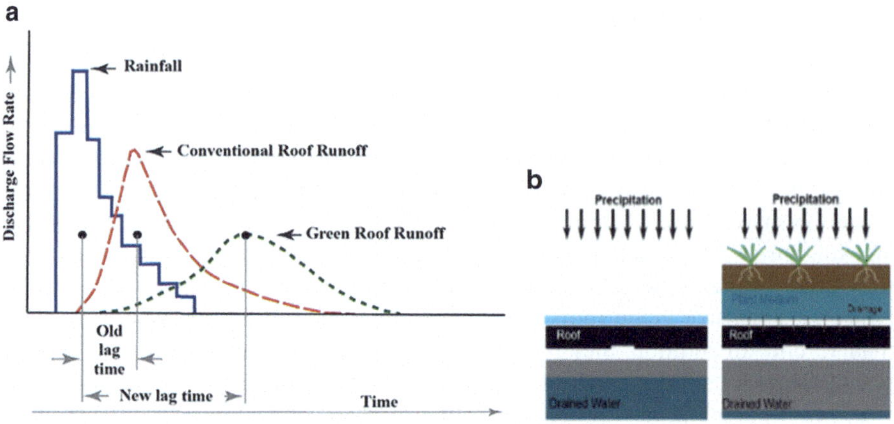

Fig. 1 Conventional and green roofs of typical rainwater and runoff hydrographs [43]

water quality. Green roof systems are the best management practices (BMP) since they catch the first cm of rainfall—the most frequently occurring rain event [32, 40–42] (Fig. 1).

The health of watersheds and aquatic ecosystems is at great risk due to the increase in stormwater volumes and associated pollution loads. Nitrogen and phosphorus, which are the major causes of eutrophication, come from excess fertilizers from agricultural fields and lawns as well as urban runoff and result in degraded water quality. Stormwater runoff brings airborne pollutants, which settle on impervious surfaces, to nearby streams and rivers in every rain. In pervious surfaces throughout the rainstorms, water replaces the air in the void spaces of the soil medium and is thus stored in soil and reduces the surface runoff. When the soil of the living green roof built on a flat surface has the maximum dryness, it serves as a water storage area at a very high capacity. In other terms, maximum water storage capacity or retention time depends on existing soil moisture content and its medium, depth, and porosity and slope of the roof and also vegetation type. The loss of stormwater of some cities is the very big deal. It could be equal to average annual household water needs of 3.6 million people of the city. If it is caught by the green roof, during the infiltration of stormwater, water could be treated, and that treatment efficiency is the function of the residence time [32, 42, 44, 45].

Rainwater harvesting (RWH) has come to the fore again due to the pollution of water resources and the distances to large settlements and global climate problems. Humans have practiced the rainwater harvest for thousands of years. Because of the many advantages of rainwater harvesting, it is now gaining popularity not only in water-restricted areas but also in the general public with conservation consciousness. This water is simple and cheap. Once the resulting rainwater has been treated, it is a reliable source of water reinforcement for all urban uses, including drinking, washing, and water consumption and irrigation. However, care must be taken in urban centers with air pollution [46].

Rainwater in places without air pollution is relatively clean and soft, has a pH value of about 5.6, and does not contain disinfection by-products, salts, minerals, and other natural and anthropogenic pollutants found in the surface water. On the other hand, in areas with air pollution from traffic and power plants, rainwater may be acidic, making plants, soils, and infrastructures corrosive and damaging with a pH as low as 3.0. In addition, if the rain has high acidity, it should be noted that roof coverings such as asphalt shingles and zinc or copper gutters and landing pipes are a source of pollution. For this reason, it is necessary to include air pollution controls for the rain collection programs to be effective.

Mitigation of Urban Heat Island Effect

The urban heat island (UHI) is a phenomenon whereby temperature levels in urban areas are higher than in the surrounding rural settings. UHI are associated with high ambient temperature effects, changes in sedimentation patterns, air pollution at the climate extremes, and the effects of air pollution in urban areas.

The mitigation of the urban heat island (UHI) effect is an important point for the sustainable urban development. UHI is a fact caused by the increase of urbanization process, that is, growth of urban areas (urbanization), structural and land cover changes together with an increase of air pollution and anthropogenic heat sources, development of specific air circulation patterns (e.g., street canyons), and other factors. The city growth has changed the nature of surfaces and reduced the presence of vegetation, building structures, and materials trapping solar radiation during the day. It caused the case of the significant temperature differences between urban and rural areas [47–49]. The impact of UHI is a reality for every urban area, regardless of its size and geographical location, but this manifestation is significantly greater for megacities, especially in hot climate regions.

Recently, anthropogenic activities in urban areas have become increasingly prominent. Excess urbanization growth results from more human activity and land changes, creating undesirable effects such as urban heat island (UHI) and a drastic difference in the temperature of the urban than the surrounding rural areas. In addition, global warming also contributes to increased UHI effects, potentially reduced precipitation, and thus aggravated effects, and summer heat load will most likely lead to increased heat waves in many areas and will affect higher densities and longer durability.

The most cost-effective strategy for mitigating the urban heat island effect in cities is the green structure or green roof. Topography, building geometry and dimensions, density, facade and roofing materials, the albedo of surfaces, urban canyons, transportation, pollution, greenery integration, etc., are the many local effects that have a great impact on the temperature profile of any area in cities. The quantitatively reflective behavior of a surface can be defined as albedo parameter that represents the total hemispherical reflection of a surface integrated over the solar spectrum. The conventional roof has an average albedo of 0.30. To increase the albedo of residential and industrial building roofs and consequently mitigate the UHI can be possible when coverings such as cool roof paintings are applied, and vegetative roofs reduce heat in three ways: by evapotranspiration, by reflecting the sun because of the higher albedo of the leaves, and by blocking the solar radiation [50].

The most cost-effective strategy for the reduction of heat island effect in the city is the emergence of greenery in cities. Green areas can be a remedy by cooling the environment in the process of evapotranspiration in which large amounts of sunlight can be transformed into latent heat. Furthermore, green areas supply a nice urban environment and have a positive effect on the comfort level of the users, reducing air temperature. During daytime trees and planting provide shading; the nights are characterized by the cooling of trees and plant coverings, green roofs, and urban microclimates. The cooling effect varies because of the higher albedo of the vegetative roofs and the evapotranspiration effect. In urban scale, green roofs have a potential to reduce the average air temperature between 0.3 and 3.0 K [47, 50].

Vegetative areas act as moisture sources for evapotranspiration (evaporative cooling), and this helps to lower city temperature [51]. Therefore, the green roof reduces the surface temperature of the roof surface and the ambient air temperature. Thus, it raises human comfort and at the same time saves energy for the buildings. Vegetation and greenery reduce the impact of urban warming [24, 51–55].

Energy Reduction

Population and infrastructure growth, modernization, and urbanization are the factors that affect the world in various energy and environmental challenges. Green roof technology is a passive energy-efficient technology used to improve the sustainability of buildings. In addition, it serves both mitigations of environmental problems and enhancement of aesthetic and architectural qualities of buildings in urban areas [56].

There are differences in annual or seasonal energy consumption between buildings with green and conventional roofs. Green roofs are positively affecting the energy consumption of a building by improving its thermal performance; even though it varies depending on the daily and seasonal weather. Poorly insulated roofs, like many existing buildings, will overheat underneath during the summer months and lead to overheating demand during the winter months. By amendment of the conventional roof to green roofs, both air conditioning and heating use reduce. In summer, the amount of solar energy used for evaporation from plants and the soil reduces the amount of energy absorbed by the roof membrane and leads to a reduction in the cooling load. In the winter months, a green roof can be added to the insulation qualities of the roof. Nevertheless, the thermal performance is highly dependent on the amount of water retained in the green roof substrate and in a humid winter climate. On hot conditions, the heat on the naked roof material that accumulates during the day continues to reenter the building during the night. In the same study, it was determined that the green roof provided less heat during the day, and therefore the nighttime temperature effect was much less. It has been found that after the sunset, the temperature of the ambient air on the vegetation of the green roof decreases considerably, and the ambient air continues to cool down during the night [51].

The green roof is able to reduce the energy consumption and to improve the internal comfort where the spring and summer seasons in sites where the climatology is characterized by high temperature and irradiance values during the day [57–59]. In a typical building, the heat load that can be encountered in the summer months is mostly due to the heat gain from the roof.

Air Pollution Reduction

City air contains high levels of pollutants, and they generally have an adverse effect on human health. More than 1 million premature deaths are thought to be linked to air pollution in developing countries around the world according to the World Health Organization (2002) [26]. It can be foreseen that these deaths will probably increase when the increasingly crowded and polluted cities are considered. Controlling the sources of air pollutants is the traditional air pollution management program that is focused on. However, this strategy does not prevent the threat posed by uncontrolled polluters. For this reason, new approaches are needed. The uses of urban vegetation are one of them. Vegetation can diminish air pollution owing to a dry deposition process and microclimate effects [26].

By changing the cooling of the albedo and evapotranspiration of the vegetation-covered urban surfaces formed by trees and other plants, microclimates can be positively influenced, thus indirectly reduce the consumption of electricity for heating or cooling purposes. The decrease in the consumption of electricity means that pollutant emissions also decrease. At reduced ambient temperatures, photochemical reactions slow down, resulting in less secondary air pollutants [60].

It is not easy to find enough space to build vegetation in a densely populated city. For example, the percentage of impervious area in New York is as high as 64% and as high as 94% in districts such as Midtown Manhattan. However, the use of green roofs in large cities can be a method to increase vegetation in these areas [61, 62].

In a 4000 m^2 field study in Singapore, it was found that the particle and SO_2 levels in the air above the green roof decreased by 6% and 37%, respectively [61]. Another study proved that some certain air pollutants can reduce by green roofs [62].

Conclusion

Rapid growth of the world population and, accordingly, the expansion of urban areas as well as global warming and climate change are creating many problems in urban life. It is expected that the population living in the cities will increase from

50% to 66% by 2050. Therefore, the sustainability of urban life will be a much more critical issue in the future. In sustainable urban life, it is expected that people living there will have a high-quality living environment without consuming too much natural resources. From this point of view, it is inevitable to seek appropriate solutions for the problems that will be experienced here for the sustainability of city life.

Green roofs are one of the green infrastructures and can be applied to city buildings appropriately. Instead of the conventional roof surface, which has no additional functions in the cities, green roofs that have vegetation can be applied. Green roofs provide many physical, environmental, and economic advantages for urban life sustainability and mitigate the effects of global warming.

The most main benefits of green roof application in the urban areas are as follows: regulate indoor and outdoor temperatures, thus reducing of energy consumption, mitigate the effect of locally generated heat islands, supply stormwater management and rainwater harvesting, and reduce air pollutants. Besides, it gives additional benefits such as the recreational area for people, a place for urban agriculture, and noise reduction.

References

1. The World Population Prospects (2015) Revision Key Findings and Advance Tables, United Nations New York, 29 July 2015, New York. You can find from http://www.un.org/en/development/desa/publications/world-population-prospects-2015-revision.html
2. Estrada F, Botzen WJW, Tol RSJ (2017) A global economic assessment of city policies to reduce climate change impacts. Nat Clim Chang 7(6):403–406
3. Lee JH, Hancock MG, Hu MC (2014) Towards an effective framework for building smart cities: lessons from Seoul and San Francisco. Technol Forecast Soc Chang 89:80–99
4. Cook-Patton SC, Bauerle TL (2012) Potential benefits of plant diversity on vegetated roofs: a literature review. J Environ Manag 106:85–92
5. Carter T, Keeler A (2008) Life-cycle cost-benefit analysis of extensive vegetated roof systems. J Environ Manag 87(3):350–363
6. Koop SHA, van Leeuwen CJ (2016) The challenges of water, waste and climate change in cities. Environ Dev Sustainability 19(2):385–418
7. Arnell NW (2004) Climate change and global water resources: SRES emissions and socio-economic scenarios. Glob Environ Chang 14(1):31–52
8. Misra AK (2014) Climate change and challenges of water and food security. Int J Sust Built Environ 3(1):153–165
9. Cerón-Palma I, Sanyé-Mengual E, Oliver-Solà J, Montero J, Rieradevall J (2012) Barriers and opportunities regarding the implementation of rooftop Eco.Greenhouses (RTEG) in Mediterranean cities of Europe. J Urban Technol 19(4):87–103
10. Addanki SC, Venkataraman H (2017) Greening the economy: a review of urban sustainability measures for developing new cities. Sust Cities Soc 32:1–8
11. Peng LLH, Jim CY (2015) Economic evaluation of green-roof environmental benefits in the context of climate change: the case of Hong Kong. Urban For Urban Green 14(3):554–561
12. Dale VH, Efroymson RA, Kline KL (2011) The land use—climate change—energy nexus. Landsc Ecol 26(6):755–773
13. Anand P, Seetharam K (2011) Climate change and living cities: Global problems with local solutions. Climate Change and Sustainable Urban Development in Africa and Asia: 21–35. ScholarBank@NUS Repository. https://doi.org/10.1007/978-90-481-9867-2_2

14. Ahvenniemi H, Huovila A, Pinto-Seppä I, Airaksine M (2017) What are the differences between sustainable and smart cities? Cities 60:234–245
15. Barrionuevo JM, Berrone P, Ricart Costa JE (2012) Smart cities, sustainable progress: opportunities for urban development. IESE Insight 14:50–57
16. Chan ALS (2011) Developing future hourly weather files for studying the impact of climate change on building energy performance in Hong Kong. Energy Build 43(10):2860–2868
17. IPCC (2007) Fresh water resources and their management climate change 2007; Impact, adaptation and vulnerability. Contribution of working group-II to the fourth assessment report of the intergovernmental panel on climate change. Cambridge University Press: Cambridge, UK.
18. Dhillon RS, Wuehlisch GV (2013) Mitigation of global warming through renewable biomass. Biomass Bioenergy 48:75–89
19. Bhuvandas N, Timbadiya PV, Patel PL, Porey PD (2014) Review of downscaling methods in climate change and their role in hydrological studies. Int J Environ Chem Ecol Geol Geophys Eng 8(10):648–653
20. Anguluri R, Narayanan P (2017) Role of green space in urban planning: outlook towards smart cities. Urban For Urban Green 25:58–65
21. Oberndorfer E, Lundholm J, Bass B, Coffman RR, Doshi H, Dunnett N, Gaffin S, Köhler M, Liu KKY, Rowe B (2007) Green roofs as urban ecosystems: ecological structures, functions, and services. Bioscience 57(10):823–833
22. Berardi U, GhaffarianHoseini A (2014) State-of-the-art analysis of the environmental benefits of green roofs. Appl Energy 115:411–428
23. Getter KL, Rowe DB, Andresen JA (2007) Quantifying the effect of slope on extensive green roof stormwater retention. Ecol Eng 31(4):225–231
24. Kolokotsa D, Santamouris M, Zerefos SC (2013) Green and cool roofs' urban heat island mitigation potential in European climates for office buildings under free floating conditions. Sol Energy 95:118–130
25. Francis LFM, Jensen MB (2017) Benefits of green roofs: a systematic review of the evidence for three ecosystem services. Urban For Urban Green 28:167–176
26. Rowe DB (2011) Green roofs as a means of pollution abatement. Environ Pollut 159(8–9): 2100–2110
27. Vijayaraghavan K (2016) Green roofs: a critical review on the role of components, benefits, limitations and trends. Renew Sust Energ Rev 57:740–752
28. Mahmoud AS, Asif M, Hassanain MA, Babsail MO, Sanni-Anibire MO (2017) Energy and economic evaluation of green roofs for residential buildings in hot-humid climates. Buildings 7(4):30. https://doi.org/10.3390/buildings7020030
29. Carter T, Jackson CR (2007) Vegetated roofs for stormwater management at multiple spatial scales. Landscape Urban Plann 80(1–2):84–94
30. Mentens JD, Hermy M (2006) Green roofs as a tool for solving the rainwater runoff problem in the urbanized 21st century? Landscape Urban Plann 77(3):217–226
31. Teemusk A, Mander Ü (2007) Rainwater runoff quantity and quality performance from a greenroof: the effects of short-term events. Ecol Eng 30(3):271–277
32. Czemiel Berndtsson J (2010) Green roof performance towards management of runoff water quantity and quality: a review. Ecol Eng 36(4):351–360
33. Petit-Boix A, Llorach-Massana P, Sanjuan D, Sierra-Pérez J, Vinyes G, Elisabet Gabarrell Durany X (2017) Application of life cycle thinking towards sustainable cities: a review. J Clean Prod 166:939–951
34. Mickovski SB, Buss K, McKenzie BM, Sokmener B (2013) Laboratory study on the potential use of recycled inert construction waste material in the substrate mix for extensive green roofs. Ecol Eng 61:706–714
35. Saadatian O, Sopian K, Salleh E, Lim CH, Riffat S, Saadatian E, Toudeshki A, Sulaiman MY (2013) A review of energy aspects of green roofs. Renew Sust Energ Rev 23:155–168
36. Yao X, Wang Z, Wang H (2015) Impact of urbanization and land-use change on surface climate in middle and lower reaches of the Yangtze River, 1988–2008. Adv Meteorol 2015:1–10

37. Meng T, Hsu D, Wadzuk B (2017) Green and smart: perspectives of city and Water Agency Officials in Pennsylvania toward adopting new infrastructure technologies for stormwater management. J Sust Water Built Environ 3(2):05017001
38. Hoang L, Fenner RA (2015) System interactions of stormwater management using sustainable urban drainage systems and green infrastructure. Urban Water J 13(7):739–758
39. Samant TM (2015) Green roofs pertaining to storm water management in urban areas: greening the city with green roofs. Iarjset 2(10):71–77
40. Whittinghill LJ, Hsueh D, Culligan P, Plunz R (2016) Stormwater performance of a full scale rooftop farm: runoff water quality. Ecol Eng 91:195–206
41. Emilsson T, Berndtsson J, Mattsson JE, Rolf K (2007) Effect of using conventional and controlled release fertiliser on nutrient runoff from various vegetated roof systems. Ecol Eng 29(3): 260–271
42. Chini C, Canning JF, Schreiber KL, Peschel JM, Stillwell A (2017) The green experiment: cities, green stormwater infrastructure, and sustainability. Sustainability 9(1):105
43. Sutton RK, Rodie SN, Shelton DP (2014) Stormwater management: green roof basics. NebGuide, G2244
44. Gregoire BG, Clausen JC (2011) Effect of a modular extensive green roof on stormwater runoff and water quality. Ecol Eng 37(6):963–969
45. Razzaghmanesh M, Beecham S, Kazemi F (2014) Impact of green roofs on stormwater quality in a South Australian urban environment. Sci Total Environ 470–471:651–659
46. Campisano A, Butler D, Ward S, Burns MJ, Friedler E, DeBusk K, Fisher-Jeffes LN, Ghisi E, Rahman A, Furumai H, Han M (2017) Urban rainwater harvesting systems: research, implementation and future perspectives. Water Res 115:195–209
47. Santamouris M (2014) Cooling the cities—a review of reflective and green roof mitigation technologies to fight heat island and improve comfort in urban environments. Sol Energy 103: 682–703
48. Di Giuseppe E, D'Orazio M (2014) Assessment of the effectiveness of cool and green roofs for the mitigation of the Heat Island effect and for the improvement of thermal comfort in nearly zero energy building. Archit Sci Rev 58(2):134–143
49. Dimond K, Webb A (2017) Sustainable roof selection: environmental and contextual factors to be considered in choosing a vegetated roof or rooftop solar photovoltaic system. Sust Cities Soc Archit 35:241–249
50. Taleghani M (2018) Outdoor thermal comfort by different heat mitigation strategies—a review. Renew Sust Energ Rev 81:2011–2018. https://doi.org/10.1016/j.rser.2017.06.010
51. Wong JKW, Lau LS-K (2013) From the 'urban heat island' to the 'green island'? A preliminary investigation into the potential of retrofitting green roofs in Mongkok district of Hong Kong. Habitat Int 39:25–35
52. Ferre ALC, Thomé AMT, Scavarda AJ (2018) Sustainable urban infrastructure: a review. Resour Conserv Recycl 128:360–372
53. Lehmann S (2014) Low carbon districts: mitigating the urban heat island with green roof infrastructure. City Culture Soc 5(1):1–8
54. Li G, Zhang X, Mirzaei PA, Zhang J, Zhao Z (2018) Urban heat island effect of a typical valley city in China: responds to the global warming and rapid urbanization. Sust Cities Soc 38: 736–745
55. Leal Filho W, Echevarria I, Leyre N, Dorothea-Franziska K, Maris M, Edward A (2018) Coping with the impacts of urban heat islands. A literature based study on understanding urban heat vulnerability and the need for resilience in cities in a global climate change context. J Clean Prod 171:1140–1149
56. Jaffal I, Ouldboukhitine S-E, Belarbi R (2012) A comprehensive study of the impact of green roofs on building energy performance. Renew Energy 43:157–164
57. Fioretti R, Pallab A, Lanzab LG, Princip P (2010) Green roof energy and water related performance in the Mediterranean climate. Build Environ 45(8):1890–1904

58. Santamouris M, Cartalis C, Synnefa A, Kolokotsac D (2015) On the impact of urban heat island and global warming on the power demand and electricity consumption of buildings—a review. Energy Build 98:119–124
59. Chan ALS, Chow TT (2013) Energy and economic performance of green roof system under future climatic conditions in Hong Kong. Energy Build 64:182–198
60. Nowak DJ, Crane DE, Stevens JC (2006) Air pollution removal by urban trees and shrubs in the United States. Urban For Urban Green 4(3–4):115–123
61. Theodosiou T (2009) Green roofs in buildings: thermal and environmental behaviour. Adv Build Energy Res 3(1):271–288
62. Yang J, Yu Q, Gong P (2008) Quantifying air pollution removal by green roofs in Chicago. Atmos Environ 42(31):7266–7273

Part IV
Environmental Studies

Natural Radioactivity and Hazard Level Assessment of Cements and Cement Raw Materials

Naim Sezgin, Bektas Karakelle, Ugur Emre Temelli, and Semih Nemlioğlu

Abstract Cement is a composite material and it consists of different raw materials. The raw materials which are used in the cement production industry are commonly obtained from rocks such as limestone, gypsum, clay, and iron ore. In addition, the cement raw materials may also include natural radionuclides such as ^{226}Ra, ^{232}Th, and ^{40}K, which may have an adverse effect on human health. Hence, determination of natural radioactivity level is very important for human health safety. In this study, natural activity concentrations of ^{226}Ra, ^{232}Th, and ^{40}K are investigated in cement and cement raw materials in Turkey as a case study. In addition, eight different radiological parameters and indices were calculated from activity concentrations. The natural radioactivity due to the presence of ^{226}Ra, ^{232}Th, and ^{40}K was measured using the gamma spectrometer coupled with HPGe detector. The mean measured activity concentrations of ^{226}Ra, ^{232}Th, and ^{40}K in the raw materials were 38.14, 92.66, and 636.63 Bq kg^{-1}, respectively, with higher activity concentrations in coal for ^{226}Ra and trass for ^{232}Th and ^{40}K. Mean activity concentrations of natural radionuclides (^{226}Ra, ^{232}Th, and ^{40}K) in cement samples were found as 34.26, 58.2, and 512, respectively. The results showed that coal and fly ash are the principal contributors for the presence of ^{226}Ra activity concentration, trass and iron ore materials for the presence of ^{232}Th, and clay and trass raw materials for the presence of ^{40}K in cements.

Keywords Natural radioactivity · Radiation hazard · Cement raw materials · Cement

N. Sezgin · S. Nemlioğlu (✉)
Department of Environmental Engineering, Istanbul University-Cerrahpaşa, Istanbul, Turkey
e-mail: snemli@istanbul.edu.tr

B. Karakelle
Cekmece Nuclear Research, and Training Center, TAEK, Istanbul, Turkey

U. E. Temelli
Department of Property Protection and Security, Istanbul University-Cerrahpaşa, Istanbul, Turkey

N. Balkaya, S. Guneysu (eds.), *Recycling and Reuse Approaches for Better Sustainability*, Environmental Science and Engineering,
https://doi.org/10.1007/978-3-319-95888-0_14

Introduction

Cement is one of the leading construction materials applied across the globe. Without cement, structures such as bridges, tunnels, and residences could not be fabricated. Of these, the latter consumes the highest volume of cement due to plastering and cement bricks [1]. The construction sector has a very important part in Turkey's economy, like many other countries of the world in the last years. The growth of Turkey's economy is almost dependent on the growth of the construction sector and the urban transformation studies performed in many cities of the country. Therefore, there are many building raw material production plants like cement and cement raw materials in Turkey.

Cement, which is a composite material, derives from rocks and various amount mixes of industrial by-products. The general manufacture procedures of cement contain raw material mixing, burning, grinding, storage, and packaging. To make cement, aluminum, silicon, iron, and calcium are the vital elements. These elements exist in a form of limestone, clay, fly ash, iron ore, etc.—the chemistry of these materials determine the quality of the cement [2]. The amount of raw materials and industrial by-products determine the type of cement [3]. The raw materials used in cement production may contain traces of natural radioactive elements like ^{238}U, ^{232}Th, and ^{40}K—the amount of these radioactive elements vary depending on the geochemical nature and geological site of the expended raw material [4–8]. Because of this, cement and its parent materials may have varying radioactive levels. When raw materials with radioactive elements are used in cement production, they cause external and internal radiation; external exposure is due to gamma radiations from ^{238}U, ^{232}Th, and ^{40}K, while internal exposure is due to short-lived radiation materials produced by radon [9, 10].

Humans spend almost 80% of their time in enclosed parameter cement. Hence, to shield humans from radiation effects of construction material, it is imperative to determine the levels of radiation in cement and its parent materials. Upon establishment of these levels, guidelines and standards of the cement materials can be instituted. Therefore, this study assesses the radionuclide levels of radium (226), thorium (232), and potassium (40) in some cement materials produced in Turkey.

Materials and Method

Sample Collection and Preparation

A total of eight raw samples, two intermediate materials of cement production, and five different product cements were collected for gamma-spectrometric measurements of natural radioactivity levels in this study. Except for water, all the measured samples were in mass form either in granulated or grinded fine powder forms. Water sample was collected from tap water into a 500 mL polyethylene bottle from the cement factory production unit. All mass-formed samples were collected in polyethylene bags. Samples were all labeled, and their specifications and sources were noted on the labels. Mass-formed samples, except for the cement samples, were first pulverized, homogenized, and sieved by a sieve of 2 mm grid mesh before the measurement of activity concentration [11, 12]. Because of their homogeneous and powder form, cement samples were expended in their original state without any preceding processing (such as pulverization, homogenization, and sieving). All mass-formed samples were dried in a temperature-controlled furnace at 110 °C for 20–24 h until constant weight was obtained, ensuring complete removal of moisture from the samples. The weighted samples were kept in a plastic container of 250 cm^3 and hermetically sealed for 4 weeks. This process was performed to ensure secular equilibrium of ^{238}U and ^{232}Th in the sample with their respective daughters [13]. The natural radioactivity levels of present radionuclide in the samples were measured using a gamma-ray spectrometry with a high purity germanium (HPGe) detector [14]. All experiments were replicated twice—the means of the results are presented.

Cement mainly constitutes of clinker—clinker is formed after heating a mixture of limestone and clay between 1400 and 1500 °C. In addition, different types of cements are produced according to the other additives which are standardized EN-197-1:2011 [15]. CEM I, commonly branded as Portland cement, contains at least 95% clinker. Selected three types of CEM II such as A-W, B-M (L-W), and A-M (P-L) are produced by 80–94%, 65–79%, and 80–94% clinker with 6–20% (fly ash), 21–35% (trass and fly ash), and 12–20% (trass and limestone) additives, respectively. CEM IV/B (P-W) is produced by 45–64% clinker and 36–55% (trass and fly ash) additives, and all types of cements included minor additives such as natural gypsum (5–0%) EN-197-1:2011 [15].

Gamma-Spectrometric Measurements

The gamma levels of the sampled cement raw materials and cement samples were performed in a Canberra Inc.-manufactured Extended Range Coaxial High-Purity Germanium (HPGe) detector, and its specifications and measurement procedures were given by Altun et al. [16].

Radiological Parameters and Hazard Indices

Radium Equivalent Activity (Ra_{eq})

Radium equivalent activity (Ra_{eq}) is a term expended for adapting safety standards of radiation protection on human population [17–19]. This activity is calculated from radionuclide concentration such as ^{226}Ra, ^{232}Th, and ^{40}K in soil or other materials as given in Eq. (1). It is assumed that 370 Bq kg^{-1} of ^{226}Ra, 259 Bq kg^{-1} of ^{232}Th, and 4810 Bq kg^{-1} of ^{40}K produce the same gamma-ray dose rate.

$$\mathrm{Ra}_{\mathrm{eq}} = A_{\mathrm{Ra}} + 1.43A_{\mathrm{Th}} + 0.077A_{\mathrm{K}} \tag{1}$$

where A_{Ra}, A_{Th}, and A_{K} are the activity concentrations of ^{226}Ra, ^{232}Th, and ^{40}K, respectively, in Bq kg^{-1}.

Estimation of the Absorbed Gamma Dose Rate (D_{R})

The D_{R} in the indoor air of the gamma ray emitted by radionuclides (^{226}Ra, ^{232}Th, and ^{40}K) of cement and its parent materials was computed based on UNSCEAR (2000) [20] and the European Commission (1999) [21] guidelines. In addition, the dose conversion coefficients of a standard room were based on UNSCEAR and the European Commission standards—the size of a standard room was 4 m × 5 m × 2.8 m. The floor, ceiling, and concrete walls measured 20 cm thick with a density of 2350 kg m^{-3}. The values of D_{R} were calculated expending Eq. (2):

$$D_{\mathrm{R}}\left(\mathrm{nGy\ h^{-1}}\right) = 0.92A_{\mathrm{Ra}} + 1.1A_{\mathrm{Th}} + 0.08A_{\mathrm{K}} \tag{2}$$

where A_{Ra}, A_{Th}, and A_{K} (in Bq kg^{-1}) are the activity concentrations of radium (226), thorium (232), and potassium (40), respectively. This study adopted 84.00 nGy h^{-1} as a D_{R} reference value. This D_{R} value represents the external, terrestrial gamma radiation and the world's average population-weighted D_{R} [20].

Annual Effective Dose Equivalent (AEDE)

AEDE is the ratio of absorbed amount in air to effective quantity received by adults. The following Eq. (3) given by UNSCEAR (2000) was expended for the annual effective dose equivalent calculation. To estimate the AEDE, a conversion coefficient and the outdoor occupancy factor of 0.7 Sv Gy^{-1} and 0.2 were used, respectively, as shown in the equation below.

$$\mathrm{AEDE}\left(\mu Sv\ \mathrm{year}^{-1}\right) = [D_{\mathrm{R}}\left(\mathrm{nGy\ h^{-1}}\right) \times 8760\left(\mathrm{h}^{-1}\right) \times 0.7\ \left(\mathrm{Sv\ Gy^{-1}}\right) \times 0.2 \times 10^{-3} \tag{3}$$

where D_{R} (nGy h^{-1}) is estimated using Eq. (5), average worldwide value of AEDE as 480 μSv y^{-1} [20].

Annual Gonadal Dose Equivalent (AGDE)

Radiation can cause different negative effects such as death or mutation of all living cells or a whole organ [22]. According to UNSCEAR (2000) [20], the bone marrow, the bone surface cells, the thyroid, the lungs, and the gonads are among the organs that are much affected by radiations. Therefore, determining the annual gonadal dosage equivalent (AGDE) is very important—it is estimated using Eq. (4) [23, 24].

$$\text{AGDE}\left(\mu Sv\ \text{year}^{-1}\right) = 3.09A_{\text{Ra}} + 4.18A_{\text{Th}} + 0.3147A_{\text{K}} \tag{4}$$

Excess Lifetime Cancer Risk (ELCR)

The prolonged contact of radiation from natural radioactivity in the soil, especially settlement areas, can cause adverse effects such as cancer. Some regulatory agencies expend a quantitative risk index assessment procedure to identify ELCR [25]. This value is calculated according to the likelihood of cancer in a population of individuals for a certain lifetime expending projected intakes, exposures, chemical-specific dosage, and response data. ELCR is estimated using Eq. (5), as quoted by Ramasamy et al. [26, 27]:

$$\text{ELCR} = \text{AEDE} \times \text{DL} \times \text{RF} \tag{5}$$

where AEDE is the yearly effective dosage equivalent, DL is 70 years of life duration, and RF is fatal cancer risk per Sievert (0.05 Sv^{-1}) [28].

Gamma Index (I_γ)

The European Commission (1999) [21] suggested an index named as the gamma index (I_γ) in order to provide the guidelines of the European Commission for building materials usage. According to the European Commission (1999), the exemption criterion of gamma dosage is 0.3 mSv y^{-1}, while the upper limit criterion stands at 1 mSv y^{-1}. The upper limit has been applied by numerous countries as a control limit [29]. The gamma index, I_γ, was defined in Eq. (6):

$$I_\gamma = \frac{A_{\text{Ra}}}{300\ \text{Bq kg}^{-1}} + \frac{A_{\text{Th}}}{200\ \text{Bq kg}^{-1}} + \frac{A_{\text{K}}}{3000\ \text{Bq kg}^{-1}} \tag{6}$$

where A_{Ra}, A_{Th}, and A_{K} are the activity concentrations of ^{226}Ra, ^{232}Th, and ^{40}K, respectively, in Bq kg^{-1}.

External Radiation Hazard (H_{ex})

External radiation hazard, H_{ex}, has been accepted as a limit value of unity in order to be nonhazardous. It is given in Eq. (7):

$$H_{ex} = \frac{A_{Ra}}{370\ \text{Bq kg}^{-1}} + \frac{A_{Th}}{259\ \text{Bq kg}^{-1}} + \frac{A_{K}}{4810\ \text{Bq kg}^{-1}} \quad (7)$$

where A_{Ra}, A_{Th}, and A_K are the activity concentrations of ^{226}Ra, ^{232}Th, and ^{40}K, respectively, in Bq kg^{-1}.

Internal Radiation Hazard (H_{in})

Internal radiation hazard, H_{in}, is shown in Eq. (8). A value ≤1 is favored [30].

$$H_{in} = \frac{A_{Ra}}{185\ \text{Bq kg}^{-1}} + \frac{A_{Th}}{259\ \text{Bq kg}^{-1}} + \frac{A_{K}}{4810\ \text{Bq kg}^{-1}} \quad (8)$$

where A_{Ra}, A_{Th}, and A_K are the activity concentrations of ^{226}Ra, ^{232}Th, and ^{40}K, respectively, in Bq kg^{-1}.

Results and Discussion

The measured activity concentrations of ^{226}Ra, ^{232}Th, and ^{40}K of the studied samples are shown in Table 1. Water and gypsum have the lowest activity concentrations of raw materials for ^{226}Ra, ^{232}Th, and ^{40}K. The highest activity concentrations were captured in raw materials: coal (95.00 Bq kg^{-1}) for ^{226}Ra and trass (270.00 Bq kg^{-1} and 1990.00 Bq kg^{-1}) for ^{232}Th and ^{40}K, respectively. The studied clinker sample has completely lower activity concentrations (16.60 Bq kg^{-1}, 30.00 Bq kg^{-1}, and 380.00 Bq kg^{-1}, respectively) than the lowest activity concentration values of the cement samples of this study for ^{226}Ra, ^{232}Th, and ^{40}K.

The least activity value of ^{226}Ra in all cement samples was found in CEM II/A-M (P-L) (26.30 Bq kg^{-1}), while the maximum activity value of ^{226}Ra was in CEM II/B-M (L-W)—47.0 Bq kg^{-1} (Table 1). The value of ^{226}Ra activity in CEM II/B-M (L-W) was still lower than both Turkey and Makedonia originated same type of cement samples in the literature (Table 2).

The highest activity values for ^{232}Th (87.0 Bq kg^{-1}) and ^{40}K (670.0 Bq kg^{-1}) were found in CEM IV/B (P-W), which has higher values than the other Turkish cement sample activity concentrations for these two radionuclides in the literature (Table 2). The lowest activity levels for ^{232}Th and ^{40}K (39.0 Bq kg^{-1} and 430.0 Bq kg^{-1}, respectively) were reached for CEM I. These results show that the contents of radium

Table 1 Activity concentrations of cement raw materials and cement

Material type	Activity concentration (Bq kg^{-1} ± relative error) ^{226}Ra	^{232}Th	^{40}K
Raw materials			
Fly ash	87.00 ± 8.70	133.00 ± 18.62	1090.00 ± 87.20
Coal	95.00 ± 10.45	72.00 ± 18.00	170.00 ± 85.00
Iron ore	26.90 ± 2.96	148.00 ± 17.76	310.00 ± 34.10
Gypsum	9.90 ± 1.09	<MDA	<MDA
Clay	23.70 ± 2.61	106.00 ± 13.78	1470.00 ± 102.90
Limestone	15.60 ± 1.72	12.30 ± 2.83	63.00 ± 14.49
Trass	42.00 ± 4.20	270.00 ± 29.70	1990.00 ± 139.30
Water	5.00 ± 0.55	<MDA	<MDA
Raw materials mean	38.14	92.66	636.63
Intermediate materials			
Raw meal	37.00 ± 4.07	40.00 ± 8.40	320.00 ± 48.00
Clinker	16.60 ± 1.83	30.00 ± 4.80	380.00 ± 34.20
Int. materials mean	26.80	35.00	350.00
Cement types			
CEM I	30.00 ± 3.30	39.00 ± 7.41	430.00 ± 47.30
CEM II/A-W	34.00 ± 3.74	46.00 ± 7.82	460.00 ± 50.60
CEM II/B-M (L-W)	47.00 ± 4.70	71.00 ± 11.36	550.00 ± 55.00
CEM II/A-M (P-L)	26.30 ± 2.89	48.00 ± 7.68	450.00 ± 45.00
CEM IV/B (P-W)	34.00 ± 3.40	87.00 ± 12.18	670.00 ± 60.30
Mean	34.26	58.20	512.00

(226), thorium (232), and potassium (40) depend on raw materials used, geological site, and geochemical nature. More or less, most parts of cements consist of a clinker. On the other hand, because of their higher activity concentrations, fly ash and/or trass should be taken into account as main origins of high amount of activity concentrations of measured radionuclides in the studied cement samples.

Radium equivalent activity, Ra_{eq}, levels of the raw materials and the studied cements were given in Table 3; 370 Bq kg^{-1} was used as a specification reference limit to all samples [12]. Ra_{eq} values of the raw materials were between 5.00 and 581.33 Bq kg^{-1}. With 581.33 Bq kg^{-1} value, only the trass sample was higher than the limit value among the raw materials. Meanwhile, fly ash was critical with its near critical value, 361.12 Bq kg^{-1}. The intermediate materials were far from the limit value as given in Table 3. All the investigated cement samples had the least Ra_{eq} ranging from 118.88 to 210.00 Bq kg^{-1}; these values are lower than their mild Ra_{eq} levels.

Absorbed gamma dose rates (D_R) in indoor air were scattered between 4.60 and 494.84 nGy h^{-1} for the raw materials, 78.67 and 103.64 nGy h^{-1} for the intermediate materials, and 104.90 and 180.58 nGy h^{-1} for the cement samples (Table 3). Most of the raw material values of D_R were higher than the world average value 84.00 nGy h^{-1}. Raw meal was higher, but the clinker was lower than the world average.

Table 2 Comparison of ^{226}Ra, ^{232}Th, and ^{40}K activity concentrations of some cement raw materials and cement

Raw materials	^{226}Ra	^{232}Th	^{40}K	Ref.
Fly ash				
Turkey	87.0	133.0	1090.0	This study
Turkey	232.3	117.1	466.2	[31]
Macedonia	140.0	80.0	540.0	[32]
	85.0	129.0	786.0	
India	45.1	39.9	88.4	[33]
Bangladeshi	117.8	157.3	1463.3	[34]
Iron (ore/oxide)				
Turkey	26.9	148.0	310.0	This study
Turkey	41.6	11.4	152.6	[31]
Egypt	160.5	87.3	121.3	[35]
Saudi Arabia	37.2	28.8	44.8	[36]
Saudi Arabia	21.6	18.6	53.6	[37]
Gypsum				
Turkey	9.9	<MDA	<MDA	This study
Turkey	8.0	11.0	35.0	[38]
Turkey	10.8	3.6	44.5	[31]
Albania	11.8	5.8	66.8	[4]
Saudi Arabia	9.0	6.5	184.8	[37]
Saudi Arabia	7.7	3.3	173	[36]
Bangladeshi	58.4	91.2	1101.1	[34]
Tanzania	9.8	4.4	81	[1]
	2.6	3.0	6.3	
Pakistan	8.2	16.2	187.7	[39]
Egypt	31.7	55.2	88.7	[35]
Macedonia	5.9	1.44	11.0	[32]
Greece	6.8	<MDA	<MDA	[40]
Clay				
Turkey	23.7	106.0	1470.0	This study
Turkey	26.7	41.8	629.3	[31]
India	63.7	38.6	313.7	[29]
Saudi Arabia	15.8	13.8	70.7	[37]
Saudi Arabia	18.2	22.4	127	[36]
Tanzania	90.7	123.3	137.7	[1]
	23.4	43.2	21.2	
Pakistan	34.7	41.2	187.6	[39]
Egypt	33.7	68.9	130.7	[35]
Limestone				
Turkey	15.6	12.30	63.0	This study
Turkey	16.5	7.7	88.1	[31]
Saudi Arabia	6.2	3.0	155.5	[37]

(continued)

Table 2 (continued)

Raw materials	^{226}Ra	^{232}Th	^{40}K	Ref.
Saudi Arabia	42.8	0.9	<MDA	[36]
Tanzania	6.4	15.2	80	[1]
	25.2	1.3	13.2	
Pakistan	28.4	11.3	63.1	[39]
Egypt	19.7	39.0	61.2	[35]
Greece	6.0	6.6	101.0	[40]
Trass				
Turkey	42.0	270.0	1990.0	This study
Turkey	67.9	76.7	681.6	[31]
Clinker				
Turkey	16.60	30.00	380.0	This study
Turkey	28.3	15.9	219.0	[31]
Albania	55.5	17.0	160.3	[4]
Bangladeshi	49.8	75.7	856.4	[34]
Saudi Arabia	79.9	7.5	6.1	[36]
Pakistan	51.1	23.2	258.4	[39]
Greece	15.0	14.0	141.0	[40]
Makedonia	31.0	20.0	234.0	[32]
Cement types				
CEM I				
Turkey	30.0	39.0	430.0	This study
Turkey	34.0	13.0	208.0	[41]
Turkey	29.8	17.5	239.0	[31]
Albania	51.2	16.1	168.8	[4]
India	35.7	37.7	159.8	[29]
Saudi Arabia	11.2	10.0	117.1	[37]
Makedonia	30.0	20.0	222.0	[32]
CEM II				
Turkey CEM II/A-W	34.0	46.0	460.0	This study
CEM II/B-M(L-W)	47.0	71.0	550.0	
CEM II/A-M(P-L)	26.3	48.0	450.0	
Turkey CEM II	51.0	18.0	221.0	[41]
Turkey CEM II/A-LL	22.4	12.6	157.1	[31]
Albania CEM II/A-LL	51.0	16.5	150.4	[4]
CEM II/B-LL	46.2	12.0	133.7	
Makedonia CEM II/A-M	45.0	29.0	272.0	[32]
CEM II/B-M	50.0	34.0	295.0	
CEM IV				
Turkey	34.0	87.0	670.0	This study
Turkey	45.0	26.0	352.0	[41]

Table 3 Radiological parameters and hazard indices of cement raw materials and cement samples

Material type	Ra_{Eq} (Bq kg^{-1})	Dose rate (D_R) (nGy h^{-1})	AEDE (μSv y^{-1})	AGDE (μSv y^{-1})	ELCR	I_γ	H_{ex}	H_{in}
Raw materials								
Fly ash	361.12	313.54	384.53	1167.03	1.35E−03	1.32	0.98	1.21
Coal	211.05	180.20	221.00	647.89	7.73E−04	0.73	0.57	0.83
Iron ore	262.41	212.35	260.42	799.10	9.11E−04	0.93	0.71	0.78
Gypsum	9.90	9.11	11.17	30.59	3.91E−05	0.03	0.03	0.05
Clay	288.47	256.00	313.96	977.89	1.10E−03	1.10	0.78	0.84
Limestone	38.04	32.92	40.38	119.40	1.41E−04	0.13	0.10	0.14
Trass	581.33	494.84	606.87	1883.24	2.12E−03	2.15	1.57	1.68
Water	5.00	4.60	5.64	15.45	1.97E−05	0.02	0.01	0.03
Raw materials mean	219.67	187.95	230.50	705.07	8.07E−04	0.80	0.59	0.70
Intermediate materials								
Raw meal	118.84	103.64	127.10	382.01	4.45E−04	0.43	0.32	0.42
Clinker	88.76	78.67	96.48	296.01	3.38E−04	0.33	0.24	0.28
Int. materials mean	103.80	91.16	111.79	339.01	3.91E−04	0.38	0.28	0.35
Cement types								
CEM I	118.88	104.90	128.65	390.74	4.50E−04	0.44	0.32	0.40
CEM II/A-W	135.20	118.68	145.55	441.78	5.09E−04	0.50	0.37	0.46
CEM II/B-M (L-W)	190.88	165.34	202.77	614.71	7.10E−04	0.70	0.52	0.64
CEM II/A-M (P-L)	129.59	113.00	138.58	423.21	4.85E−04	0.48	0.35	0.42
CEM IV/B (P-W)	210.00	180.58	221.46	679.10	7.75E−04	0.77	0.57	0.66
Mean	156.91	136.50	167.40	509.91	5.86E−04	0.58	0.42	0.52
	370.00[a]	84.00[a]	480.00[a]		2.90E−04	1.00[b]	1.00	1.00

[a]UNSCEAR (2000) [20] world average
[b]Council Directive (2014) [42]

However, all D_R values of the cement samples were more or less higher than the world average 84.00 nGy h^{-1}.

AEDE values were calculated in the ranges of 6.64–606.87 μSv y^{-1}, 96.48–127.10 μSv y^{-1}, 128.65–221.46 μSv y^{-1} for the raw and the intermediate product materials and the studied cement samples, respectively, as given in Table 3. Compared to the world average reference value of 480.00 μSv y^{-1}, all AEDE readings were below this value—except trass.

Annual gonadal dose equivalent (AGDE) values varied between 15.45 and 1883.24 μSv y^{-1}, 296.01 and 382.01 μSv y^{-1}, and 390.74 and 679.10 μSv y^{-1}, for the raw and the intermediate product materials and the cement samples, respectively (Table 3). The average value of AGDE is 509.91 μSv y^{-1} for the studied cements.

Excess lifetime cancer risk (ELCR) values were obtained between 1.97×10^{-5} and 2.12×10^{-3}, 3.38×10^{-4} and 4.45×10^{-4}, and 4.50×10^{-4} and 7.75×10^{-4}, for the raw and the intermediate product materials and the cement samples, respectively, as given in Table 3. Only water and gypsum remained lower than the limit reference value, 2.90×10^{-4}. It should be considered that according to their raw and intermediate materials' ELCR levels, all the studied cement sample values were higher than the limit level.

The gamma index (I_γ) values were calculated between 0.02 and 2.15, 0.33 and 0.43, and 0.44 and 0.77, for the raw and the intermediate product materials and the cement samples, respectively, as shown in Table 3. Fly ash, clay, and trass were higher than the limit value, 1.00. All the calculated samples for the raw and intermediate materials were stayed lower levels for this parameter. The samples' I_γ values were below the reference point.

External radiation hazard (H_{ex}) values were found between 0.01 and 1.57, 0.24 and 0.32, and 0.32 and 0.57 for the raw and the intermediate product materials and the cement samples, respectively, as presented in Table 3. The limit reference value, 1.00, was only exceeded by trass among the raw and the intermediate materials. Fly ash drew attention on the boundary of the reference limit. Comparing the H_{ex} results of the samples, their H_{ex} values were below the standard limit.

In the latter of radiation hazard indices, internal radiation hazard, H_{in}, values were found between 0.03 and 1.68, 0.28 and 0.42, and 0.40 and 0.60 for the raw and the intermediate product materials and the cement samples, respectively, as given in Table 3. The limit reference value, 1.00, was only exceeded by trass and fly ash among the raw and the intermediate materials. All the cement samples of this study were lower than the limit value.

Conclusions

In this study, cement factory raw materials, intermediate materials, and five different types of cement samples were investigated to determine the activity concentrations of their natural radionuclides and evaluated using some important radiation parameters, such as radiation indices. Gamma-ray spectrometry method was used to determine the activity concentration. Results indicated an average of 34.26 Bq kg^{-1}, 58.20 Bq kg^{-1}, and 512.00 Bq kg^{-1} in ^{226}Ra, ^{232}Th, and ^{40}K, respectively. Radium equivalent activity (Ra_{eq}) levels of the investigated cements varied between 118.88 and 210.00 Bq kg^{-1}, which were below the allowable concentration of 370 Bq kg^{-1}. The D_R for the selected samples were between 104.90 and 180.58 nGy h^{-1}, which were higher than the world average value of 84.00 nGy h^{-1}. Annual effective dose equivalent (AEDE) values for the cement samples were in the range of 128.65–221.46 μSv y^{-1}, which were all lower than the world average reference value 480.00 μSv y^{-1}. Annual gonadal dose equivalent (AGDE) values varied between 390.74 and 679.10 μSv y^{-1}, and the mean value was 509.91 μSv y^{-1} for the cement samples. Excess lifetime cancer risk (ELCR) values of the studied cements were between 4.50×10^{-4} and 7.75×10^{-4}. All the studied cement sample values were higher than the limit level of 2.90×10^{-4}. The gamma index (I_γ) values of the cement samples were between 0.44 and 0.77. All the cement samples' I_γ values were below the standard limit value, 1.00. External radiation hazard index (H_{ex}) values of the samples were in a range of 0.32 and 0.57. The internal radiation hazard index (H_{in}) values of the investigated cements were found between 0.40 and 0.60, which were lower than the limit value, 1.00.

This study shows that the raw materials could be effective on the natural radiation levels and amounts of the related parameters. Especially, levels of radionuclides of trass and fly ash are important among the raw materials of the investigated cements. Coal as fuel of cement production has also importance because of high amount of radionuclide content. Consequently, natural radioactivity levels of the product cements could be mainly affected by the raw materials related to their geological locations. When trass and/or fly ash added into the cement content, they could be effective constituents of natural radioactivity levels of the produced cements. Coal should be taken into account regarding the natural radioactivity evaluations. It can be stated that in order to decrease the natural radioactivity level of cements, raw material and fuel should be selected from lower radioactive originated locations and sources, considering the economic conditions, as well.

References

1. Amasi AI, Mtei KM, Nathan IJ, Jodlowski P, Dinh CN (2014) Natural radioactivity in Tanzania cements and their raw materials. Res Environ Earth Sci 6(10):469–474
2. British Geological Survey (2005). Mineral Profile: Cement Raw Materials. UK, British Geological Survey, Natural Environment Research Council, Office of the Deputy Prime Minister, p.20. https://www.bgs.ac.uk/downloads/start.cfm?id=1408
3. European Commission, Integrated Pollution Prevention and Control (IPPC) (2001) Reference document on best available techniques in the cement and lime manufacturing industries. https://theconstructor.org/building/manufacture-of-cement/13709. Accessed 15 Apr 2018
4. Shala F, Xhixha G, Kac M¸ Xhixha E, Hasani F, Xhixha E, Shyti M, Kuqi DS, Prifti D, Qafleshi M (2017) Natural radioactivity in cements and raw materials used in Albanian cement industry. Environ Earth Sci 76, 670
5. Yuce G, Ugurluoglu D, Dilaver AT, Eser T, Sayin M, Donmez M, Ozcelik S, Aydin F (2009) The effects of lithology on water pollution: natural radioactivity and trace elements in water resources of Eskisehir Region (Turkey). Water Air Soil Pollut 202(1–4):69–89
6. Callegari I, Bezzon GP, Broggini C, Buso GP, Caciolli A, Carmignani L, Colonna T, Fiorentini G, Guastaldi E, Kac M, Xhixha E, Mantovani F, Massa G, Menegazzo R, Mou L, Pirro A, Rossi Alvarez C, Strati V, Xhixha G, Zanon A (2013) Total natural radioactivity map of Tuscany (Italy). J Maps 9(3):438–443
7. Strati V, Baldoncini M, Bezzon GP, Broggini C, Buso GP, Caciolli A, Callegari I, Carmignani L, Colonna T, Fiorentini G, Guastaldi E, Kaceli Xhixha M, Mantovani F, Menegazzo R, Mou L, Rossi Alvarez C, Xhixha G, Zanon A (2014) Total natural radioactivity map of Veneto (Italy). J Maps 11(4):545–551
8. Kaceli Xhixha M, Albèri M, Baldoncini M, Bezzon GP, Buso GP, Callegari I, Casini L, Cuccuru S, Fiorentini G, Guastaldi E, Mantovani F, Mou L, Oggiano G, Puccini A, Rossi

Alvarez C, Strati V, Xhixha G, Zanon A (2016) Uranium distribution in the Variscan basement of Northeastern Sardinia. J Maps 12(5):1029–1036
9. Righi S, Bruzzi L (2006) Natural radioactivity and radon exhalation in building materials used in Italian dwellings. J Environ Radioact 88:158–170
10. Khan K, Khan HM (2001) Natural gamma-emitting radionuclides in Pakistani Portland cement. Appl Radiat Isot 54:861–865
11. Awodugba AO, Adelabu JSA, Awodele MK, Ishola GA (2007) Gamma ray activity in raw materials and the end product from the West African Portland Cement PLC, Ewekoro, South Western Nigeria. Indoor Built Environ 16(6):569–572
12. Mollah AS, Ahmad GU, Hussain SR, Rahman MM (1986) The natural radioactivity of some building materials used in Bangladesh. Health Phys 50:849–851
13. Schotzing V, Debertin K (1983) Photon emission probabilities per decay of ^{226}Ra and ^{232}Th equilibrium with their daughter products. Int J Appl Radiat Isot 34:533–538
14. Ajayi OS, Ajayi IR (1999) Survey of environmental gamma radiation levels of some areas of Ekiti and Ondo State, Southwestern Part of Nigeria. Niger J Phys 11:17–21
15. European Standard EN 197-1:2011, Cement part 1: composition, specifications and conformity criteria for common cements
16. Altun M, Sezgin N, Nemlioglu S, Karakelle B, Can N, Temelli UE (2017) Natural radioactivity and hazard-level assessment of Portland cements in Turkey. J Radioanal Nucl Chem 314:941–948
17. Matiullah AA, Rehman S, Faheem M (2004) Measurement of radioactivity in the soil of Bahawalpur division, Pakistan. Radiat Prot Dosim 112(3):443–447
18. Nageswara Rao MV, Bhatti SS, Rama Seshu P, Reddy AR (1996) Natural radioactivity in soil and radiation levels of Rajasthan. Radiat Prot Dosim 63:207–216
19. NEA-OECD (1979) Exposure to radiation from natural radioactivity in building materials. Report by NEA Group of Experts of the Nuclear Energy Agency. OECD, Paris
20. United National Scientific Committee on the Effects of Atomic Radiation (UNSCEAR) (2000) Sources and risks of ionizing radiation. Report to the general assembly with annexes. United Nations, New York, NY
21. European Commission (EC) (1999) Radiation Protection, 112. Radiological protection principles concerning the natural radioactivity of building materials. Directorate-General Environment. Nuclear Safety and Civil Protection
22. Zaim N, Atlas H (2016) Assessment of radioactivity levels and radiation hazards using gamma spectrometry in soil samples of Edirne, Turkey. J Radioanal Nucl Chem 310(3):959–967
23. Mamont-Ciesla K, Gwiazdowski B, Biernacka M, Zak A (1982) Radioactivity of building materials in Poland. In: Vohra G, Pillai KC, Sadavisan S (eds) Natural radiation environment. Halsted Press, New York, NY, p 551
24. Vohra G, Pillai KC, Sadavisan S (eds) (1982) Natural radiation environment. Halsted Press, New York, NY
25. Okeyode IC, Jibiri NN (2013) Excess lifetime cancer risks associated with the use of sediments from Ogun River, Nigeria as building material. Res J Phys 7(1):1–8
26. Ramasamy V, Suresh G, Meenakshisundaram V, Ponnusamy V (2011) Horizontal and vertical characterization of radionuclides and minerals in river sediments. Appl Radiat Isot 69:184–195
27. Ramasamy V, Sundarrajan M, Paramasivam K, Meenakshisundaram V, Suresh G (2013) Assessment of spatial distribution and radiological hazardous nature of radionuclides in high background radiation area, Kerala, India. Appl Radiat Isot 73:21–31
28. International Commission on Radiological Protection (ICRP) (2007) The Recommendations of the International Commission on Radiological Protection. ICRP Publication 103. Ann ICRP 37, pp 2–4
29. Raghu Y, Ravisankar R, Chandrasekaran A, Vijayagopal P, Venkatraman B (2017) Assessment of natural radioactivity and radiological hazards in building materials used in the Tiruvannamalai District, Tamilnadu, India, using a statistical approach. J Taibah Univ Sci 11:523–533

30. Krieger R (1981) Radioactivity of construction materials. Betonwerk Fertigteil Techn 47:468–473
31. Turhan S (2008) Assessment of the natural radioactivity and radiological hazards in Turkish cement and its raw materials. J Environ Radioact 99(2):404–414
32. Stojanovska Z, Nedelkovski D, Ristova M (2010) Natural radioactivity and human exposure by raw materials and end product from cement industry used as building materials. Radiat Meas 45(8):969–972
33. Kumar V, Ramachandran TV, Prasad R (1999) Natural radioactivity of Indian building materials and by-products. Appl Radiat Isot 51:93–96
34. Asaduzzaman K, Mannan F, Khandaker MU, Farook MS, Elkezza A, Bin Mohd Amin Y, Sharma S, Bin Abu Kassim H (2015) Assessment of natural radioactivity levels and potential radiological risks of common building materials used in Bangladeshi dwellings, assessment of natural radioactivity levels and potential radiological risks of common building materials used in Bangladeshi dwellings. PLoS One 10(10):e0140667. https://doi.org/10.1371/journal.pone.0140667
35. El-Taher A, Makhluf S, Nossair A, Halim ASA (2010) Assessment of natural radioactivity levels and radiation hazards due to cement industry. Appl Radiat Isot 68(1):169–174
36. Al-Dadi MM, Hassan HE, Sharshar T, Arida HA, Badran HM (2014) Environmental impact of some cement manufacturing plants in Saudi Arabia. J Radioanal Nucl Chem 302(3):1103–1117
37. Alashrah S, El-Taher A (2016) Gamma spectroscopic analysis and associated radiation hazards parameters of cement used in Saudi Arabia. J Environ Sci Technol 9:238–245
38. Tufan MC, Disci T (2013) Natural radioactivity measurements in building materials used in Samsun, Turkey. Radiat Prot Dosim 156(1):87–92
39. Aslam M, Gul R, Ara T, Hussain M (2012) Assessment of radiological hazards of naturally occurring radioactive materials in cement industry. Radiat Prot Dosim 151(3):483–488
40. Papaefthymiou H, Gouseti O (2008) Natural radioactivity and associated radiation hazards in building materials used in Peloponnese, Greece. Radiat Meas 43(8):1453–1457
41. Ozdis BE, Cam NF, Ozturk BC (2017) Assessment of natural radioactivity in cements used as building materials in Turkey. J Radioanal Nucl Chem 311:307–316
42. Council Directive 2013/59/Euratom of 5 Dec. 2013 laying down basic safety standards for protection against the dangers arising from exposure to ionising radiation, and repealing Directives 89/618/Euratom, 90/641/Euratom, 96/29/Euratom, 97/43/Euratom and 2003/122/Euratom. L13, 57

Ecological Footprint Calculation

Emine Elmaslar Özbaş, Selda Yiğit Hunce, Hüseyin Kurtuluş Özcan, and Atakan Öngen

Abstract Ecological footprint is a method that was created to calculate the load for nature of a specific population, and it also calculates the biological effective and water-containing areas which are required to obtain renewable sources for the use of people. The data obtained by the ecological footprint resource accounting provides guidance for strategic decisions on resource management. Many studies exist in the web area in order to calculate the ecological footprint automatically after entering the data. In this study, the questions in a web site were asked to people living in Istanbul in different age intervals. It was observed that the percentages of the components constituting the ecological footprints calculated for three different age ranges (three different age ranges between individuals aged 50 and below) varied according to the age ranges. In addition, the values of ecological footprints according to the education and income levels were examined for the male and the female individuals. For all the age ranges, it can be said that the ecological footprints of individuals with high economic income are higher than those with low economic income. Also, the ecological footprint of the males of the same income level is higher than the ecological footprints of the females.

Keywords Ecological footprint · Age · Income rate

Introduction

Human activities create ecological pressures such as pollution and resource depletion, and it leads to the changes in the ecosystems [1]. Climate change, pollution, and decrease in biodiversity can be listed as the main resulting environmental effects; however, changes in the ecosystems have several different environmental impacts in addition to the listed main impacts. Developing countries having relatively poor and

E. E. Özbaş (✉) · S. Y. Hunce · H. K. Özcan · A. Öngen
Department of Environmental Engineering, Istanbul University-Cerrahpaşa, Istanbul, Turkey
e-mail: elmaslar@istanbul.edu.tr

N. Balkaya, S. Guneysu (eds.), *Recycling and Reuse Approaches for Better Sustainability*, Environmental Science and Engineering,
https://doi.org/10.1007/978-3-319-95888-0_15

vulnerable populations are primarily affected from the results (e.g., water shortage) of that kind of environmental changes [2, 3].

In the recent past, various tools have been suggested with different characteristics that will enable us to measure the pressure created by human activities and development trends and to ensure the prosperity of the future generations [4]. EC measurement is one of the well-accepted tools among them. The term ecological footprint (EF) was firstly introduced in 1992 by Canadian ecology economist William Rees and then developed by Wackernagel [5, 6]. EF is a method that was created to calculate the load for the nature of a specific population, and it also calculates the biological effective and water-containing areas which are required to obtain renewable sources for the use of people [5–7]. The data obtained by EF resource accounting provides guidance for strategic decisions on resource management. This can play a key role in getting greater success and financial security of a country [8]. EF has been determined as a simple but effective method to identify the environmental impact [9, 10].

The Ecological Footprint Atlas [11] reveals that human demands and activities have exceeded the earth's biocapacity, and an ecological overshoot started in 1970s [3–12]. In simple terms, the EF value measures how quickly human activities consume the resources of the nature, and it is generally presented in conjunction with biocapacity. The biocapacity is a means of measuring the bio-productivity of the nature and represents the biosphere's regenerative capacity [13–18].

The major benefits of EF are the following: its calculation is not complex, it gives an idea by using the past data, it gives a chance to compare the data of a chosen sample area worldwide, and it has a standard measure by means of units of land per capita [19]. The major drawbacks of using EF to determine environmental impacts are the following: it neglects the technological (environmental friendly or not) changes between the compared ones, it is a statistical measure, it ignores the underground sources such as water, and it ignores the flows in or out of the sample area [18–20].

In this study, the questions in a web site were asked to people living in Istanbul in different age intervals to calculate the EF values. Three different age ranges between individuals aged 50 and below were evaluated to compare the EF values variation among young and mature people. The data of the female and the male individuals were interpreted separately. Also, education and income levels were examined in order to determine the effect of spending habits on EF.

Material and Method

The calculation of EF is an area of expertise that requires a detailed study. Many studies exist in the web area in order to calculate EF automatically after entering the data. In this study, the questions in the web site [21] were asked to people living in Istanbul and having different age intervals. The given answers were employed to the survey form in the same web site, so the EF of individuals surveyed were calculated.

The frequency of consumption of animal products (red meat, poultry, fish, eggs, milk) and the type and quality of consumed food (processed, packaged, imported products) were asked regarding the food footprint. Due to the need for more biological productive space in the production process compared to the plant products, animal products have been used to calculate the food footprint.

The amount of garbage produced by the individual compared to his neighbor was questioned for the calculation of the waste footprint. There was only one question in this subject. The reason is to avoid double counts by taking into account the amount of waste produced in food, housing, and transportation footprints.

The housing footprint questions were about the size of the housing in question, the quality, the number of people living in the house, and the use of electricity. The size of the residence in terms of square meters and its ratio of the number of the individuals living there were related to the amount of usable space. The more people living per square meters results in smaller EF. Mass housing, dormitory, apartments, and detached house with garden have EF values increasing, respectively. Having electricity and water network in the housing is an element that increases the EF in terms of energy consumption and the usage of biological productive area.

Transportation footprint was calculated by the data of the use of public transport, the use of special vehicles, the average length of road travel per week, the fuel consumption of the vehicle used, and the annual airline use. While the use of public vehicles reduces the EF, the use of special vehicles and airline increases the EF as the excess amount of road travel creates air and noise pollution and increases the need for asphalt road construction.

The housing, food, transportation, and waste footprints of each individual and how many planets are required for the individuals to maintain their consumption patterns were obtained based upon the automatic calculation of the program. Figure 1 shows the location of the city of Istanbul in Turkey.

Fig. 1 The location of Istanbul in Turkey [22]

Results

The EF survey results of individuals are given in Table 1 according to 16–20, 21–35, and 36–50 age intervals. The EF values of the age intervals are presented in Table 2 according to the income rates. The values in the "travel" subcategory have reached to high levels for the individuals in the 16–20 age range, most of whom were students. Owing to the fact that the travel time cannot be changed in eco-hint options, it is found that the EF values were not decreased adequately. When the EF based on the income rates of the individuals in the 16–20 age range was examined, it was seen that EF value increased with the increase in the income rates.

In the individuals in the 21–35 age group, the majority of whom consisting of students, a reduction in travelling rates was not determined resulting from the living requirements of Istanbul. In the other subcategory, it was found that the technological equipment purchasing rates had an increasing effect. It was seen that the high rates of technological equipment purchasing had an increasing effect on EF especially for the participators in the 30s age range. When the EFs of the female and the male in the 21–35 age range were compared based on financial situation, it was observed that the value of EF increased with the increase in the income rate.

When the EFs of individuals in the 36–50 age range were examined, it was seen that the EF hints were not useful for this age group who were constantly renewing their homes and often buying furniture and technological equipment. It was also seen that the EF value increased with the increase in the income rates in this group too, as the others. The total EF values for 2013 are given in Table 3 for some countries.

Table 1 The variation of EF values and the contribution of consumption elements forming EF

Age	Food (%)	Travel (%)	Home (%)	Others (%)	Carbon footprint (ton/year)	Ecological footprint (number of planet Earths)
16–20	20.3	28.7	21.5	29.6	12.554	3.07
21–35	21.1	29.1	17.5	32.3	10.778	2.716
36–50	22.1	22.2	19.6	36.1	13.41	3.034

Table 2 EF values of different age intervals according to income rates

Age	Income rate (TL)	Ecological footprint (number of planet Earths)	Age	Income rate (TL)	Ecological footprint (number of planet Earths)	Age	Income rate (TL)	Ecological footprint (number of planet Earths)
16–20	400–1200	2.23	21–35	400–1200	2.56	36–50	400–1200	1.98
	1200–2000	2.93		1200–2000	2.59		1200–2000	2.3
	>2000	3.86		>2000	2.82		>2000	4.22

Table 3 Total EF values for some countries [23]

Country	Total ecological footprint (2013) (×10^5 global hectares)
Dominica	1.62 (the smallest total EF in 2013)
China	50096.54 (the largest total EF in 2013)
United States of America	27245.96
Australia	2048.38
Canada	3084.62
United Kingdom	3229.76
Indonesia	3637.59
Italy	2693.33
Turkey	2430.73
Spain	1871.29
Iran	2317.62

In the total EF assessment, Turkey has the 15th place in the list of countries having the largest total EF with the 2430.73 × 10^5 GH value (according to the assessment held in 2013) [23].

The value of EF which was 0.73 number of planet Earths (6984.54 × 10^6 global hectares) in 1961 increased to 1.68 number of planet Earths (2060.28 × 10^6 global hectares) in 2013. As stated in the related literature, the EF value which was 0.48 number of planet Earths (4228 × 10^4 global hectares) in 1961 for Turkey has increased to 1.87 number of planet Earths (2430.73 × 10^5 global hectares) in 2013 [24]. According to the biocapacity values, it is seen that the global biocapacity value was 3.12 global hectares per person in 1961 and it increased to 1.71 global hectares per person value in 2013. For Turkey, it is stated that the biocapacity value of 2.64 global hectares per person in 1961 was decreased to the value of 1.47 global hectares per person in 2013 [24]. When the trend between 1961 and 2013 was examined, it can be seen that the worldwide value of EFs has rapidly increased in comparison with the biocapacity and ecological deficit has emerged. It can be said that the ecological reserve has decreased rapidly. The similar trend seems to be valid also for Turkey.

When the EF was assessed according to the different age groups, the EF value of individuals in the 16–20 age group (3.07 number of planet Earths) in 2013 was observed as approximately two times higher than the Turkey's average EF (1.87 number of planet Earths) value (Table 1). When the EF values for the age groups are evaluated in general, the average EF for the sample of this study was calculated as 2.94 number of planet Earths. This value is still quite above the 1.87 which is the average number of planet Earths of Turkey.

When the income level and age groups were evaluated together, the average EF value was found as 4.22 number of planet Earths for the 36–50 age group having an income over 2000 TL. The average EF value for the same age group having an income of 400–1200 TL was determined as 1.98 number of planet Earths (Table 2). This is the lowest EF value in terms of the combined assessments of the income level and the age group. These values are also over the average of Turkey (based on 2013 values).

Table 4 EF values (number of planet Earths) based on income rate and gender

Income level	400–1200 TL	1200–2000 TL	>2000 TL
Female	2.68	2.54	3.34
Male	2.32	2.90	3.92

Table 5 EF values (number of planet Earths) depending on the level of education and gender

Education level	Primary school	Secondary school	High school	University
Female	2.35	2.36	2.64	3.21
Male	1.98	2.27	2.87	3.63

When the income rate is considered, it is seen that the EF values of the females were higher than those of the males (2.68 number of planet Earths for females, 2.32 number of planet Earths for males) for the people having 400–1200 TL income. The average EF values calculated for the other two income levels were higher for the males than the females (Table 4).

According to the values of EFs based on education levels, it was determined that EF values also increased by the level of education (Table 5). It was seen that the values of EF were found higher for the female individuals for the surveyed group having only primary and secondary school graduation. The EF values for the primary school graduates were 2.35 number of planet Earths for the females and 1.98 number of planet Earths for the males; also for the secondary school graduates, the values were 2.36 and 2.27 number of planet Earths for the females and the males, respectively. The males graduated from high school and university had higher EF values than the females of the same educational level. The reasons for the increment in the EF values by the level of education can be listed as follows: the level of income of the people generally ascends by the level of education, people having higher income rates tend to spend more money especially for the imported luxury products, the possibilities of possessing cars and carrying out their daily travel with these vehicles are increasing as the purchasing power rises, and fewer people live in larger houses as the level of income increases.

When the results were evaluated, the EF values of the people living in Istanbul were found quite higher than the overall average EF values of Turkey. Turkey's population in 2016, which is the year that this study was carried out, was 78,945,645, and the population of Istanbul in the same year was 14,804,116. In addition, the population density of Istanbul was 2849 capita/km^2 [25]. Approximately 20% of Turkey's population lives in Istanbul. In a city with such a high population, transportation is a big problem, and people spend a lot of time in traffic. According to Table 1, it can be seen that the travel-induced EF constitutes a large part of the total EF value. In addition, Istanbul has the highest gross national product among the other cities of Turkey [25]. Considering all these data, it can be said that the EF value of Istanbul is higher than the average value of Turkey, as expected.

Conclusion

In this study, three different age intervals were chosen, and the results of these age intervals were discussed according to the income rates. The EF test provides an awareness about the destructive and impoverishing effects of the liberal economic policies that are rapidly depleting the planet's resources and forcing the carrying capacity.

According to the results obtained from this study, the values of EFs differ depending on the income level and age groups. It has been seen that the EF value is increased by the increase in the economic income for all age groups surveyed in this study.

Another result obtained from this study was that the EF values of people living in Istanbul were found to be notably higher than the mean EF value of Turkey. This situation can be explained by the fact that the population density of Istanbul is higher than the others, the hours spend in traffic is longer than the others due to the heavy traffic, and the income levels of the people living in Istanbul are higher.

In addition, the results show that there is a difference between the values of EFs of the female and the male of the same income and education level. Although the EF values increase with the increase in the income level for both genders as a general trend, the EF of the males of the same income level was higher than the EF of the females. Actively driving of the males was more than the females, and traveling longer distances during the day can be considered as a possible cause of this trend. When the data was examined based on the education levels, it was seen that the EF values of the individuals also increased as the education levels increased. It is related with the increase in the economic income by the level of education and the increase in the consumption habits depending on the income and the education level.

References

1. Steffen W, Sanderson A, Tyson P, Jäger J, Matson PA, Moore III B, Oldfield F, Richardson K, Schellnuber H, Turner II BL (2005) Global change and the earth system: a planet under pressure. Springer, Berlin
2. Field C, Barros V, Dokken D, Mach K, Mastrandrea M, Bilir T, Chatterjee M, Ebi K, Estrada Y, Genova R (2014) IPCC, 2014: Climate change 2014: Impacts, adaptation, and vulnerability. Part A: Global and sectoral aspects. Contribution of working Group II to the fifth assessment report of the ıntergovernmental panel on climate change. Cambridge University Press, Cambridge
3. Rudolpha A, Figge L (2017) Determinants of ecological footprints: what is the role of globalization? Ecol Indic 81:348–361
4. Niccolucci V, Tiezzi E, Pulselli FM, Capineri C (2012) Biocapacity vs ecological footprint of world regions: a geopolitical interpretation. Ecol Indic 16:23–30
5. Huiqin L, Linchun H (2011) Evaluation on sustainable development of scenic zone based on tourism ecological footprint: case study of yellow crane tower in Hubei Province, China. Energy Procedia 5:145–151

6. Begum RA, Pereira JJ, Jaafar AH, Al-Amin AQ (2009) An empirical assessment of ecological footprint calculations for Malaysia. Resour Conserv Recycl 53:582–587
7. Akman Y, Ketenoğlu O, Kurt L, Yiğit N (2012) Ekolojik Sentez. Palme Publihing, Ankara
8. De Rothschild D, Dicks L, Woodford C, Scott M, Dennis-Bryan K, Dipper F (2008) Dünya Önemlidir. Tudem Publishing, Ankara, Turkey
9. Cagiao J, Gomez B, Dom enech JL, Mainar SG, Lanza HG (2011) Calculation of the corporate carbon footprint of the cement industry by the application of MC3 methodology. Ecol Indic 11:1526–1540
10. Cucek L, Klemes JJ, Kravanja Z (2012) A review of footprint analysis tools for monitoring impacts on sustainability. J Clean Prod 34:9–20
11. Ewing B, Moore D, Gold finger S, Oursler A, Reed A, Wackernagel M (2010) Ecological Footprint Atlas, Global Footprint Network, Oakland
12. WWF (2014) Living planet report 2014: species and spaces, people and places, introduction page9, http://assets.worldwildlife.org/publications/723/files/original/WWF-LPR2014
13. Bastianoni S, Niccolucci V, Pulselli RM, Marchettini N (2012) Indicator and indicandum: sustainable way vs prevailing conditions in the ecological footprint. Ecol Indic 16:47–50
14. Karl HE (2004) Actual land demand of Austria 1926–2000: a variation on ecological footprint assessments. Land Use Policy 21(3):247–259
15. Wackernagel M, Rees WE (1996) Our ecological footprint: reducing human impact on the earth. New Society Publishers, Gabriola Island, BC
16. Monfreda C, Wackernagel M, Deumling D (2004) Establishing national natural capital accounts based on detailed ecological footprint and biological capacity assessments. Land Use Policy 21(3):231–246
17. Geng Y, Zhang L, Chen X, Xue B, Fujita T, Dong H (2014) Urban ecological footprint analysis: a comparative study between Shenyang in China and Kawasaki in Japan. J Clean Prod 75(14):130–142
18. Peng W, Wang X, Li X, He C (2018) Sustainability evaluation based on the emergy ecological footprint method: a case study of Qingdao, China, from 2004 to 2014. Ecol Indic 85:1249–1261
19. Moffatt I (2000) Commentary forum: the ecological footprint ecological footprints and sustainable development. Ecol Econ 32:359–362
20. Südaş HD, Özeltürkay EY (2015) Analyzing the thoughts of ecological footprints of university students: a preliminary research on Turkish students. Procedia Soc Behav Sci 175:176–184
21. Ekolojik ayak izim. http://ekolojikayakizim.org. Accessed 10 Jan 2018
22. Wikiwand (2018) Istanbul city in Turkey map. https://www.google.com.tr/url?sa=i&rct=j&q=&esrc=s&source=images&cd=&cad=rja&uact=8&ved=2ahUKEwiOvt7msObaAhUIuBQKHdW5BVwQjRx6BAgBEAU&url=http%3A%2F%2Fwww.wikiwand.com%2Ftr%2F%25C4%25B0stanbul%2527da_2007_T%25C3%25BCrkiye_genel_se%25C3%25A7imleri&psig=AOvVaw3_Mr0WgzD8ma-wWmf11Md1&ust=1525328267827036. Accessed 25 Apr 2018
23. Footprint data. https://data.world/footprint/nfa-201-edition. Accessed 12 Jan 2018
24. Global Footprint Network (2018) https://www.footprintnetwork.org. Accessed 16 Jan 2018
25. Turke Istatistical Institute (2018) https://www.tuik.gov.tr. Accessed 05 Jan 2018

Black Carbon Aerosols in Urban Air: Sources, Concentrations, and Climate Change

Burcu Onat, Ülkü Alver Şahin, and Burcu Uzun

Abstract Black carbon (BC) is one of the key atmospheric aerosol components of locomotive air quality and climate change. BC has the capacity to absorb light across all visible wavelengths; it reradiates solar radiation as heat, inducing a climate warming effect. Due to its shorter lifetime in the atmosphere compared to carbon dioxide, it is an excellent target for emission reductions. BC is a specific marker of primary combustion of fossil fuel and biomass. BC concentrations in urban areas are variable; in developed countries, motorized transport vehicles are considered to be the most important source of BC, whereas in developing countries, biomass burning may be important. The interest of policy makers in BC was rise up owing to arising evidence on health effects and the impact of BC on global warming. The authors present here a general knowledge on BC to clarify its sources, concentration levels, effect on climate change, and adverse health effect.

Keywords Black carbon · Sources · Urban areas · Traffic

Introduction

Black carbon is sourced from combustion processes such as fossil fuel burn, gasoline and diesel engines, and coal-fired power plants [1]. BC is a component of fine particles that is generated from the incomplete combustion. The type of combustion and fuels affects the BC production. The significant amounts of BC are produced by diesel combustion, while small amount of BC occurs with the combustion of natural gas. Depending on the combustion efficiency and fuel type, BC is co-emitted with organic carbon (OC) in the different ratios [2–5]. BC is different from other forms of carbon compounds included in atmospheric aerosols for it is insoluble in water and in organic solvents and it is refractory [2, 6]. The solar radiation can be absorbed

B. Onat · Ü. A. Şahin (✉) · B. Uzun
Department of Environmental Engineering, Istanbul University-Cerrahpaşa, Istanbul, Turkey
e-mail: ulkualver@istanbul.edu.tr

N. Balkaya, S. Guneysu (eds.), *Recycling and Reuse Approaches for Better Sustainability*, Environmental Science and Engineering,
https://doi.org/10.1007/978-3-319-95888-0_16

by BC at all wavelengths [7]. BC is always spread to atmosphere along with inorganic gases and OC [2, 8]. OC, one of the carbon fractions in particulate matter (PM), has cooling effect different from BC [3, 4]. BC is one of the most important pollutants which contribute significantly to global warming after carbon dioxide (CO_2) [8, 9]. In consequence of the adverse health effects and to be significant contributor to global warming of BC, government agencies have advanced various strategies for decreasing BC emissions [10, 11]. Government reports of European Commission countries and United States (USA) emphasized the importance of having the good-quality data on BC emission sources to apply efficient air quality policies [10, 12, 13]. To improve air quality and to develop mitigation strategies for BC emissions require an advanced understanding of BC's sources in urban areas, spatial patterns, and the effect of transport.

Black Carbon Sources

Black carbon is a primary pollutant spreaded from uncompleted combustion and is connected with a kindly of different major sources [14] such as transport, electricity generation, residential heating, cooking, and biomass burning such as agricultural burning and wildfires [2, 10, 15].

The total global BC emissions are guessed to be approximately 7600 Gg (8.4 million tons) for 2000. Figure 1 shows that the major BC emissions in the world are within the four major categories: these are open biomass burning (including wildfires), transport, domestic, and industry. The greatest amounts of BC are produced in the South and Central America, parts of Africa and Asia. The use of solid fuels for heating in developing countries (such as Asian and African) accounts for 60–80% of total energy-related BC emissions [2]. BC resources in developing countries are quite dissimilar than in the United States [16]. For the United States, the major BC emissions are inside of the four main (52%), domestic/residential (3.6%), and energy (6.8%) [15, 16].

Wildfires cause significant biomass BC emissions [17, 18], typically in the summer months in Canada [19]. Wood combustion for domestic heating has been estimated as the second largest BC source in Canada after transportation [20]. Wildfires were estimated as main contributor (83%) of the total BC in Russia [5].

Because of the adverse health effects, the BC source apportionment studies have been focused on the BC fraction of atmospheric particulate matter. The major preferred methods used for BC source apportionment are the aethalometer model, radiocarbon method, and macro-tracer method [12, 21–23]. These methods allow to be obtained the BC emissions within two categories as fossil fuel combustion and biomass burning. The studies conducted in Europe showed that the major BC source was fossil fuel combustion, and also biomass burning was a significant contributor

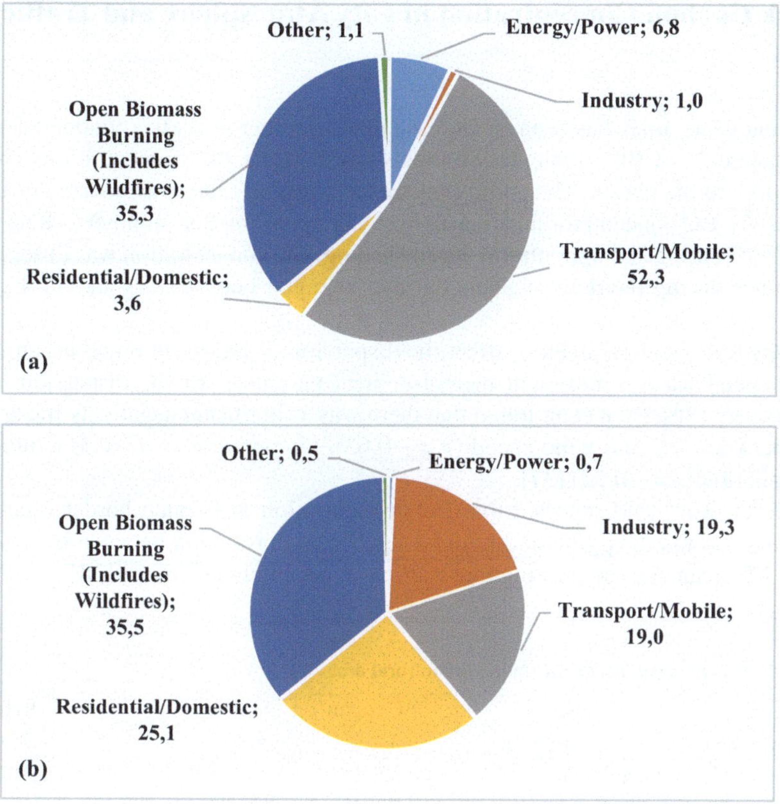

Fig. 1 Black carbon emission inventories for (**a**) the United States (year 2005 data) [16] and (**b**) global (year 2000 data) [15]

in time at rural areas. The forecast of more clarified BC source contribution can be possible with the chemical mass balance (CMB) and positive matrix factorization (PMF) models which are valid methods among US studies. In developing countries such as China and India, lots of BC source apportionment studies have been conducted [24–26]. Considering the anthropogenic BC emissions in the global scale, China is the largest BC emitter because of the results of urbanization and economic growth [27, 28]. In Beijing, coal combustion and biomass burning were the main sources of BC, accounting for 61–83% [29].

Diesel engines are in change of more than 90% of BC emissions sourced from road transportation; even diesel engines are responsible for 70% of total BC emissions in some countries [30, 31]. The diesel vehicles have been guessed to be the most important BC source in Canada [20]. Traffic emissions are predominant source of carbonaceous aerosol in Madrid [32–34].

Black Carbon Concentration in City Atmosphere and Traffic Sites

In recent years, there has been an increase in the number of studies conducted on the regional scale of BC (Table 1). Studies show that the BC concentrations change according to the region. The anthropogenic activities are the main sources of BC in cities [35]. BC concentration increases both during winter and summer as a result of domestic heating and agricultural waste burning. BC concentration was observed to be higher during daytime in Shanghai and changed between 0.3 and 11.4 μg/m^3 [36].

Meteorological parameters affect the dispersion of BC in rural and urban areas. Wind speed was an important meteorological parameter to BC dispersion in the atmosphere [36]. Bhat et al. found that there was a significant negatively linear relation between BC and wind speed ($r = -0.63$), temperature ($r = -0.51$), and total precipitation ($r = -0.55$) [37].

Shores et al. [38] reported that BC concentration at Mexico border cities was between 0.8 and 2.2 μg/m^3 and found a significant correlation between BC and CO at the Mexican sites suggesting that traffic is a predominant source [38].

Table 1 Previous studies on BC in the urban/rural area

Location	Site type	BC (μg/m^3)	Season	Reference
Beijing/China	Suburban	9.3	Annual	[35]
New York/United States	Urban	0.52–0.76 4.0–10.0	Annual Daily	[44]
Otay mesa/United States	Rural	0.8	June–July	[38]
Parque Morelos/Mexico	Urban	2.2	June–July	[38]
El Trompo/Mexico	Urban	1.7	June–July	[38]
Beijing/China	Urban	12.3 (summer) 17.9 (winter)	August–December	[39]
Istanbul/Turkey	Traffic area	6.05–9.92	Winter	[42]
Istanbul/Turkey	Traffic area	6.06/6.96	Summer/winter	[43]
Shanghai/China	Urban	2.3	Annual	[36]
Beijing/China	Urban	1.3–6.1	January	[40]
Himalaya/India	Urban	6.0	Annual	[37]
Karachi/Pakistan	Urban	2.2–12.5	January 2007–June 2016	[41]
Madrid/Spain	Traffic urban area	3.7	Annual	[8]
	Rural area	2.6	Annual	
Beijing/China	Urban	5.3	Hourly	[29]

Table 2 BC exposure concentrations in travel modes

Location	BC (μg/m³)					Reference
	Car	Bus	Subway	Bicycle	Walk	
Arnhem, Netherlands	8.2 (diesel) 9.3 (gasoline)	9.0 (diesel) 5.1 (electric)	–	5.3 (low traffic) 6.6 (high traffic)	–	[50]
Barcelona/Spain	16.7	7.0	–	8.5	5.7	[51]
Flanders/ Belgium	6.4	6.6	5.1	3.2–3.6	3.2–3.6	[48]
Thessaloniki/ Greece	Route 1 9.0 (windows open) 4.2 (windows closed)	Route 1 7.4	–	Route 1 10.5	–	[56]
Shanghai/China	–	7.3	9.4	6.6	5.6	[57]
Barcelona/Spain	–	5.5	7.0	–	4.4–9.6	[58]
Queensland/ Australia	4.4 (windows open) 1.7 (windows closed)	2.4	–	1.0	1.2	[59]
London/England	4.4 (GM[a])	5.6 (GM[a])	–	–	3.1 (GM[a])	[60]
Sacramento/ California	0.50	0.95	0.25 (light rail)	0.71	–	[61]

[a]Geometric means

BC concentrations may differ according to the seasons. Bhat et al. [37] found that the annual average concentration of BC in northwestern Himalaya was 6.0 μg/m³ and BC concentration measurement was highest in autumn and lowest in spring. They found that the BC emissions sourced from biomass burning during autumn, spring, and winter seasons were equal to BC emissions sourced from fossil fuel and biomass burning during summer season [37]. In Ontario, Canada, the BC emissions sourced from vehicular traffic were determined to be major source of BC and the highest BC emission levels observed in the highway site [14]. Becerril-Valle et al. found that the BC concentrations were 3.7 μg/m³, 2.3 μg/m³, and 2.6 μg/m³ at the traffic urban site, at the urban background, and in the rural area, respectively [8].

Several studies have been conducted on BC aerosol in Beijing, focusing on its temporal variation, size distribution, and source apportionment (Table 2) [29, 35, 39, 40]. In Beijing, the mean BC concentrations were 17.9 μg/m³ in winter and 12.3 μg/m³ in summer near road site [39], ranging from 6.1 μg/m³ [40] to 35.33 μg/m³ [29] during the hazy conditions and 1.3 μg/m³ during non-haze periods [40].

BC concentrations in the metropolitan cities are variable. In Karachi BC concentration ranged from 2.2 to 12 μg/m^3 [41]. In different traffic areas of Istanbul, BC concentrations were observed between 6.05 and 9.92 μg/m^3 [42] and between 6.06 and 6.96 μg/m^3 [43]. In New York, BC daily mean concentration ranged from 4.0 to 10.0 μg/m^3, while the annual average BC was between 0.52 and 0.76 μg/m^3 during 2007–2010 period [44]. BC should not be thought as an only local pollutant during the arrangement of the policy decisions about air quality in future [14]. The forward trajectory analyses showed that any regional air quality management strategies should be arranged taking into account BC contributions from the border area [37, 38, 45].

In developed countries, motorized transport and mainly diesel vehicles are considered to be the most significant sources of BC, whereas in developing countries, biomass burning may be important [46–48].

People can be exposed to high concentrations of BC during travelling in transport vehicles [49–52]. In high vehicle density urban areas, commuting is considered as one of the high-exposure periods among various daily activities [53]. Personal monitoring studies showed that the transport activity can be responsible for quite a large part of integrated personal exposure to combustion-related pollutants [54, 55].

Measurements of personal exposure to BC have been reported recently in different transport modes [48, 50, 51, 56–61] (Table 2). The travel modes (i.e., the transport system, technology, or energy source), characteristics of the path traveled (i.e., street configuration, micrometeorology, or traffic loads), and commuting time might influence the BC exposure [62]. Car and bus commuters are exposed to higher levels of BC than commuters cycling and walking [48, 50, 57, 59, 60], but the lowest BC concentrations were observed in the electric buses [50, 59]. Also, the ventilation mode of vehicle is a significant factor that affects the in-vehicle BC concentrations. In some studies, the measurements were made in car for two situations: windows open and windows closed. When the windows were open, people were exposed to higher BC concentration [51, 56, 59].

In subway, the average BC concentrations ranged from 5.1 to 9.4 μg/m^3 [48, 57, 58]. Some factors may affect pollutant exposure levels in subway wagons such as service time, ventilation, passenger numbers, platform screen doors, driving conditions, etc. [63].

Black Carbon and Climate Change

Black carbon (BC) aerosols play a major role in regional as well as global climate by interacting with the solar radiation, modifying the cloud properties (affect cloud droplet number concentrations and cloud microphysical properties), decreasing precipitation, and reducing the snow albedo [7, 64–66].

BC has the capacity to absorb light across all visible wavelengths [6, 67, 68], so it reradiates solar radiation as heat, inducing a climate warming effect [69]. Because of these effects and its shorter lifetime in the atmosphere compared to carbon dioxide, it is an excellent target for emission reductions [70]. In addition, the BC particle deposits in the Arctic cause a decrease of the Earth's ability to reflect the solar radiation and hastening melt [1].

The knowledge of the relative contributions of the different sources of BC is required to investigate their respective climate impacts. Assessing these impacts remains difficult, however, compounded by the role of BC in cloud formation, which is in turn dependent upon its size and internal mixing with other chemical species [14, 71, 72]. Globally averaged direct radiative forcing due to BC is estimated as 0.9 W/m^2 (0.4–1.2 W/m^2) by Ramanathan and Carmichael (2008) and 0.60 (with uncertainty of −61 to +70%) W/m^2 by IPCC (2013) [7, 73, 74].

Bond et al. estimated the industrial-era radiative forcing of BC as +1.1 W/m^2 (+0.17, +2.1), while Myhre et al. estimated the direct radiative forcing (RF) over 1750–2010 of +0.4 (+0.05, +0.8) W/m^2 for fossil fuel and biofuel BC and +0.2 (+0.03 + 0.4) for biomass burning BC [2, 75]. UNEP and WMO (2011) have estimated that implementation of proposed BC and CH_4 control measures by 2030 could prevent up to 0.5 °C of additional warming by 2050 [8].

The global emission of BC in 2007 was estimated to be ~8.7 Tg/year, and of this 1.6 Tg/year was from industry, 2.8 Tg/year from residential/commercial sectors, and 1.5 Tg/year from transportation [76]. The Fifth Assessment Report of the Intergovernmental Panel on Climate Change (IPCC) suggests that the direct radiative forcing of BC aerosol is +0.6 W/m^2: +0.4 W/m^2 of that is due to emissions from the burning of fossil fuels and biofuels [7]. Bond et al. estimated a larger BC direct radiative forcing effect of +0.71 W/m^2, with +0.51 W/m^2 from fossil and biofuel emissions and + 0.2 W/m^2 from biomass burning [2, 77].

Health Effects of Black Carbon

As a constituent of $PM_{2.5}$, BC is responsible for respiratory and cardiovascular diseases [78–81]. However, there are limited monitoring data on BC for epidemiologic analyses [78], especially in developing countries [82]. BC is a better indicator of air pollution health effects than ambient $PM_{2.5}$ based on more consistent and stronger associations from epidemiological studies [12, 83, 84].

BC is also of concern from a human health perspective, with epidemiological evidence linking exposure to BC with cardiopulmonary hospital admissions and mortality [84]. BC is associated with lung cancer, cardiopulmonary, cardiovascular, and all-cause mortality [83, 84].

Health outcomes associated with BC include cardiovascular effects [85–87], respiratory effects [88, 89], and mortality [90]. Some of these effects can be demonstrated at ambient concentrations below 1 $\mu g/m^3$ [85, 86, 91].

A number of studies have reported a strong association between BC and road traffic [92–94] and biomass burning [95] emissions. While BC aerosols are not the only cause of adverse health effects due to particles, they are a major factor, specially the ultrafine BC. Indeed, the recent WHO report concludes that "combustion-derived aerosols are particularly significant in terms of their health effects" [96]. Recently, emissions from diesel vehicles have received significant attention. Diesel vehicles are a source of particulate and black carbon (BC) emissions. BC, a product of incomplete fuel combustion, has adverse effects on human health, including respiratory diseases and lung cancer [97]. Recent epidemiological results for BC suggest that health effects per mass may be up to 10 times higher than PM_{10} [80].

References

1. EPA (2018) https://www.epa.gov/air-research/black-carbon-research. Accessed 20 Mar 2018
2. Bond TC, Doherty SJ, Fahey DW, Forster PM, Berntsen T, DeAngelo BJ, Flanner MG, Ghan S, Karcher B, Koch D, Kinne S, Kondo Y, Quinn PK, Sarofim MC, Schultz MG, Schulz M, Venkataraman C, Zhang H, Zhang S, Bellouin N, Guttikunda SK, Hopke PK, Jacobson MZ, Kaiser JW, Klimont Z, Lohmann U, Schwarz JP, Shindell D, Storelvmo T, Warren SG, Zender CS (2013) Bounding the role of black carbon in the climate system: a scientific assessment. J Geophys Res 118:5380–5552
3. Sand M, Berntsen TK, von Salzen K, Flanner MG, Langner J, Victor DG (2016) Response of arctic temperature to changes in emissions of short-lived climate forcers. Nat Clim Chang 6:286–289
4. Stohl A, Aamaas B, Amann M, Baker LH, Bellouin N, Berntsen TK, Boucher O, Cherian R, Collins W, Daskalakis N, Dusinska M, Eckhardt S, Fuglestvedt JS, Harju M, Heyes C, Hodnebrog Ø, Hao J, Im U, Kanakidou M, Klimont Z, Kupiainen K, Law KS, Lund MT, Maas R, MacIntosh CR, Myhre G, Myriokefalitakis S, Olivie D, Quaas J, Quennehen B, Raut JC, Rumbold ST, Samset BH, Schulz M, Seland Ø, Shine KP, Skeie RB, Wang S, Yttri KE, Zhu T (2015) Evaluating the climate and air quality impacts of short-lived pollutants. Atmos Chem Phys 15:10529–10566
5. Evans M, Kholod N, Kuklinski T, Denysenko A, Smith SJ, Staniszewski A, Bond TC (2017) Black carbon emissions in Russia: a critical review. Atmos Environ 163:9–21
6. Petzold A, Ogren JA, Fiebig M, Laj P, Li SM, Baltensperger U, Holzer-Popp T, Kinne S, Pappalardo G, Sugimoto N, Wehrli C, Wiedensohler A, Zhang X (2013) Recommendations for reporting "black carbon" measurements. Atmos Chem Phys 13:8365–8379
7. IPCC-Intergovernmental Panel on Climate Change (2013) Climate change 2013: the physical science basis, report of working Group I to the fifth assessment of the IPCC. Cambridge University Press, Cambridge
8. Becerril-Valle M, Coz E, Prévôt ASH, Močnik G, Pandis SN, de la Campa AS, Artíñano B (2017) Characterization of atmospheric black carbon and co-pollutants in urban and rural areas of Spain. Atmos Environ 169:36–53
9. UNEP-CCAC (2014),Time to act to reduce short-lived climate pollutants, Judith Maréchal and GRID-Arendal (Ed.), ISBN: 978-82-7701-130-1.
10. US EPA, March 2012. Report to congress on black carbon, p 388
11. UNEP/WMO (2011), Integrated Assessment of Black Carbon and Tropospheric Ozone, Bart Ullstein (Ed.), Banson, UK, ISBN: 92-807-3141-6.
12. Briggs NL, Long CM (2016) Critical review of black carbon and elemental carbon source apportionment in Europe and the United States. Atmos Environ 144:409–427
13. Belis CA, Larsen BR, Amato F, El Haddad I, Favez O, Harrison RM, Hopke PK, Nava S, Paatero P, Prevot A, Quass U, Vecchi R, Viana M (2014) European guide on air pollution source apportionment with receptor models, p 92. https://doi.org/10.2788/9307. Report EUR 26080 EN

14. Healy RM, Sofowote U, Su Y, Debosz J, Noble M, Jeong CH, Jones K (2017) Ambient measurements and source apportionment of fossil fuel and biomass burning black carbon in Ontario. Atmos Environ 161:34–47
15. Lamarque JF, Bond TC, Eyring V, Granier C, Heil A, Klimont Z, Lee D, Liousse C, Mieville A, Owen B, Schultz MG, Shindell D, Smith SJ, Stehfest E, Van Aardenne J, Cooper OR, Kainuma M, Mahowald N, McConnell JR, Naik V, Riahi K, VanVuuren DP (2010) Historical(1850-2000) gridded anthropogenic and biomass burning emissions of reactive gases and aerosols: methodology and application. Atmos Chem Phys 10(15):7017–7039. https://doi.org/10.5194/acp-10-7017-2010
16. EPA (2012) https://www.epa.gov/air-research/black-carbon-research. Accessed 5 Apr 2018
17. Van der Werf GR, Randerson JT, Giglio L, Collatz GJ, Kasibhatla PS, Arellano AF Jr (2006) Interannual variability in global biomass burning emissions from 1997 to 2004. Atmos Chem Phys 6:3423–3441
18. Schultz MG, Heil A, Hoelzemann JJ, Spessa A, Thonicke K, Goldammer JG, Held AC, Pereira JMC, van het Bolscher M (2008) Global wildland fire emissions from 1960 to 2000. Glob Biogeochem Cycles 22:1–17
19. Randerson JT, Liu H, Flanner MG, Chambers SD, Jin Y, Hess PG, Pfister G, Mack MC, Treseder KK, Welp LR, Chapin FS, Harde JW, Goulden ML, Lyons E, Neff JC, Schuur EAG, Zender CS (2006) The impact of Boreal forest fire on climate warming. Science 314:1130–1132
20. Council, Arctic, (2015) Enhanced black carbon and methane emissions reductions: an arctic council framework for action. Iqaluit 2015a SAO Report to Ministers, Annex 4.
21. Favez O, El Haddad I, Piot C, Boreave A, Abidi E, Marchand N (2010) Intercomparison of source apportionment models for the estimation of wood burning aerosols during wintertime in an Alpine city (Grenoble, France). Atmos Chem Phys 10(12):5295–5314
22. Larsen BR, Gilardoni S, Stenstrom K, Niedzialek J, Jimenez J, Belis CA (2012) Sources for PM air pollution in the Po Plain, Italy: II. Probabilistic uncertainty characterization and sensitivity analysis of secondary and primary sources. Atmos Environ 50:203–213
23. Belis CA, Karagulian F, Larsen BR, Hopke PK (2013) Critical review and metaanalysis of ambient particulate matter source apportionment using receptor models in Europe. Atmos Environ 69:94–108
24. Chen B, Andersson A, Lee M, Kirillova EN, Xiao Q, Krusa M, Shi M, Hu K, Lu Z, Streets DG, Du K, Gustafsson O (2013) Source forensics of black carbon aerosols from China. Environ Sci Technol 47(16):9102–9108. https://doi.org/10.1021/es401599r
25. Liu J, Li J, Zhang Y, Liu D, Ding P, Shen C, Shen K, He Q, Ding X, Wang X, Chen D, Szidat S, Zhang G (2014) Source apportionment using radiocarbon and organic tracers for PM2.5 carbonaceous aerosols in Guangzhou, South China: contrasting local- and regional-scale haze events. Environ Sci Technol 48(20):12002–12011. https://doi.org/10.1021/es503102w
26. Gustafsson O, Krusa M, Zencak Z, Sheesley RJ, Granat L, Engstrom E, Praveen PS, Rao PS, Leck C, Rodhe H (2009) Brown clouds over South Asia: biomass or fossil fuel combustion? Science 323(5913). https://doi.org/10.1126/science.1164857
27. Wang R, Tao S, Balkanski Y, Ciais P, Boucher O, Liu J, Piao S, Shen H, Vuolo MR, Valari M (2014) Exposure to ambient black carbon derived from a unique inventory and high-resolution model. Proc Natl Acad Sci 111:2459–2463
28. Zhang Q, Streets DG, Carmichael GR, He K, Huo H, Kannari A, Klimont Z, Park I, Reddy S, Fu J (2009) Asian emissions in 2006 for the NASA INTEX-Bmission. Atmos Chem Phys 9:5131–5153
29. Liu Y, Yan C, Zheng M (2018) Source apportionment of black carbon during winter in Beijing. Sci Total Environ 618:531–541
30. Kupiainen K, Klimont Z (2007) Primary emissions of fine carbonaceous particles in Europe. Atmos Environ 41:2156–2170
31. Amann M, Klimont Z, Wagner F (2013) Regional and global emissions of air pollutants: recent trends and future scenarios. In: Gadgil A, Liverman DM (eds) Annual review of environment and resources, p 38

32. Plaza J, Artí~nano B, Salvador P, Gomez-Moreno FJ, Pujadas M, Pio CA (2011) Short-term secondary organic carbon estimations with a modified OC/EC primary ratio method at a suburban site in Madrid (Spain). Atmos Environ 45:2496–2506
33. Salvador P, Artí~nano B, Alonso DG, Querol X, Alastuey A (2004) Identification and characterisation of sources of PM10 in Madrid (Spain) by statistical methods. Atmos Environ 38:435–447
34. Salvador P, Artíñano B, Viana M, Alastuey A, Querol X (2012) Evaluation of the changes in the Madrid metropolitan area influencing air quality: analysis of 1999–2008 temporal trend of particulate matter. Atmos Environ 57:175–185
35. Gao R, Niu S, Zhang H, Guo J, Meng D, Ma J (2007) A comparative study on black carbon aerosol observations in regions of Beijing and Lhasa in 2006. Proc SPIE Int Soc Opt Eng 6679:66791T–66791T-8
36. Zha S, Cheng T, Tao J, Zhang R, Chen J, Zhang Y, Du J (2014) Characteristics and relevant remote sources of black carbon aerosol in Shanghai. Atmos Environ 135:159–171
37. Bhat MA, Romshoo SA, Beig G (2017) Aerosol black carbon at an urban site-Srinagar, Northwestern Himalaya, India: seasonality, sources, meteorology and radiative forcing. Atmos Environ 165:336–348
38. Shores CA, Klapmeyer ME, Quadros ME, Marr LC (2013) Sources and transport of black carbon at the California–Mexico border. Atmos Environ 70:490–499
39. Song S, Wu Y, Xu J, Ohara T, Hasegawa S, Li J (2013) Black carbon at a roadside site in Beijing: temporal variations and relationships with carbon monoxide and particle number size distribution. Atmos Environ 77:213–221
40. Wang Q, Huang RJ, Cao J, Tie X, Shen Z, Zhao S, Zhou Y (2016) Contribution of regional transport to the black carbon aerosol during winter haze period in Beijing. Atmos Environ 132:11–18
41. Bibi S, Alam K, Chishtie F, Bibi H, Rahman S (2017) Temporal variation of black carbon concentration using aethalometer observations and its relationships with meteorological variables in Karachi, Pakistan. J Atmos Sol Terr Phys 157:67–77
42. Ozdemir H, Pozzoli L, Kindap T, Demir G, Mertoglu B, Mihalopoulos N, Unal A (2014) Spatial and temporal analysis of black carbon aerosols in Istanbul megacity. Sci Total Environ 473:451–458
43. Onat B, Şahin AÜ, Mangır N, Doğan M (2017) Black carbon and particulate matter concentration at urban traffic site in İstanbul, 8th atmospheric sciences symposium, ATMOS2017, 01–04 Nov 2017, Istanbul, Turkey
44. Wang Y, Hopke PK, Rattigan OV, Chalupa DC, Utell MJ (2012) Multiple-year black carbon measurements and source apportionment using delta-c in Rochester, New York. J Air Waste Manage Assoc 62(8):880–887
45. Li C, Bosch C, Kang S, Andersson A, Chen P, Zhang Q (2016) Sources of black carbon to the Himalayan-Tibetan Plateau glaciers. Nat Commun 7:12574
46. Highwood EJ, Kinnersley RP (2006) When smoke gets in our eyes: the multiple impacts of atmospheric black carbon on climate, air quality and health. Environ Int 32:560–566
47. Kirchstetter TW, Aguiar J, Tonse S, Fairley D, Novakov T (2008) Black carbon concentrations and diesel vehicle emission factors derived from coefficient of haze measurements in California: 1967e2003. Atmos Environ 42:480–491
48. Dons E, Panis LI, Van Poppel M, Theunis J, Wets G (2012) Personal exposure to black carbon in transport microenvironments. Atmos Environ 55:392–398
49. Suarez L, Mesias S, Iglesias V, Silva C, Caceres DD, Ruiz-Rudolph P (2014) Personal exposure to particulate matter in commuters using different transport modes (bus, bicycle, car and subway) in an assigned route in downtown Santiago, Chile. Environ Sci Process Impacts 16(6):1309–1131
50. Zuurbier M, Hoek G, Oldenwening M, Lenters V, Meliefste K, van den Hazel P (2010) Commuters' exposure to particulate matter air pollution is affected by mode of transport, fuel type, and route. Environ Health Perspect 118:783–789

51. De Nazelle A, Fruin A, Westerdahl S, Martinez D, Ripol D, Kubesch N, Nieuwenhuijsen M (2012) A travel mode comparison of commuters' exposures to air pollutants in Barcelona. Atmos Environ 59:151–159
52. Ragettli MS, Corradi E, Braun-Fahrlander C, Schindler C, de Nazelle A, Jerrett M, Ducret-Stich RE, Künzli N, Phuleria HC (2013) Commuter exposure to ultrafine particles in different urban locations, transportation modes and routes. Atmos Environ 77:376–384
53. Duci A, Chaloulakou A, Spyrellis N (2003) Exposure to carbon monoxide in the Athens urban area during commuting. Sci Total Environ 309:47–58
54. Dons E, Int Panis L, Van Poppel M, Theunis J, Willems H, Torfs R (2011) Impact of time activity patterns on personal exposure to black carbon. Atmos Environ 45:3594–3602
55. Fruin SA, Winer AM, Rodes CE (2004) Black carbon concentrations in California vehicles and estimation of in-vehicle diesel exhaust particulate matter exposures. Atmos Environ 38(25):4123–4133
56. Vouitsis I, Taimisto P, Kelessis A, Samaras Z (2017) Microenvironment particle measurements in Thessaloniki, Greece. Urban Clim 10(4):608–620
57. Li B, Lei XN, Xiu GL, Gao CY, Gao S, Qian NS (2015) Personal exposure to black carbon during commuting in peak and off -peak hours in Shanghai. Sci Total Environ 524–525:237–245
58. Moreno T, Reche C, Rivas I, Minguillon MC, Martins V, Vargas C, Buonanno G, Parga J, Pandolfi M, Brines M, Ealo M, Fonseca AS, Amato F, Sosa G, Capdevila M, De Miguel E, Querol X, Gibbons W (2015) Urban air quality comparison for bus, tram, subway and pedestrian commutes in Barcelona. Environ Res 142:495–510
59. Williams RD, Knibbs LD (2016) Daily personal exposure to black carbon: a pilot study. Atmos Environ 132:296–299
60. Rivas I, Kumar P, Hagen-Zanker A, Andrade FM, Slovic DA, Pritchard PJ, Geurs TK (2017) Determinants of black carbon, particle mass and number concentrations in London transport microenvironments. Atmos Environ 161:247–262
61. Ham W, Vijayan A, Schulte N, Herner JD (2017) Commuter exposure to $PM_{2.5}$, BC and UFP in six common transport microenvironment in Sacramento, California. Atmos Environ 167:335–345
62. Hertel O, Hvidberg M, Ketzel M, Storm L, Stausgaard L (2008) A proper choice of route significantly reduces air pollution exposure - a study on bicycle and bus trips on urban streets. Sci Total Environ 389:58–70
63. Xu B, Hao J (2017) Air quality inside subway metro indoor environment worldwide: a review. Environ Int 107:33–46
64. Lohmann U, Feichter J (2005) Global indirect aerosol effects: a review. Atmos Chem Phys 5:715–737
65. Flanner MG, Zender CS, Hess PG, Mahowald NM, Painter TH, Ramanathan V, Rasch PJ (2009) Springtime warming and reduced snow cover from carbonaceous particles. Atmos Chem Phys 9:2481–2497
66. Chen LWA, Doddridge BG, Dickerson RR, Chow JC, Mueller PK, Quinn J, Butler WA (2001) Seasonal variations in elemental carbon aerosol, carbon monoxide and sulfur dioxide: implications for sources. Geophys Res Lett 28(9):1711–1714
67. Goldberg ED (1985) Black carbon in the environment. John Wiley & Sons Ltd., New York
68. Horvath H (1993) Atmospheric light absorption a review. Atmos Environ Part A Gen Top 27:293–317
69. Jacobson MZ (2000) A physically based treatment of elemental carbon optics: implications for global direct forcing of aerosols. Geophys Res Lett 27:217–220
70. Jacobson MZ (2010) Short-term effects of controlling fossil-fuel soot, biofuel soot and gases, and methane on climate, arctic ice, and air pollution health. J Geophys Res Atmos 115:D14209
71. Cubison MJ, Ervens B, Feingold G, Docherty KS, Ulbrich IM, Shields L, Prather K, Hering S, Jimenez JL (2008) The influence of chemical composition and mixing state of Los Angeles urban aerosol on CCN number and cloud properties. Atmos Chem Phys 8:5649–5667

72. Fierce L, Bond TC, Bauer SE, Mena F, Riemer N (2016) Black carbon absorption at the global scale is affected by particle-scale diversity in composition. Nat Commun 7:12361
73. Ramanathan V, Carmichael G (2008) Global and regional climate changes due to black carbon. Nat Geosci 1:221–227
74. Girach IA, Nair VS, Babu SS, Nair PR (2014) Black carbon and carbon monoxide over Bay of Bengal during W_ICARB: source characteristics. Atmos Environ 94:508–517
75. Myhre G, Shindell D, Breon FM, Collins W, Fuglestvedt J, Huang J, Koch D, Lamarque JF, Lee D, Mendoza B, Nakajima T, Roboc A, Stephens G, Takemura T, Zhang H (2013) Anthropogenic and natural radiative forcing. In: Stocker TF, Qin D, Plattner GK, Tignor M, Allen SK, Boschung J, Nauels A, Xia Y, Bex V, Midgley PM (eds) Climate change 2013: the physical science basis. Contribution of working Group I to the fifth assessment report of the intergovernmental panel on climate change. Cambridge University Press, Cambridge
76. Wang R (2015) Global emissions of black carbon from 1960 to 2007, global emission inventory and atmospheric transport of black carbon. Springer, New York, pp 115–129
77. Wang Q, Huang RJ, Zhao Z, Zhang N, Wang Y, Ni H, Zhang J (2016) Size distribution and mixing state of refractory black carbon aerosol from a coastal city in South China. Atmos Res 181:163–171
78. Geng F, Hua J, Mu Z, Peng L, Xu X, Chen R, Kan H (2013) Differentiating the associations of black carbon and fine particle with daily mortality in a Chinese city. Environ Res 120:27–32
79. Jansen KL, Larson TV, Koenig JQ, Mar TF, Fields C, Stewart J, Lippmann M (2005) Associations between health effects and particulate matter and black carbon in subjects with respiratory disease. Environ Health Perspect 113:1741–1746
80. Janssen NA, Hoek G, Simic-Lawson M, Fischer P, van Bree L, ten Brink H, Keuken M, Atkinson RW, Anderson HR, Brunekreef B, Cassee FR (2011) Black carbon as an additional indicator of the adverse health effects of airborne particles compared with PM_{10} and $PM_{2.5}$. Environ Health Perspect 119:1691–1699
81. Straif K, Keil U, Taeger D, Holthenrich D, Sun Y, Bungers M, Weiland SK (2000) Exposure to nitrosamines, carbon black, asbestos, and talc and mortality from stomach, lung, and laryngeal cancer in a cohort of rubber workers. Am J Epidemiol 152:297–306
82. Chen Y, Schleicher N, Fricker M, Cen K, Liu XL, Kaminski U, Norra S (2016) Long-term variation of black carbon and PM2. 5 in Beijing, China with respect to meteorological conditions and governmental measures. Environ Pollut 212:269–278
83. Grahame TJ, Klemm R, Schlesinger RB (2014) Public health and components of particulate matter: the changing assessment of black carbon. J Air Waste Manag Assoc 64(6):620–660. https://doi.org/10.1080/10962247.2014.912692
84. Janssen NAH, Gerlofs-Nijland ME, Lanki T, Salonen RO, Cassee F, Hoek G, Fischer P, Brunekreef B, Krzyzanowski M (2012) Health effects of black carbon. World Health Organization (WHO), Geneva, p 96
85. Adar SD, Gold DR, Coull BA, Schwartz J, Stone PH, Suh HH (2007) Focused exposures to airborne traffic particles and heart rate variability in the elderly. Epidemiology 18(1):95–103
86. McCracken J, Baccarelli A, Hoxha M, Dioni L, Melly S, Coull B, Suh H, Vokonas P, Schwartz J (2010) Annual ambient black carbon associated with shorter telomeres in elderly men: veterans affairs normative aging study. Environ Health Perspect 118(11):1564–1570
87. Wellenius GA, Burger MR, Coull BA, Schwartz J, Suh HH, Koutrakis P, Schlaug G, Gold DR, Mitteleman MA (2012) Ambient air pollution and the risk of acute ischemic stroke. Arch Intern Med 172(3):229–234
88. Lin W, Huang W, Zhu T, Hu M, Brunekreef B, Zhang Y, Liu X, Cheng H, Gehring U, Li C, Tang X (2011) Acute respiratory inflammation in children and black carbon in ambient air before and during the 2008 Beijing Olympics. Environ Health Perspect 119(10):1507–1512
89. Patel MM, Chillrud SN, Correa JC, Hazi Y, Feinberg M, Deepti KC, Prakash S, Ross MJ, Levy D, Kinney PL (2010) Traffic-related particulate matter and acute respiratory symptoms among New York City area adolescents. Environ Health Perspect 118(9):1338–1343

90. Gan WQ, Koehoorn M, Davies HW, Demers P, Tamburic L, Brauer M (2011) Long-term exposure to traffic-related air pollution and the risk of coronary heart disease hospitalization and mortality. Environ Health Perspect 119(4):501–507
91. Dons E, Temmerman P, Van Poppel M, Bellemans T, Wets G, Int Panis L (2013) Street characteristics and traffic factors determining road users' exposure to black carbon. Sci Total Environ 447(1):72–79
92. Hamilton RS, Mansfield TA (1991) Airborne particulate elemental carbon: its sources, transport and contribution to dark smoke and soiling. Atmos Environ 25:715–723
93. Watson JG, Chow JC, Lowenthal DH, Pritchett LC, Frazier CA, Neuroth GR, Robbins R (1994) Differences in the carbon composition of source profiles for diesel- and gasoline powered vehicles. Atmos Environ 28(15):2493–2505
94. Pakkanen TA, Kerminen VM, Ojanena CH, Hillamo RE, Aarnio P, Koskentalo T (2000) Atmospheric black carbon in Helsinki. Atmos Environ 34:1497–1506
95. Sandradewi J, Prevot ASH, Weingartner E, Schmidhauser R, Gysel M, Baltensperger U (2008) A study of wood burning and traffic aerosols in an Alpine valley using a multi-wavelength aethalometer. Atmos Environ 42:101–112
96. Reche C, Querol X, Alastuey A, Viana M, Pey J, Moreno T, Dall'Osto M (2011) New considerations for PM, black carbon and particle number concentration for air quality monitoring across different European cities. Atmos Chem Phys 11(13):6207–6227
97. Fann N, Lamson A, Anenberg S, Wesson K, Risley D, Hubbell B (2012) Estimating the national public health burden associated with exposure to ambient $PM_{2.5}$ and ozone. Risk Anal 32(1):81–95. https://doi.org/10.1111/j.1539-6924.2011.01630

Climate Change and Greenhouse Gases in Turkey

Ülkü Alver Şahin, Burcu Onat, and Coşkun Ayvaz

Abstract Within the course of significant changes in the history of humankind, the Industrial Revolution, together with the developments in science and technology, resulted in increases in the world population, which led to increases in anthropogenic greenhouse gas emissions and concerns for human welfare. As a result, the composition of the atmosphere has changed, creating a change in the global climate. According to the Fourth Assessment Report of the Intergovernmental Panel on Climate Change, this change will severely diminish the impact of the struggles for sustainable development. Turkey is located in the Mediterranean Basin, which will be the region affected most from climate change, the first impact of which will be severe drought. Climate change will significantly influence water, agriculture, and energy sectors. Because of steady population growth and intensive industrialization, total greenhouse gas (GHG) emissions in Turkey steadily increased from 1990 through 2015. The energy sector in Turkey accounted for 67.8% of total GHG emissions, followed by industrial processes at 15.7%, agriculture at 10.8%, and waste at 5.7%. According to the modeling results covering between 2016 and 2099 in Turkey, especially in summer, an increase in the mean temperature between 1 and 4 °C is expected throughout different areas of the country.

Keywords Climate change · Greenhouse gases · CO_2 · Turkey

Introduction

The first studies on climate change were done 150 years ago. In 1859, John Tyndall announced that changes in water vapor and carbon dioxide (CO_2) would lead to climate change. In 1896, global warming directly caused by anthropogenic CO_2 was calculated by Svante Arrhenius; however, in 1938, Guy Stewart Callendar introduced for the first time that the earth was gradually warming. Approximately 30 years

Ü. A. Şahin · B. Onat (✉) · C. Ayvaz
Department of Environmental Engineering, Istanbul University-Cerrahpaşa, Istanbul, Turkey
e-mail: bonat@istanbul.edu.tr

N. Balkaya, S. Guneysu (eds.), *Recycling and Reuse Approaches for Better Sustainability*, Environmental Science and Engineering,
https://doi.org/10.1007/978-3-319-95888-0_17

later, in the beginning of 1970s, it was discovered that artificial chlorofluorocarbons (CFCs) were depleting the stratospheric ozone (O_3) layer and that CFCs, methane (CH_4), and O_3 were causing the greenhouse effect. In 1977, scientists asserted that the most crucial problem of the next century would be the global climate changes and their aftermath. In 1987, several countries signed the Montreal Protocol on Substances that Deplete the Ozone Layer, which set restrictions on the use of O_3-depleting gases. By the 1990s, it had been discovered that atmospheric aerosol has a cooling effect. Based on this, it was accepted that greenhouse gases (GHGs) are positively correlated, and atmospheric aerosols are negatively correlated, with a rise in the global temperature [1].

Greenhouse Effect and Greenhouse Gases

Radiation from the sun to the earth comes in very short wavelengths within the visible spectrum and is also a factor that affects the earth's climate. The change in the sun–earth radiation balance can happen in the following three different ways: (1) alterations in earth's orbit or sun, which consequently lead to changes in the amount of the radiation that reaches the earth; (2) changes in the reflection of solar radiation, known as "Albedo" (factors such as cloud cover, atmospheric particulates, and plant cover); and (3) changes in the back-reflected long-wave radiation from the earth to space, such as the changes resulting from alterations in GHG concentrations. These changes affect earth's climate both directly and indirectly.

To establish a radiation balance, the earth must reflect as much energy back into the space that it absorbs from the sun; however, because the earth is colder than the sun, the energy reflected from it will be within the infrared region of the spectrum, which cannot be seen by the humans. In an ideal situation in which the amount of radiation absorbed from the sun is reflected equally back to the space, the temperature of the earth's surface should be −19 °C; nevertheless, some parts of thermal radiation reflected by the continents and water is retained by the atmosphere and clouds and then returned to the earth. This event heats the earth's surface and raises the temperature to 14 °C [2]. The phenomenon that causes the atmosphere to be much warmer than it should be from only absorbed solar radiation is known as the "greenhouse effect." The greenhouse effect increases the ambient temperature of the earth's atmosphere by ~33 °C.

The increase in the temperature of the earth's surface (i.e., the greenhouse effect) is a result of GHGs. These GHGs, such as atmospheric CO_2, water vapor (H_2O), CH_4, and nitrous oxide (N_2O), absorb the radiation reflected from the earth's surface, and their molecules heat up. These warmer molecules collide with the other molecules in the air, which leads to warming of the other parts of the atmosphere. In addition, after these GHGs warm, they begin to radiate. Some of this radiation escapes into the space, but some returns to the earth. This radiation, in addition to solar radiation, creates the greenhouse effect. In the Annex A of the Kyoto Protocol, the gases defined as GHGs are CO_2, CH_4, N_2O, hydrofluorocarbons (HFCs), perfluorocarbons (PFCs), sulfur hexafluoride (SF_6), and nitrogen trifluoride (NF_3).

Together, CO_2, water vapor, and clouds contribute to 90% of the greenhouse effect; the other GHGs are responsible for the remaining 10% [3].

Organic carbon, deposited as fossils for millions of years, is mined and used as fuel (coal, petroleum, and natural gas). Globally, 80% of the anthropogenic CO_2 in the atmosphere comes from the transportation or industrial emissions. The remaining 20% is from deforestation and biomass burning. Approximately 900 m^2 of forest can deposit 25 ton of carbon; however, one-half of the forests on earth were destroyed in the second half of the twentieth century. In addition, because of the carbonate raw material used in some industries, such as cement and manufactured ceramic, more CO_2 is emitted into the atmosphere. Each year, the amount of CO_2 emitted into the atmosphere by humans corresponds to 6.5 × 109 ton. The amount that the CO_2 sinks can absorb is 3.3 × 109 ton/year (1.7 ± 0.5 ocean/atmosphere, 1.4 ± 0.7 land/atmosphere) [4]; therefore, atmospheric CO_2 concentration has continuously increased and, in the last 150 years, has exceeded 400 ppm.

The half-life of atmospheric CO_2 is predicted to be between 50 and 200 years; therefore, even if CO_2 emissions are completely eliminated, the time required to reduce CO_2 levels to that in the pre-industrial development era (i.e., ~280 ppm) would be between 100 and 300 years [5]. Fossil fuels are not renewable energy sources, and the reserves within the earth will last for only another 40–200 years. If the use of fossil fuels continues until all of the reserves are depleted, the atmospheric CO_2 levels over the next century are predicted to be four times the concentration of those today.

CH_4 gas is formed when microbes decompose organic matter under anaerobic conditions. This gas is generated both naturally in the environment and by human activities. Its concentration in the atmosphere is ~1.7 ppm [5]. Wetlands, swamps, some coastal sediments, oceans, and plants are the most significant sources of CH_4 and are responsible for 20% of the earth's CH_4 generation. Decomposing landfills, burning fossil fuels, and agricultural activities are examples of human activities that generate CH_4. Among all agricultural activities, animal husbandry, such as raising sheep and cattle, and rice planting and processing are the highest sources of CH_4. Since 1750, the concentration of CH_4 in the atmosphere has increased by 150%, and since 1960, there has been a sixfold increase. On earth, the total CH_4 emissions are calculated to be 598 Tg/year, with approximately 60% of this amount being anthropogenic [2].

N_2O, another type GHG, shows lower emissions than other GHGs. The natural sources of N_2O are the oceans and the oxidation of ammonia in the atmosphere and soil. In particular, tropical soils emit a considerable amount of N_2O into the atmosphere. The anthropogenic sources comprise, for example, farming activities (e.g., fertilizer use, husbandry), burning fossil fuels, and some industrial activities. In particular, the use of nitrogen fertilizer and burning of coal increase N_2O emissions. In addition, nylon and nitric acid production emit N_2O. Compared to the pre-Industrial Revolution, its atmospheric concentration has increased 16% and continues to increase by ~0.25% annually. The tropospheric concentration of N_2O is close to 312 ppb; however, the concentration during the Industrial Revolution was 275 ppb [5]. Total global N_2O emissions are calculated as 14.7 Tg/year, and ~6.9 Tg/year are emitted from anthropogenic sources.

Halocarbons are composed of carbon/sulfur and/or hydrogen and at least one of the chlorine, fluorine, bromine, or iodine series. Chlorofluorocarbons (CFCs), hydrochlorofluorocarbons (HCFCs), hydrofluorocarbons (HFCs), and sulfur hexafluoride (SF_6) are classified as being in this group of molecules. Some molecules, such as SF_6, can last ~3200 years in the atmosphere and are referred to as "immortal molecules." Before industrialization, the atmospheric concentrations of ethyl bromide and methyl bromide organohalide compounds were very low. Nevertheless, technological advancements in the chemical industry are responsible for the immense amount of the anthropogenic gases emitted into the atmosphere in the second half of the twentieth century. These gases are used in refrigerators and air conditioners as coolants, in sprays, and as industrial cleaning solvents. When they are emitted into the atmosphere, they deplete stratospheric O_3. At the same time, CFCs and HCFCs are GHGs. With the Montreal Protocol, the use of these compounds was restricted and was completely stop in 2030; however, since their atmospheric lifetimes are between 60 and 100 years and their concentration has built up continuously since the 1960s, they are considered to be important GHGs [4].

Radiative forcing is generally measured as the heat transfer on the unit spherical area in the uppermost part of the atmosphere given in W/m^2 units. In other words, if a factor has positive radiative force, in the end, the energy of the surface–atmosphere system will increase, resulting in warming and vice versa. Some of the radiative forcing values from human activity-related parameters are given in Fig. 1. All GHGs have high radiative forcing values, yet the most effective is CO_2.

With the intensification of industrialization, GHGs have increased considerably, and, as a result, the total radiative forcing became 2.45 W/m^2. This amount corresponds to 1% of the total amount of incoming solar heat, yet it is also equal to the energy generated by 1.8 × 109 ton petroleum in just 1 min, and more than 100-fold of world's energy required for heating [6].

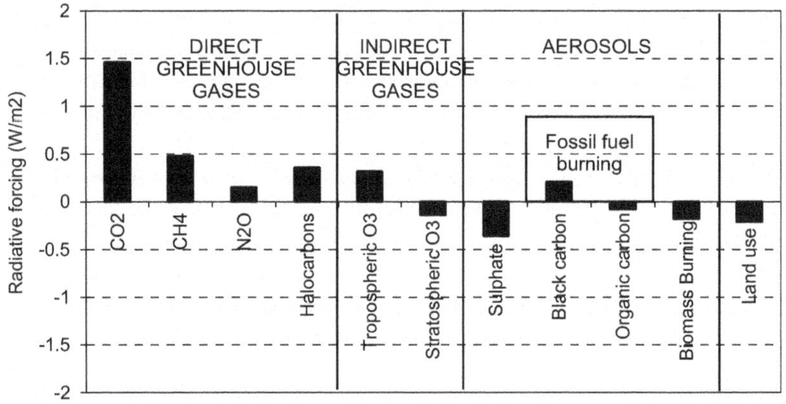

Fig. 1 Radiative forcing after the 1700s resulted from the effective factors on climate change [7]

Table 1 Greenhouse gases, concentrations, lifetime in atmosphere, and global warming potential (GWP) for 100 years [7]

Greenhouse gas	Formula	Concentration in atmosphere (ppt)		Atmospheric lifetime (years)	GWP for 100 years
		1750	1998		
Carbon dioxide	CO_2	278,000	358,000	(variable)	1
Methane	CH_4	700	1745	12.2 ± 3	23
Nitrous oxide	N_2O	275	311	120	296
HFC-22	CHF_2Cl	0	132	11.9	1700
Perfluoro-methane	CF_4	40	80	>50,000	5700
CFC-12	CCl_2F_2	0	533	100	10,600
Perfluoro-ethane	C_2F_6	0	3	10,000	11,900
Sulfur hexafluoride	SF_6	0	4.2	3200	22,200

Each GHG's effect on global warming is different. The potential warming of GHG is defined as the global warming potential (GWP) and determined according to a reference GHG CO_2 [7]. Moreover, the residence times of these gases in the atmosphere are very effective factors of their concentrations and global warming potentials (Table 1). Each GHG has a characteristic molecular property in its greenhouse effect or global warming potential. For example, one CFC molecule has between a 6000- and 7000-fold greater impact, and one CH_4 has 21-fold greater impact, than one CO_2 molecule. On the other hand, the CO_2 concentration in the atmosphere is greater. In general, 60% of human-related warming comes from CO_2 gases, 15–20% from CH_4, and the remaining from other GHGs (N_2O, CFC, and O_3) [6].

Climate Change

Climate change is defined as follows: in addition to the natural climate change in cyclic and observable time intervals, changes in the climate result from direct or indirect human impact which alters the composition of the atmosphere [8].

Climate change is nothing new, and there have been times when the CO_2 levels were much higher or much lower than today (e.g., Antarctica was once a tropical paradise and the equator region was once covered with ice). Nonetheless, since humans first appeared and for about 10,000 years since that time, the climate became more stabilized [9].

Anthropological CO_2 and changes in the composition of the atmosphere must be taken together to determine the amount of CO_2 in the atmosphere and its impact on climate. The temperature on earth has been regularly recorded since 1850. Figure 2 shows the recorded changes in the temperature and CO_2 concentration after the beginning of the industrial era. These data clearly show a similar trend between the amount of GHGs and changes in the global temperature.

From 1800 to 1870, which is regarded as the first Industrial Revolution, an increase in population increased the use of fossil fuel and the rate of GHG emissions. Ice measurements from 1850 to 1870 illustrated that the mean global temperature was 13.6 °C and atmospheric CO_2 concentration was 290 ppm. In the second Industrial Revolution between 1870 and 1910, fertilizers and other chemicals were being used regularly. According to the report of the Intergovernmental Panel on Climate Change (IPCC), between 1750 and 2011, cumulative anthropogenic CO_2 emissions into the atmosphere were 2040 ± 310 $GtCO_2$. Approximately

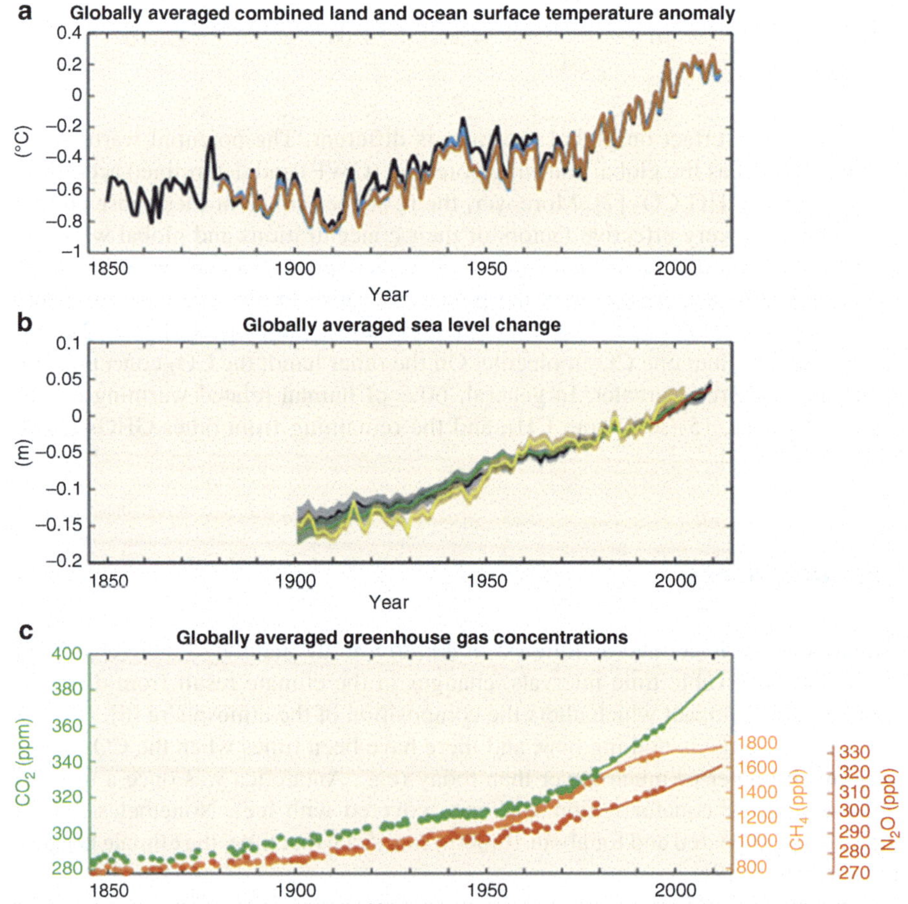

Fig. 2 Annual changes in the earth's land and sea surface temperatures (**a**), sea level (**b**), and atmospheric greenhouse gases (**c**) [10]

40% of these emissions have remained in the atmosphere (880 ± 35 $GtCO_2$); the remainder was removed from the atmosphere and stored on land (in the plants and soils) and in the oceans [10].

The increase in GHG concentrations within the last 200 years is considered to be primarily responsible for the increases in the global surface temperatures in the twentieth century. From the end of the nineteenth century, the mean surface temperature increased by 0.8 ± 0.2 °C [10]. Land and sea temperature records showed that this rate of increase is not constant (Fig. 2). Global surface temperatures increased by 0.35 °C between 1910 and 1940; however, there was a cooling period after this increase, and a 0.1 °C decrease was observed. From 1970 to the present time, there has been an increase of 0.55 °C. These temperature changes can be considered trivial; however, even if not immediately problematic, when this change takes place globally, the consequences can be inevitable and extremely dire. For example, just a few degrees' difference in the mean global temperature separates us from that of the Ice Age [9]. When the temperature records spanning ~160 years are investigated, we observed that our last 12 years were the hottest (Fig. 2).

Effects of Climate Change

Even though the significance of global climate change was not understood until recently, many people are aware that if nothing is done to prevent it, the consequences will be worse. From research data, it has been shown that the world will confront much more significant and serious dangers 20–30 years after climate changes. For example, it takes a long time to heat large water masses, such as seas and oceans. In the beginning of summer, the water is a bit colder than during the middle and the last parts of summer. The same logic holds true for global warming; therefore, the effect of extra absorbed heat will be apparent 20–30 years later. This also means that the current changes in our carbon use will have an effect 20–30 years later. Storms, water scarcity, heat waves, monsoons, the destruction of the coral reefs, food shortages, deforestation, and sea level rise are expected to be more apparent and frequent. On earth, the expected changes at the continent level [9] related to global warming in 10–30 years are provided in Table 2.

Climate Change in Turkey

Turkey became a party to the United Nations Framework Convention on Climate Change (UNFCCC) in May 2004 and to the Kyoto Protocol on February 5, 2009. The Climate Change Coordination Committee was established in 2010 in Turkey and renamed as the Climate Change and Air Management Cooperation Committee in 2013 to determine a countrywide policy and related strategies to address climate change. The GHG emissions per capita in Turkey are very low compared to that in countries belonging to the Organization for Economic Co-operation and

Table 2 Consequences of climate change at the continent level [9]

	Water scarcity	Food production	Diseases	Floods	Heat waves and fires	Coral reefs and glaciers
Africa	75–250 million people are under risk until 2020	Yield reductions up to 50% until 2020	Increase in malaria in eastern mountainous regions of Africa	High risk at coastal sites	Increase in forest fires due to droughts and increased temperatures	Annihilation of coral reefs in Red Sea and East Africa is expected
Asia	120 million–1.2 billion people are under risk until 2020	Yield reductions up to 2.5–10 % until 2020, 49 million people are in the verge of hunger	Due to the flood and increased temperatures, rise in cholera and typhus outbreaks	2.5 million people are under the risk of floodings, and even more people will face sea level rise and hurricane-related problems	In the South and East Asia, increase in high temperature-related deaths and decrease in low temperature-related deaths	Until 2035, complete disappearance of Himalayan glaciers and 30% reduction in coral reefs
Australia, New Zealand	Until 2030 20% reduction in water resources, less snow cover in south and east parts	Reduction in the fish amount of tropical islands around corals, decrease in products in east and south of Australia and east of New Zealand	0.1–0.3 million more people will suffer from contagious fever until 2020	Submersion risks at coasts and small islands due to the sea level rise	4–25% rise in the frequency of forest fires in east and south parts of Australia and twofold increase in hot temperature-related deaths	58–81% destruction of the corals is expected until 2030

Europe	Reduction in the water resources of south parts up to 23%, increase in drought and fires in the vicinity of Mediterranean	Increase in yield in north, decrease in south	No change is expected	Increased number of storms in North Atlantic, sea level rise, severe storms, and rains especially effective on coastal and internal regions of the continent, increase in floodings	More frequent heat waves, which are first seen in 2003 and claimed 30,000 lives, decrease in the cold weather-related deaths	Decrease in the snow cover, disappearance of the considerable amount of the glaciers on top of high mountains
Middle and South America	7–77 million people are under at risk until 2020	Reduction in rice yield, until 2020 5 million people are under at risk of hunger	Risk of malaria is expected to expand through south	Flood risks at the regions near the sea level like Argentine, El Salvador	Increase in the fires in South America, which also include Amazon rainforests	Near to complete melting of tropical glaciers, disappearance of corals in Caribbean Sea
North America and polar regions	Reduction is expected in already in-use water resources	Reduction in quality and yield especially in grape production	Lyme illness, which is transmitted with ticks, is expected to advance 200 km north	Increase in floods and tropical storms especially in coasts of Mexico Bay	Rise in the risk of fires owing to the increased droughts	Melting of the considerable amount of glaciers in Alaska and Arctic region until 2030

Development (OECD) and European Union; however, when this emission per capita of Turkey is multiplied by 70 million people, Turkey becomes the thirteenth highest GHG emitter in the world, which is why it is important for Turkey to take steps to reduce emissions. In 2010, the Ministry of Environment and Urbanization, General Directorate of Environmental Management, Climate Change, prepared the National Climate Change Strategy and National Climate Change Action Plan and in 2011 created the Climate Change Adaptation Strategy and Action Plan [11]. Turkey has also participated in the Paris Agreement, which aimed to hold the mean global temperature increase to <2 °C and was scheduled to go into effect beginning in 2020 and presented its road map for the struggle against climate change. As a fast-developing economy, Turkey planned to decrease forecasted GHG emissions approximately <21% in 2030.

Since 2004, the Turkish Statistical Institute has calculated the country's GHG emissions using the IPCC guide [12]. As a result, in 2007, the Climate Change and Air Management Coordination Cooperation Committee prepared the First National Communication on Climate Change, which is one of the most significant commitments Turkey has made within the scope of UNFCCC [13]. Finally, in 2016, the Sixth National Communication of Turkey was prepared and published [14].

Turkey's National Greenhouse Gases Inventory comprises CO_2, CH_4, N_2O, fluorinated gases (F-gases), NOx, non-methane volatile organic compounds, and CO emissions that result from the sources illustrated in Fig. 3. Turkey's total GHG emissions in terms of CO_2 equivalency in 2006 and 2013 were 331.8 Mtons and 459.10 Mtons, respectively [14]. Between 1990 and 2013, GHG emissions continuously increased, and between 2000 and 2013, emissions increased by 47.7%. From 1990 to 2013, GHG emissions per capita increased from 3.96 ton CO_2-e to 6.04 ton CO_2-e; however, although this value was below the 2012 average of 12.47 ton CO_2-e for OECD countries, it was above the world's average of 4.88 ton CO_2-e. Turkey's CO_2 emissions were 79.15% of the total GHG emissions. The remaining percentages of the total GHG gases were CH_4 (14.33%), N_2O (5.06%), and F-gases (1.45%).

The sources of total emissions in Turkey can be categorized as 67.8% from the energy sector, 15.7% from industrial processes and product usage, 10.8% from agricultural activities, and 5.7% from the waste sector. The GHG emissions from the energy sector are from fuel combustion; 67.8% of the energy sector is shared with 24.8% by the energy industry, 15% by transportation, 13.6% by industry, and 12.8% by fuel combustion in households, services, and agriculture/forestry/fishery subsectors [14].

Climate change varies from region to region. Turkey's long-term temperature anomalies show a general trend that is quite similar worldwide; however, there are some remarkable differences. Global temperature increases, for example, began in the 1980s, yet it began in Turkey only after 1990 (Fig. 4). This temperature increase is seen mostly in summer in Turkey. For example, between 1970 and 2000, there was an increase in summer temperatures of 1.5 °C.

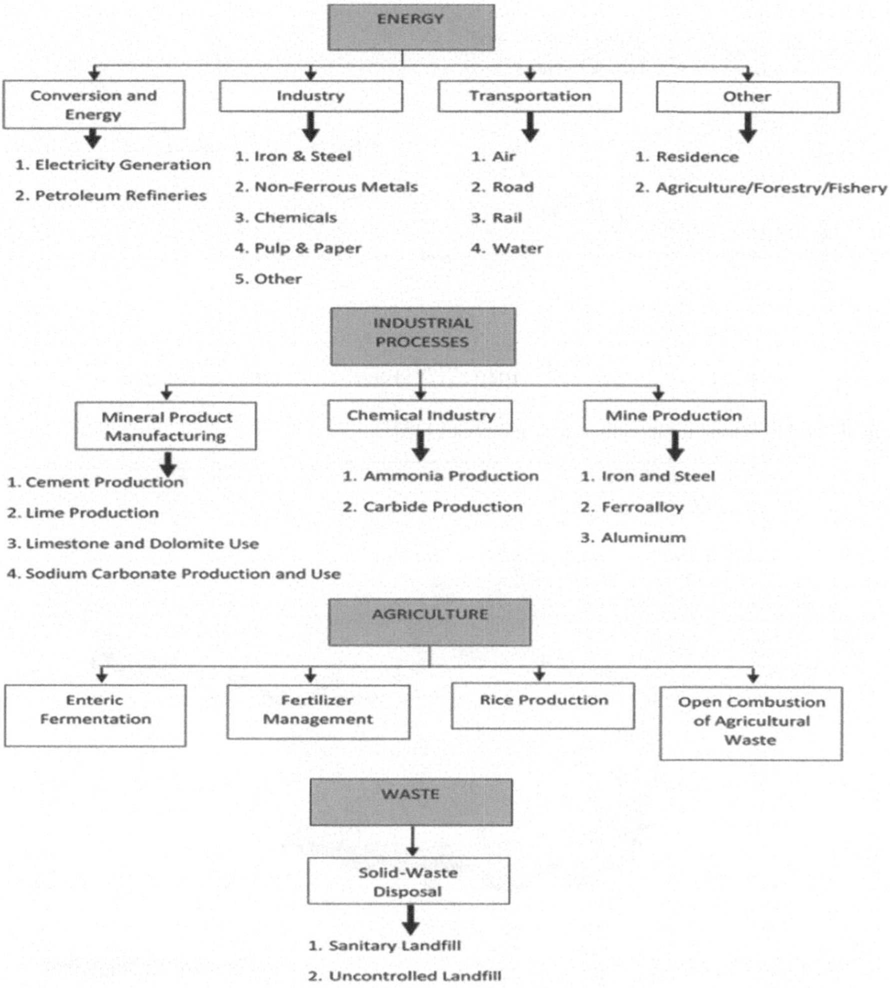

Fig. 3 Sources of greenhouse gases from the National Greenhouse Gases Inventory in Turkey [12]

According to IPCC, the annual mean temperatures in Turkey will increase in the future years by an average of 2.5–4.0 °C and will increase by 4 °C in the Aegean and Eastern Anatolia regions. This situation has also been reported in one study in Turkey. According to this modeling study done by Şen in 2013 [15], south Turkey will face a critical threat of drought, and in the Aegean, Mediterranean, and Southeast Anatolia regions, winter precipitation will be reduced by 20–50% (Fig. 5).

According to the results of the model of Demircan et al. [16], after 2071, the winter temperature will increase by ~2 °C and in spring and autumn by ~3 °C. In

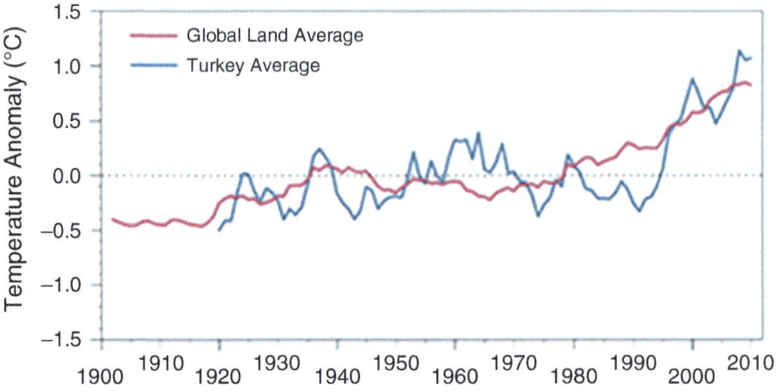

Fig. 4 Turkey's long term temperature anomalies [15]

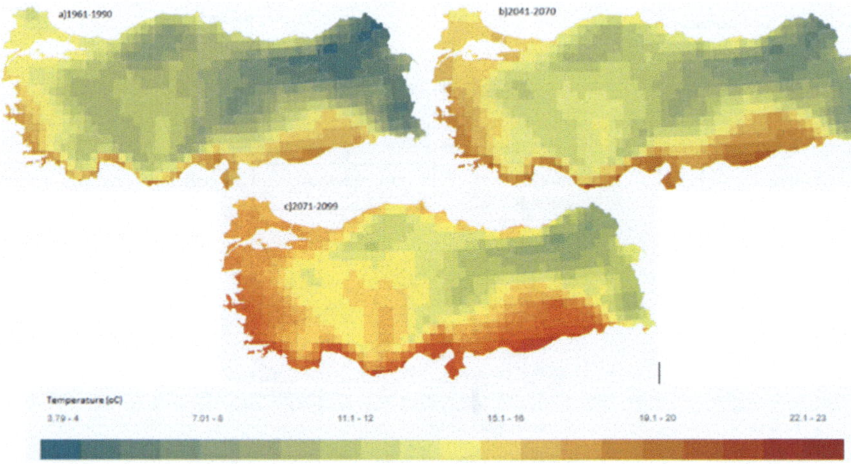

Fig. 5 Turkey's predicted mean annual temperature distribution in 1961–1990, 2041–2070, and 2071–2099 based on modeling results [17]

summer, temperatures will increase by up to 4 °C in the Southeast Anatolia region and coastal areas of the Aegean region. On the other hand, the flood risk will increase in the northern regions. In particular, the potential effects of global warming are declining water resources, forest fires, drought, desertification, and ecological degradation. Studies define Turkey's three main risks regarding climate change as follows: (1) reduced precipitation in the interior areas of Anatolia and southern watersheds, (2) increased temperatures that will alter precipitation in winter from snow to rain, and (3) excessive amounts of precipitation, especially in the northern and western coastal regions of Anatolia.

Conclusion

The first emphasis on climate change was made by Arrhenius, saying that "... if the amount of atmospheric CO_2 increases two-fold, the mean world temperature will rise 5.7°C..." in history [18]. Today's modeling studies confirm this situation. Also according to models for Turkey, the annual mean temperatures will increase in the future years by an average of 2.5–4.0 °C [15].

To meet its commitments as a party to the Kyoto Protocol, Turkey must revise its national policies. As such, the first activity would be to revise the relevant regulations to make them more effective in reducing GHG emissions from industry, motor vehicles, and heating. This could be accomplished by providing new infrastructures to create heat using less energy, using fewer fuel-consuming vehicles for long-range transportation, using less energy-consuming technological systems in the industry, providing incentives for renewable energy consumption and mechanisms, and considering environmental protection as a fundamental concept in the transportation and industry sectors.

References

1. Weart SR (2008) The discovery of global warming. Harvard University Press, Cambridge
2. Solomon S, Qin D, Manning M, Chen Z, Marquis M, Averty KB, Tignor M, Miller HL (2007) IPCC (Intergovernmental Panel on Climate Change) the physical science basis. Contribution of Working Group I to the fourth assessment report of the Intergovernmental Panel on Climate Change. Cambridge University Press, Cambridge
3. Kadıoğlu M (2001) Global climate change and Turkey. Guncel, Istanbul
4. Houghton JT, Ding Y, Griggs DJ, Noguer M, Linden PJ, Dai X, Maskell K, Johnson CA (2001) IPCC (Intergovernmental Panel on Climate Change), climate change 2001: the scientific basis. Cambridge University Press, Cambridge
5. Metcalfe S, Derwent C (2005) In: Matthews J (ed) Atmospheric pollution and environmental change. Oxford University Press, New York
6. Hardy JT (2003) Climate change: causes, effects, and solutions. Wiley, Chichester
7. Seinfeld JH, Pandis SN (2006) Climate and chemical composition of the atmosphere. In: Atmospheric chemistry and physics; from air pollution to climate change, Chap 23. Wiley, Chichester
8. UNFCC (2003) The United Nations Framework Convention on Climate Change, Official Journal numbered 25320, 18 Dec 2003
9. Walker G, King SD (2008) The hot topic-how to tackle global warming and still keep the lights on. Bloomsbury, London
10. Pachauri RK, Meyer LA (2014) IPCC, 2014: climate change 2014: synthesis report. Contribution of Working Groups I, II and III to the fifth assessment report of the Intergovernmental Panel on Climate Change. IPCC, Geneva, Switzerland
11. NCCAP (National Climate Change Action Plan) (2011) Republic of Turkey 2011–2023, Ankara, Turkey
12. Turkey Statistical Institute (TSI) (2007) SG National Emissions Inventory report, Ankara
13. Apak G, Ubay B (2007) Turkey First National Communication on Climate Change, Ministry of Environment and Forestry, Ankara

14. 6NDR-National Declaration Report (2016) Ministry of Environment and Urbanization, General Directorate of Environmental Management
15. Sen OL (2013) A holistic view of climate change and its impacts in Turkey, Mercator-IPC Report, SU-Istanbul Policy Center. http://ipc.sabanciuniv.edu/wp-content/uploads/2013/12/IklimDegisikligiRaporWEB.16.12.13.pdf. Accessed 15 Mar 2018
16. Demircan M, Gürkani H, Eskioğlu O, Arabacı H, Coşkun M (2017) Climate change projections for Turkey: three models and two scenarios. Turk J Water Sci Manag 1(1):23
17. Şen ÖL (2018) Clımate change in Turkey, as part of the MERCATOR-IPC Fellowshıp Program. http://climatechangeinturkey.com/mapping-climate-change-temperature.html. Accessed 21 Mar 2018
18. Arrhenius S (1896) On the influence of carbonic acid in the air upon the temperature of the ground. Philos Mag J Sci 41(251):237–276

Ecological Risk Assessment of Heavy Metal Pollution in Urban Soils: A Case Study

Naim Sezgin, Semih Nemlioğlu, and Nilgün Balkaya

Abstract In this study, geo-accumulation index (I_{geo}), enrichment factor (EF), potential ecological risk (E_r), and risk index (RI) of five metals (Cu, Pb, Zn, Hg, and Cd) and Sc (as reference element for enrichment factor calculation) in soil samples were applied for assessing the pollution pattern and possible sources of heavy metals. Soil samples were collected from 27 different locations in Avcilar in the European part of Istanbul, Turkey. The mean I_{geo} values of Cu, Pb, Zn, Hg, and Cd were found between 0.12 and 0.96, which means that the soil contamination levels of these metals were uncontaminated to moderately contaminated. The EF mean values of Cu, Pb, Zn, Hg, and Cd were 1.70, 1.30, 1.30, 2.29, and 1.26, respectively. These EF results indicated that Cu, Pb, Zn, and Cd in the study area soil were affected by natural sources, whereas Hg was mainly caused by nature and partly influenced by human activities. The order of the mean values of E_r in the soil samples was in the following descending order of Hg > Cd > Cu > Pb > Zn. Most of the soil samples showed low and moderate ecological risks.

Keywords Soil pollution · Geo-accumulation index · Enrichment factor · Ecological risk index · Metal contamination

Introduction

Soil is a fundamental and nonrenewable natural resource. It hosts goods and services vital to the ecosystems and human life. Our crops, feed, and fuel are obtained from the soil. Soil filters and cleans huge amount of water each year. It is a major storehouse for carbon and helps regulate emissions of various greenhouse gases such as carbon dioxide [1]. However, soil is a fragile structure. Therefore, damages

N. Sezgin (✉) · S. Nemlioğlu · N. Balkaya
Department of Environmental Engineering,
Istanbul University-Cerrahpaşa, Istanbul, Turkey
e-mail: nsezgin@istanbul.edu.tr

N. Balkaya, S. Guneysu (eds.), *Recycling and Reuse Approaches for Better Sustainability*, Environmental Science and Engineering,
https://doi.org/10.1007/978-3-319-95888-0_18

can occur in soils because of accidents or improper management. Contamination of soil by chemicals is one of these damages [2].

Sustainable development goals stated by the United Nations in the 2030 Agenda for Sustainable Development identified the need to reduce the number of deaths and illnesses from hazardous chemicals and soil pollution and contamination by 2030 (target 3.9) and to reduce the release of chemicals and all wastes to air, water, and soil in order to minimize their adverse impacts on human health and the environment by 2020 (target 12.4). To restore degraded soils and improve soil health by 2030 (target 15.3) is also another Sustainable Development Goal [3].

As stated by Ashraf et al. [4], petroleum hydrocarbons, heavy metals, pesticides, and solvents are the most common chemicals causing soil pollution. Heavy metal accumulation may take place from some natural processes or human activities that increase heavy metal accumulation in the soil environment [5]. When heavy metals reach to the soil and water, then they can accumulate as nonbiodegradable forms. Thus, they can cause adverse effects for living organisms in the receiving environment. Because of the dynamic equilibria between the soil components, heavy metals may occur in different phases such as the solid phase (immobilized contaminants) or dissolved in soil. The final form of heavy metals is the most dangerous type because of the fact that their ions are mobile, so they can leach and be absorbed by the living organisms [6].

The increasing heavy metal concentrations in urban surface soil can cause toxic effects for the human and the other living organisms [7, 8]. Mobile and/or available forms of heavy metals in soil or even in food chain may enter the human body through ingestion, digestion, and inhalation systems [9]. Implications for transport, exposure, environmental availability, and human health and ecological risks of metals in urban soils were described in many literatures and even summarized with illustrations as given by Luo et al. [10]. As stated by Jiang et al. [11], it is necessary to evaluate human health risks from heavy metal exposure due to soil pollution and identify the contamination sources to improve the soil environment and protect the human health.

In this study, concentrations of Cu, Pb, Zn, Hg, and Cd were measured in 27 soil samples, which were collected from Avcilar-Istanbul in Turkey. The selected area of this study is near industrial districts, construction and demolition activities, thermal power plant, port, and the D-100 highway which has heavy traffic. In addition, Istanbul University Avcilar Campus, which hosts approximately 22,000 people (student, teaching staff, and administrative staff), is also located in the selected area. In the present study, the concentrations of heavy metals were identified in soil samples obtained from Avcilar-Istanbul. Geo-accumulation index (I_{geo}), enrichment factor (EF), ecological risk (E_r), and risk index (RI) were calculated in order to assess the soil pollution and toxicity due to the selected major and heavy metal elements.

Materials and Methods

Study Area, Soil Sampling, and Chemical Analysis

Avcilar is a district of Istanbul, Turkey, located in the European side of the city, just to the West of the Kucukcekmece inlet of the Sea of Marmara. Avcilar has a population of about 500,000 people. Besides, there are more than 250 industrial companies (small and big companies) and two main highways [D-100 and E-80 (TEM)]. The study area was selected between the residential areas of Avcilar district, D100 highway, medium- and small-scale industrial facilities, and Kucukcekmece Lake shoreline. It also includes a university campus (Istanbul University Avcilar Campus) and green fields. There are 5 faculties, 2 vocational schools, 2 institutes, 1 dormitory, 5 other centers, and approximately 22,000 people (student, teaching staff, and administrative staff) in Avcilar Campus. The campus has mainly two types of areas: (1)

Fig. 1 Sampling points in the study area. *Source* Google Earth

main study and living area and (2) green fields. The campus is near the lakeside of Kucukcekmece Lake. Aerial image of the study area is given in Fig. 1. The colored points on the map indicate the soil sampling places in the study area. The sampling places were located within 28°43′13.51″E to 28°44′10.53″E longitude and 40°59′1.54″N to 40°59′47.00″N latitude.

Twenty-seven topsoil samples (0–20 cm) were collected from different sample points of the study area as indicated in Fig. 1. Each sample was collected and then placed into polyethylene bags for transport to the laboratory. Sampling locations were recorded in the form of global positioning system coordinates. After air-drying, all soil samples were sieved through a 0.5 mm (30 mesh BS) stainless steel sieve to remove plant roots, debris, glasses, and other materials. 0.5 g of subsamples was digested using 3 mL of 3:1:2 $HCl:HNO_3:H_2O$ at 95 C for 1 h and diluted with 10 mL water. Then, the samples' metal concentrations as Cu, Pb, Zn, Hg, Cd, and Sc were measured using inductively coupled plasma mass spectrometry (ICP-MS, Perkin-Elmer ELAN 9000) in the accredited Acme Analytical Laboratory in Vancouver, Canada. Quality controls involved routine analyses of standards and duplicates. Approximately 5% of the samples were analyzed as internal and external control samples; the variability was found to be <10% for all elements analyzed.

Geo-accumulation Index (I_{geo})

The geo-accumulation index (I_{geo}), which was suggested by Muller, has been usefully applied for the determination of soil contamination in the last decades [12, 13]. The geo-accumulation index can be calculated by Eq. (1):

$$I_{geo} = \log_2 \left(\frac{C_n}{1.5 B_n} \right) \tag{1}$$

where C_n is the measured concentration of elements in the soil samples and B_n is the geochemical reference value. The constant 1.5 is used to account for the possible variations in the reference values as they are affected by natural fluctuations and anthropogenic influence [14, 15]. Muller's classification for the assessment of the level of contamination is given in Table 1.

Enrichment Factors and Ecological Risks

The metal contamination in soils from anthropogenic sources can be obtained by the enrichment factor (EF) given in Eq. (2) [10]. The EF of an element in a studied sample is based on the standardization of a measured element against a reference element. The reference element is often characterized by low occurrence variability,

Table 1 I_{geo} classifications, categories of EF values, potential ecological risk categories of E_r and RI values

I_{geo} classifications	
I_{geo} value	*Class – Soil quality*
$I_{geo} \leq 0$	0 – Uncontaminated
$0 < I_{geo} \leq 1$	1 – Uncontaminated to moderately contaminated
$1 < I_{geo} \leq 2$	2 – Moderately contaminated
$2 < I_{geo} \leq 3$	3 – Moderately to heavily contaminated
$3 < I_{geo} \leq 4$	4 – Heavily contaminated
$4 < I_{geo} \leq 5$	5 – Heavily to extremely contaminated
$5 < I_{geo}$	6 – Extremely contaminated
Enrichment categories of EF values	
EF value	*Enrichment category*
EF ≤ 2	Minimal enrichment
2 < EF ≤ 5	Moderate enrichment
5 < EF ≤ 20	Significant enrichment
20 < EF ≤ 40	Very high enrichment
40 < EF	Extremely high enrichment
E_r and RI values	*Ecological risk category*
$E_r < 40$	Low potential ecological risk
$40 \leq E_r < 80$	Moderate potential ecological risk
RI < 150	Low ecological risk
150 ≤ RI < 300	Moderate ecological risk
300 ≤ RI < 600	Considerable ecological risk
600 < RI	Very high ecological risk

with the most commonly used elements being Al, Fe, Ti, Sc, Si, Sr, and K. In this study, Sc was selected as the reference element.

A value of EF close to 1 indicates a natural origin, where values greater than 10 are considered to be originated mainly from anthropogenic sources. The EF analyses are shown in Table 1 [11]. The EF of each heavy metal was calculated according to Eq. (2) [11]:

$$\mathrm{EF} = \frac{\left(\dfrac{C_x}{C_{\mathrm{ref}}}\right)_{\mathrm{sample}}}{\left(\dfrac{C_x}{C_{\mathrm{ref}}}\right)_{\mathrm{background}}} \tag{2}$$

where (C_x/C_{ref}) is the ratio of concentrations between the heavy metal and a reference metal in the sample and the background soil.

To evaluate the ecological risk index (RI) of heavy metals, the approach that was originally introduced by Hakanson [16] was used. The RI was calculated from the following Eqs. (3–5):

$$\mathrm{RI} = \sum_{m}^{i=1} E_{\mathrm{r}} \tag{3}$$

$$E_{\mathrm{r}} = T_{\mathrm{r}} \times C_{\mathrm{f}} \tag{4}$$

$$C_{\mathrm{f}} = \frac{C_{\mathrm{s}}}{C_{\mathrm{n}}} \tag{5}$$

where C_s and C_n are the heavy metal sample and background concentrations, respectively, E_r is the ecological risk of each element, and RI shows the ecological risk of multiple elements. C_f is the contamination factor. Hakanson [16] defined T_r as a "toxic-response factor" for a given substance and demonstrated this value for Cu, Pb, Zn, Hg, and Cd to be 5, 5, 1, 40, and 30, respectively. The E_r and RI values and the means of these values are given in Table 1 [17].

Result and Discussion

Metal Concentrations

The means of metal (Cu, Pb, Zn, Hg, Cd, and Sc) concentrations in the soil samples which were taken from Avcilar-Istanbul are shown in Table 2. The arithmetic means of Cu, Pb, Zn, Hg, Sc, and Cd concentrations were found to be 26.75, 21.23, 80.89, 0.03, 4.19, and 0.16 mg/kg, respectively. Spatial distribution of Cu, Pb, Zn, Hg, and Cd concentrations in the soil samples collected from the study area are given in Fig. 2. Heavy metal concentrations of nearly all soil samples taken from Avcilar were found to be higher than that of the reference soil (ZA23). It can be said that, according to these results, the existence of all heavy metals investigated in the soil samples are probably associated with anthropogenic sources. The heavy metal concentrations of the soil samples obtained from the study area in Avcilar were compared with those of the soils of Turkey cities and/or regions, mean values of the average concentrations for world soils, and the earth's crust. The results are given in Table 3.

Geo-accumulation Index (I_{geo}) Analysis Results

As shown in Table 4, the mean I_{geo} values of Cu, Pb, Zn, Hg, and Cd were found as 0.52, 0.12, 0.14, 0.96, and 0.14, respectively. These results showed that the soil contamination levels of these metals were identified as uncontaminated to moderately contaminated because the I_{geo} values were found between 0 and 1.

Table 2 Mean of metal concentrations (mg/kg) in soil samples

Sampling point	Cu	Pb	Zn	Hg	Sc	Cd
ZA1	27.30	21.00	70.00	0.04	5.70	0.20
ZA2	22.60	25.50	106.00	0.02	5.60	0.10
ZA3	25.10	23.90	79.00	0.02	3.60	0.20
ZA4	25.80	18.60	89.00	0.05	4.40	0.10
ZA4A	20.20	16.30	68.00	0.02	3.60	0.10
ZA5	18.20	19.50	54.00	0.02	3.40	0.10
ZA6	39.40	22.20	118.00	0.05	4.30	0.20
ZA7	18.90	18.20	63.00	0.01	3.60	0.10
ZA8	19.30	20.50	70.00	0.02	3.80	0.20
ZA9	18.80	14.30	95.00	0.02	3.60	0.20
ZA10	15.60	11.70	53.00	0.02	3.00	0.10
ZA11	28.20	29.30	102.00	0.03	4.20	0.30
ZA12	47.30	24.80	129.00	0.05	4.80	0.20
ZA13	29.70	25.20	85.00	0.04	6.10	0.20
ZA14	18.40	24.30	103.00	0.02	2.80	0.20
ZA15	32.20	23.80	93.00	0.02	7.60	0.20
ZA16	32.80	22.60	92.00	0.04	5.10	0.20
ZA18	20.00	19.80	68.00	0.02	4.00	0.10
ZA19	22.80	23.30	69.00	0.03	5.10	0.20
ZA20	26.00	23.80	73.00	0.01	4.80	0.40
ZA21	28.40	45.20	112.00	0.10	3.60	0.10
ZA22	97.30	23.20	107.00	0.04	3.70	0.20
ZA23 (Rf. S)	12.40	13.00	49.00	0.01	3.20	0.10
ZA24	19.40	14.80	50.00	0.02	3.70	0.10
ZA25	14.20	13.90	37.00	0.01	2.40	0.10
ZA26	19.20	13.40	57.00	0.02	2.50	0.20
ZA27	22.70	21.10	93.00	0.04	4.90	0.10
Mean	26.75	21.23	80.89	0.03	4.19	0.16

Rf. S reference soil point

Enrichment Factor Analysis Results

Soil contamination was assessed on the basis of enrichment factor (EF). The enrichment factors (EFs) of all analyzed metals were calculated for each soil sample relative to the background value of the local soil (point ZA23), with Sc chosen as the reference element. Table 4 shows the EF values and the means of heavy metals in the soil. The mean EF values of Cu, Pb, Zn, Hg, and Cd were found as 1.70, 1.30, 1.30, 2.29, and 1.26, respectively. The mean EF values of Cu, Pb, Zn, and Cd ≤ 2 showed the minimal pollution of these metals in the soil samples taken from Avcilar. The mean value of EF for Hg was found in the range of 2–5, indicating moderate pollution. These results indicated that Cu, Pb, Zn, and Cd in the soil samples were affected by natural sources, whereas Hg was mainly caused by the nature and partly

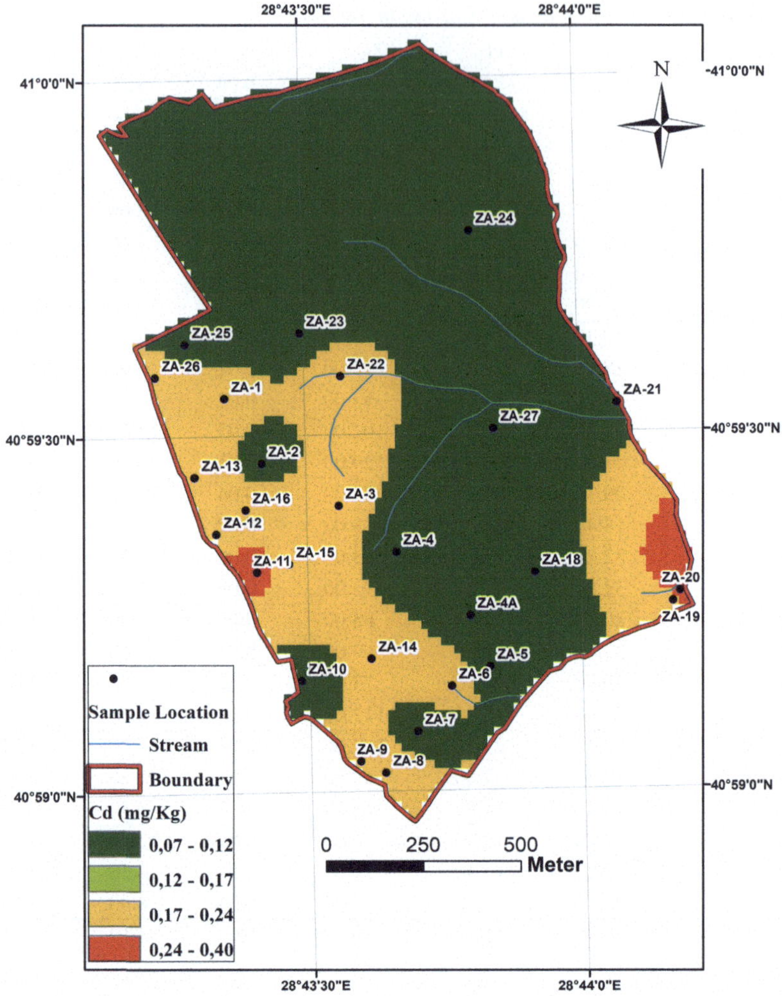

Fig. 2 Spatial distribution of Cd, Cu, Hg, Pb, and Zn concentrations

influenced by the human activities. It can be said that the samples having an enrichment factor of higher than 1.5 indicated strong human influence [28, 29]. As can be seen from Table 4, most of the sites' enrichment factors (EFs) for the heavy metals (Cu, Pb, Zn, Hg, and Cd) were higher than 1.5. According to this result (i.e., EFs > 1.5), the related sampling points were mainly located near the D100 highway subjected to heavy traffic pollution loads, in campus main roads, and rainwater surface runoff route whose upstream is located just in front of the industrial area. These highlighted points also have the same tendency regarding the I_{geo} values (i.e., $I_{geo} > 1.0$).

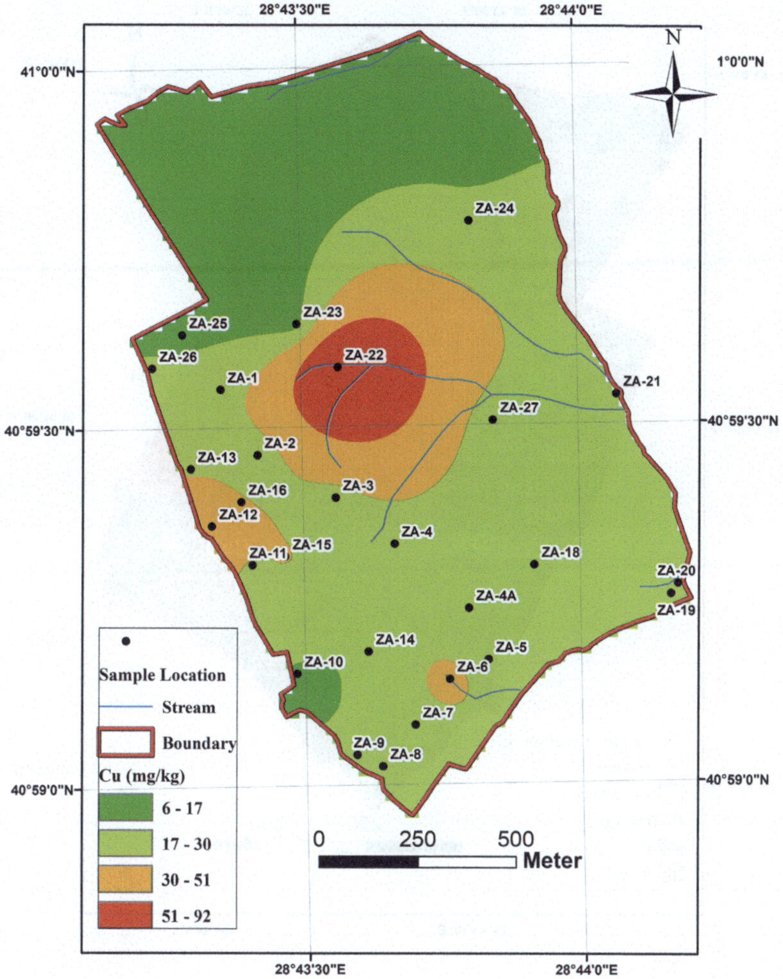

Fig. 2 (continued)

Ecological Risk Analysis Results

In this study, the ecological risk index is also calculated to assess the soil pollution and toxicity due to metal elements. As stated by Islam et al. [30], the potential ecological risk index is a comprehensive potential ecological risk index, which is the sum of E_r. It demonstrates the ecological risk caused by the overall contamination and shows the sensitivity of the biological community to the toxic substance. The ecological risk factors for each element (E_r) and the potential ecological risk index (RI) are summarized in Table 5. As can be seen from Table 5, the mean E_r of all the studied metals for all sites were in the descending order of Hg > Cd > Cu > Pb > Zn. The mean E_r values of Cu, Pb, Zn, Hg, and Cd were obtained as 10.79, 8.17, 1.65, 117.04, and 49.44, respectively. It can be said that, except for sampling point 21, the

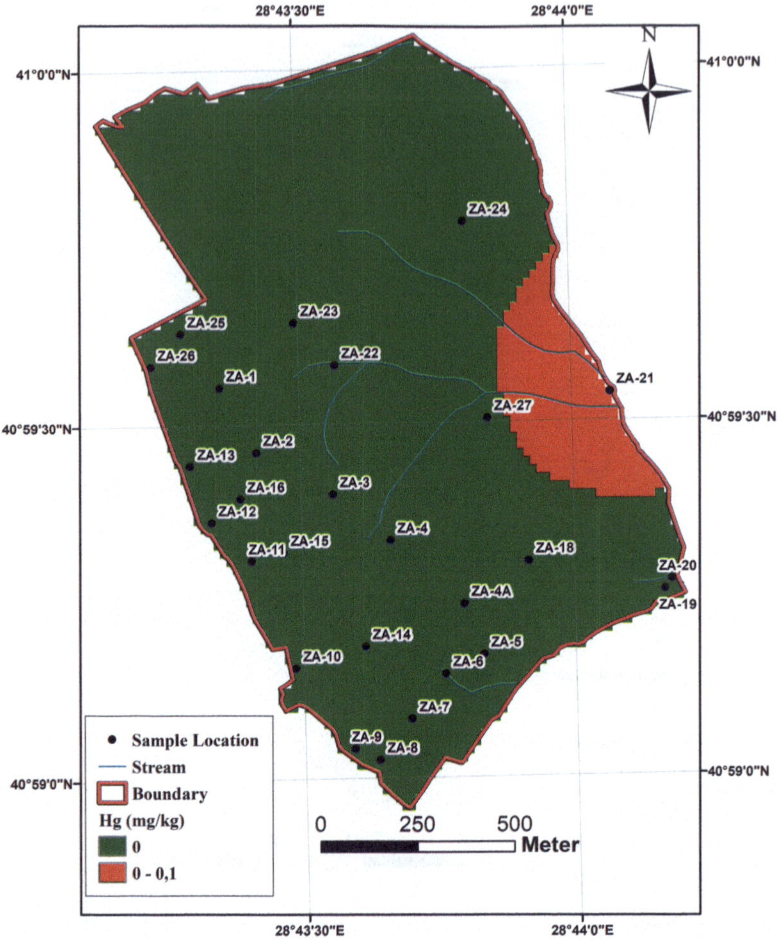

Fig. 2 (continued)

soil samples from Avcilar Campus showed low or moderate ecological risk. In view of the RI, soil from sampling point 21 (lakeside of Kucukcekmece Lake) showed considerable ecological risk, which might be due to the Kucukcekmece Lake contaminated by pollution loads from all kinds of sources such as households, industries, agriculture, forests, closed municipal solid waste dumping site, atmospheric fallout, etc.

Heavy metals originate from local sources mostly due to the industries (nonferrous industries, power plants, iron, steel, and chemical industries), agriculture (irrigation with polluted water, use of mineral fertilizers especially phosphates, contaminated manure, sewage sludge, and pesticides containing heavy metals), waste incineration, burning of fossil fuels, and traffic loads.

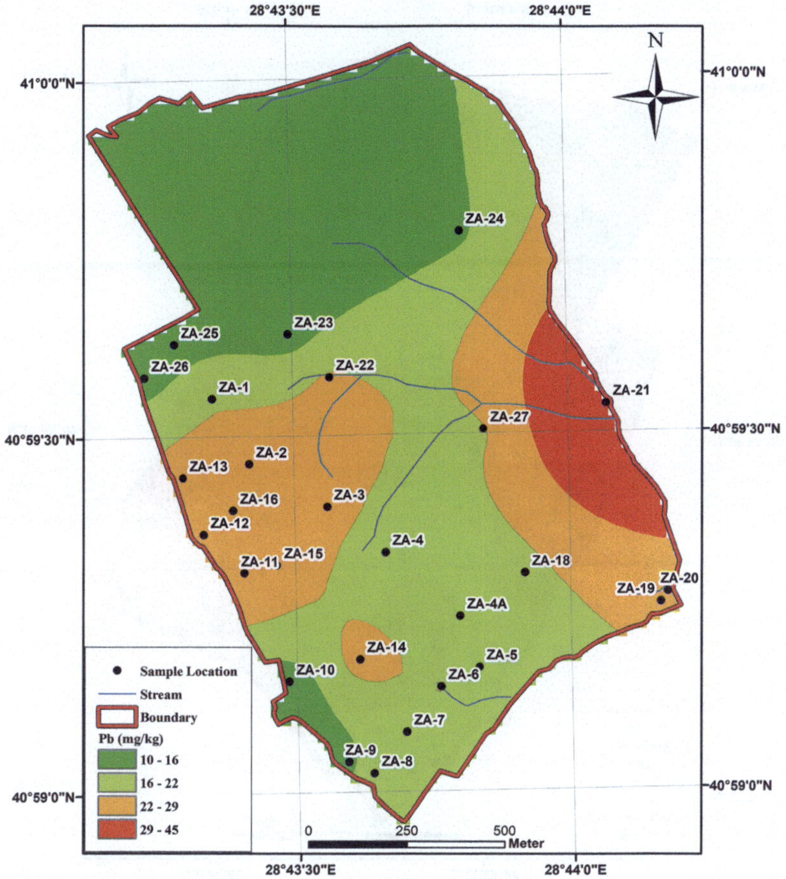

Fig. 2 (continued)

Conclusions

In the present study, the pollution status and risk assessment of heavy metals (Cu, Pb, Zn, Hg, and Cd) in the soil samples taken from Avcilar-Istanbul have been investigated. The mean values of heavy metal concentrations in the studied soil samples for Cu, Pb, Zn, Hg, Cd, and Sc were found to be 26.75, 21.23, 80.89, 0.03, 0.16, and 4.19 mg/kg, respectively. The mean I_{geo} values of Cu, Pb, Zn, Hg, and Cd in the soil samples were obtained as Hg > Cu > Zn, Cd > Pb in the descending order. According to the EF analysis, the results indicated minimal pollution for Cu, Pb, Zn, and Cd, and the sources of these metals were natural. The mean EF value for Hg showed moderate pollution, and its sources could be mainly caused by the nature and partly influenced by the human activities. The ecological risk results for Cu, Pb, Zn, Hg, and Cd were found to be moderate or low ecological risks in most sampling

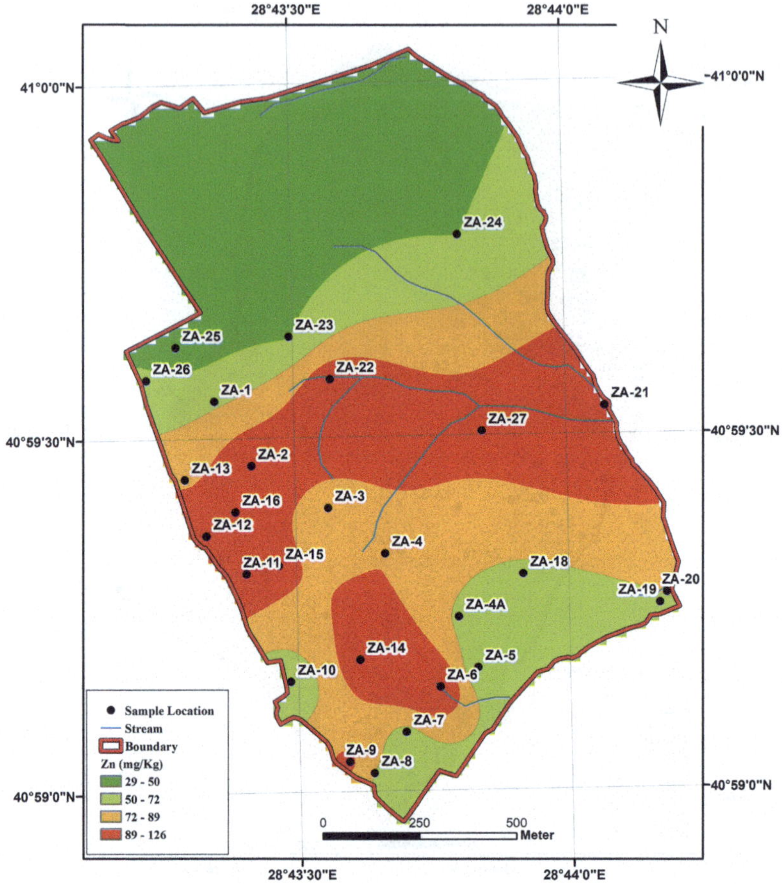

Fig. 2 (continued)

locations (except for point ZA21) of the study area. The ecological risk result of sampling point ZA21 showed that this location had considerable ecological risk. Point ZA21 had the lowest elevation level in the study area and was located just in the lakeside of Kucukcekmece Lake. Point ZA21 was a completely open location to main wind directions. Kucukcekmece Lake water can easily reach to this sampling point, and the Sea of Marmara-originated seawater can be mixed into the lake water. Industrial sewage leakage into the lake was also highly possible. Atmospheric deposition of both industrial and highway-originated particles could possibly be the heavy metal contamination risk sources for the study area as well. Particulate matter-originated heavy metal transportation via rainwater surface runoff and enrichment in the lower locations of the site were again highly possible because of the high inclination of the studied area. On the other hand, moderate ecological risk for Cu, Pb, Zn, Hg, and Cd was determined for eight sampling points (ZA1, ZA4, ZA6, ZA12, ZA13, ZA16, ZA22, and ZA27) in Avcilar. Except for points ZA22 and

Table 3 Heavy metal concentrations (mg/kg) in the soils of Turkey cities and/or regions, average concentrations for world soils, and the earth's crust

City or area of Turkey	Cu	Pb	Zn	Hg	Cd	Ref.
Istanbul-Avcilar	26.75	21.23	80.89	0.03	0.16	This study
Konya	2.36	2.65	4.50		0.02	[18]
Tunçbilek-Kütahya	19.6	21.2	57.9	0.63	0.11	[19]
Gebze-Kocaeli	95.88	246	632	0.10	4.41	[20]
Tavşanlı-Kütahya	32.6	82.6	165.3	0.22	1.5	[21]
Van	20	80	12.1		5.9	[22]
Amik plain-Hatay		5.56			0.19	[23]
Central and east parts of Black Sea region	44.73	29.27	57.51		1.26	[24]
Southern Turkey	1.03	1.21	4.94		0.02	[25]
Average concentrations for world soils	14	25	62	0.1	1.1	[26]
Earth's crust average value	39	17	67	0.08	0.1	[27]

Table 4 Geo-accumulation index (I_{geo}) analysis results and EF values

	(I_{geo}) analysis results					EF values of metals in soil samples				
Sampling point	Cu	Pb	Zn	Hg	Cd	Cu	Pb	Zn	Hg	Cd
ZA1	0.55	0.11	−0.07	1.42	0.42	1.24	0.91	0.80	2.25	1.12
ZA2	0.28	0.39	0.53	0.42	−0.58	1.04	1.12	1.24	1.14	0.57
ZA3	0.43	0.29	0.10	0.42	0.42	1.80	1.63	1.43	1.78	1.78
ZA4	0.47	−0.07	0.28	1.74	−0.58	1.51	1.04	1.32	3.64	0.73
ZA4A	0.12	−0.26	−0.11	0.42	−0.58	1.45	1.11	1.23	1.78	0.89
ZA5	−0.03	0.00	−0.44	0.42	−0.58	1.38	1.41	1.04	1.88	0.94
ZA6	1.08	0.19	0.68	1.74	0.42	2.36	1.27	1.79	3.72	1.49
ZA7	0.02	−0.10	−0.22	−0.58	−0.58	1.35	1.24	1.14	0.89	0.89
ZA8	0.05	0.07	−0.07	0.42	0.42	1.31	1.33	1.20	1.68	1.68
ZA9	0.02	−0.45	0.37	0.42	0.42	1.35	0.98	1.72	1.78	1.78
ZA10	−0.25	−0.74	−0.47	0.42	−0.74	1.34	0.96	1.15	2.13	0.96
ZA11	0.60	0.59	0.47	1.00	1.00	1.73	1.72	1.59	2.29	2.29
ZA12	1.35	0.35	0.81	1.74	0.42	2.54	1.27	1.76	3.33	1.33
ZA13	0.68	0.37	0.21	1.42	0.42	1.26	1.02	0.91	2.10	1.05
ZA14	−0.02	0.32	0.49	0.42	0.42	1.70	2.14	2.40	2.29	2.29
ZA15	0.79	0.29	0.34	0.42	0.42	1.09	0.77	0.80	0.84	0.84
ZA16	0.82	0.21	0.32	1.42	0.42	1.66	1.09	1.18	2.51	1.25
ZA18	0.10	0.02	−0.11	0.42	−0.58	1.29	1.22	1.11	1.60	0.80
ZA19	0.29	0.26	−0.09	1.00	0.42	1.15	1.12	0.88	1.88	1.25
ZA20	0.48	0.29	−0.01	−0.58	1.42	1.40	1.22	0.99	0.67	2.67
ZA21	0.61	1.21	0.61	2.74	−0.58	2.04	3.09	2.03	8.89	0.89
ZA22	2.39	0.25	0.54	1.42	0.42	6.79	1.54	1.89	3.46	1.73
ZA23	−0.58	−0.58	−0.58	−0.58	−0.74	1.00	1.00	1.00	1.00	0.90
ZA24	0.06	−0.40	−0.56	0.42	−0.74	1.35	0.98	0.88	1.73	0.78
ZA25	−0.39	−0.49	−0.99	−0.58	−0.74	1.53	1.43	1.01	1.33	1.20
ZA26	0.05	−0.54	−0.37	0.42	0.42	1.98	1.32	1.49	2.56	2.56
ZA27	0.29	0.11	0.34	1.42	−0.74	1.20	1.06	1.24	2.61	0.59
Mean	0.52	0.12	0.14	0.96	0.14	1.70	1.30	1.30	2.29	1.26

Table 5 Ecological risk factors (E_r) and the potential ecological risk index (RI) of Cu, Pb, Zn, Hg, and Cd in soil samples

	E_r						
	Cu	Pb	Zn	Hg	Cd		
T_r	5	5	1	40	30	RI	Ecological risk
ZA1	11.01	8.08	1.43	160.00	60.00	240.51	Moderate ecological risk
ZA2	9.11	9.81	2.16	80.00	30.00	131.08	Low ecological risk
ZA3	10.12	9.19	1.61	80.00	60.00	160.93	Low ecological risk
ZA4	10.40	7.15	1.82	200.00	30.00	249.37	Moderate ecological risk
ZA4A	8.15	6.27	1.39	80.00	30.00	125.80	Low ecological risk
ZA5	7.34	7.50	1.10	80.00	30.00	125.94	Low ecological risk
ZA6	15.89	8.54	2.41	200.00	60.00	286.83	Moderate ecological risk
ZA7	7.62	7.00	1.29	40.00	30.00	85.91	Low ecological risk
ZA8	7.78	7.88	1.43	80.00	60.00	157.10	Low ecological risk
ZA9	7.58	5.50	1.94	80.00	60.00	155.02	Low ecological risk
ZA10	6.29	4.50	1.08	80.00	27.00	118.87	Low ecological risk
ZA11	11.37	11.27	2.08	120.00	90.00	234.72	Low ecological risk
ZA12	19.07	9.54	2.63	200.00	60.00	291.24	Moderate ecological risk
ZA13	11.98	9.69	1.73	160.00	60.00	243.40	Moderate ecological risk
ZA14	7.42	9.35	2.10	80.00	60.00	158.87	Low ecological risk
ZA15	12.98	9.15	1.90	80.00	60.00	164.04	Low ecological risk
ZA16	13.23	8.69	1.88	160.00	60.00	243.80	Moderate ecological risk
ZA18	8.06	7.62	1.39	80.00	30.00	127.07	Low ecological risk
ZA19	9.19	8.96	1.41	120.00	60.00	199.56	Low ecological risk
ZA20	10.48	9.15	1.49	40.00	120.00	181.13	Low ecological risk
ZA21	11.45	17.38	2.29	400.00	30.00	461.12	Considerable ecological risk
ZA22	39.23	8.92	2.18	160.00	60.00	270.34	Moderate ecological risk
ZA23	5.00	5.00	1.00	40.00	27.00	78.00	Low ecological risk
ZA24	7.82	5.69	1.02	80.00	27.00	121.54	Low ecological risk
ZA25	5.73	5.35	0.76	40.00	27.00	78.83	Low ecological risk
ZA26	7.74	5.15	1.16	80.00	60.00	154.06	Low ecological risk
ZA27	9.15	8.12	1.90	160.00	27.00	206.17	Moderate ecological risk
Mean	10.79	8.17	1.65	117.04	49.44	187.08	–

ZA27, six of these sampling points (ZA1, ZA4, ZA6, ZA12, ZA13, and ZA16) were located in the living area of the campus. It is possible to increase ecological risk of Cu, Pb, Zn, Hg, and Cd in the near future according to the increasing levels of the present industrial facilities' future activities. Because of this reason, possibly increased levels of ecological risks for these five metals should considerably be taken into account to protect human life and the environment in the study area of Avcilar-Istanbul.

References

1. FAO (2017) Voluntary guidelines for sustainable soil management. Food and Agriculture Organization of the United Nations, Rome
2. Hodson ME (2010) The need for sustainable soil remediation. Elements 6:363–368
3. http://www.un.org/en/development/desa/population/migration/generalassembly/docs/globalcompact/A_RES_70_1_E.pdf. Accessed 15 Apr 2018
4. Ashraf MA, Maah MJ, Yusoff I (2014) Chapter from the book environmental risk assessment of soil contamination. Soil contamination, risk assessment and remediation. IntechOpen. p 3–56
5. Mazurek R, Kowalska J, Gąsiorek M, Zadrożny P, Józefowska A, Zaleski T, Kępka W, Tymczuk M, Orłowska K (2017) Assessment of heavy metals contamination in surface layers of Roztocze National Park forest soils (SE Poland) by indices of pollution. Chemosphere 168:839–850
6. Vareda JP, Valente AJM, Durães L (2016) Heavy metals in Iberian soils: removal by current adsorbents/amendments and prospective for aerogels. Adv Colloid Interf Sci 237:28–42
7. Chandrasekaran A, Ravisankar R, Harikrishnan N, Satapathy KK, Prasad MVR, Kanagasabapathy KV (2015) Multivariate statistical analysis of heavy metal concentration in soils of Yelagiri Hills, Tamil Nadu. India-spectroscopical approach. Spectrochim Acta 137:589–600
8. Yuswir NS, Praveena SM, Aris AZ, Ismail SNS, Hashim Z (2015) Health risk assessment of heavy metal in urban surface soil (Klang District, Malaysia). Bull Environ Contam Toxicol 95:80–89
9. Luo X, Yu S, Li X (2012) The mobility, bioavailability, and human bioaccessibility of trace metals in urban soils of Hong Kong. Appl Geochem 27:995–1004
10. Luo XS, Yu S, Zhu YG, Li XD (2012) Trace metal contamination in urban soils of China. Sci Total Environ 421–422:17–30
11. Jiang Y, Chao S, Liu J, Yang Y, Chen Y, Zhang A, Cao H (2017) Source apportionment and health risk assessment of heavy metals in soil for a township in Jiangsu Province, China. Chemosphere 168:1658–1668
12. Barbieri M (2016) The importance of enrichment factor (EF) and geoaccumulation index (Igeo) to evaluate the soil contamination. J Geol Geophys 5(237):1–4
13. Ololade IA (2014) An assessment of heavy metal contamination in soils within auto-mechanic workshops using enrichment and contamination factors with geoaccumulation indexes. J Environ Prot 5(11):970–982
14. Aiman U, Mahmood A, Waheeda S, Malik RN (2016) Enrichment, geo-accumulation and risk surveillance of toxic metals for different environmental compartments from Mehmood Booti dumping site, Lahore city, Pakistan. Chemosphere 144:2229–2237
15. Xu JL, Jing BB, Zhang KX, Cui YC, Malkinson D, Kopel D, Song K, Da LJ (2017) Heavy metal contamination of soil and tree-ring in urban forest around highway in Shanghai, China. Human Ecol Risk Assess Int J 23(7):1745–1762
16. Hakanson L (1980) An ecological risk index for aquatic pollution-control – a sedimentological approach. Water Res 14:975–1001
17. Jiao X, Teng Y, Zhan Y, Wu J, Lin X (2015) Soil heavy metal pollution and risk assessment in Shenyang Industrial District, Northeast China. PLoS One 10(5). https://doi.org/10.1371/journal.pone.0127736
18. Onder S, Dursun S, Gezgin S, Demirbas A (2007) Determination of heavy metal pollution in grass and soil of city Centre green areas (Konya, Turkey). Pol J Environ Stud 16(1):145–154
19. Özkul C (2016) Heavy metal contamination in soils around the Tunçbilek thermal power plant (Kütahya, Turkey). Environ Monit Assess 188:284–295
20. Abanuz GY (2011) Heavy metal contamination of surface soil around Gebze industrial area, Turkey. Microchem J 99:82–92

21. Arık F, Yaldız T (2010) Heavy metal determination and pollution of the soil and plants of southeast Tavşanlı (Kütahya,Turkey). Clean (Weinh) 38(11):1017–1030
22. Türkdoğan MK, Kilicel F, Kara K, Tuncer I, Uygan I (2002) Heavy metals in soil, vegetables and fruits in the endemic upper gastrointestinal cancer region of Turkey. Environ Toxicol Pharmacol 13:175–179
23. Karanlık S, Agca N, Yalçın M (2011) Spatial distribution of heavy metals content in soils of Amik Plain (Hatay, Turkey). Environ Monit Assess 173:181–191
24. Ozyazici MA, Dengiz O, Ozyazici G (2017) Spatial distribution of heavy metals density in cultivated soils of central and east parts of Black Sea region in Turkey. Eurasian J Soil Sci 6(3):197–205
25. Agca N (2015) Spatial distribution of heavy metal content in soils around an industrial area in Southern Turkey. Arab J Geosci 8:1111–1123
26. Alloway BJ (2013) Sources of heavy metals and metalloids in soils. In: Alloway BJ (ed) Heavy metals in soils: trace metals and metalloids in soils and their bioavailability, environmental pollution, vol 22. Springer, Dordrecht. https://doi.org/10.1007/978-94-007-4470-7_2
27. Taylor SR, McLennan SM (1995) The geochemical evolution of the continental crust. Rev Geophys 33:165–241
28. Rashed MN (2010) Monitoring of contaminated toxic and heavy metals, from mine tailings through age accumulation, in soil and some wild plants at Southeast Egypt. J Hazard Mater 178:739–746
29. Zhang J, Liu CL (2002) Riverine composition and estuarine geochemistry of particulate metals in China-weathering features, anthropogenic impact and chemical fluxes. Estuar Coastal Shelf Sci 54:1051–1070
30. Islam MS, Ahmed MK, Habibullah-Al-Mamun M, Masunaga S (2015) Potential ecological risk of hazardous elements in different land-use urban soils of Bangladesh. Sci Total Environ 512–513:94–102

Part V
Wastewater Management

Microbial Growth Kinetics of an Anaerobic Acidogenic Bioreactor

Yalçın Aşkın Öktem

Abstract Kinetic evaluation of an acidogenic completely stirred tank reactor (CSTR) treating a chemical synthesis-based pharmaceutical wastewater was investigated. For kinetic modeling of the acidogenic reactor (Monod, Contois, Grau, et al.), first-order kinetic models were used. These models have high correlation coefficients in this study. Monod models have shown high correlation coefficients (99%) which were used in describing the process kinetics of the anaerobic acidogenic reactor. Kinetic studies showed that sludge yield (Y) and decay rate (K_d) were 0.25 mg VSS/mg COD and 0.79 1/day, respectively.

Keywords Acidification · Degradation kinetic · Monod · Anaerobic · CSTR

Introduction

Pharmaceutical wastewaters are characterized by high dissolved organics and changeable amount of salts [1, 2]. The high organic content of chemical synthesis-based pharmaceutical wastewater makes it especially suitable for processes based on anaerobic biotechnology [3]. Anaerobic digestion performance, both in terms of removal efficiency and energy recovery, a proper model, as well as dependable kinetic and stoichiometric parameters are required. In the literature, there is a general compromise on considering the hydrolysis as it is a kinetically limiting step of the all process [4]. Microbial kinetics plays a considerable role in the setup and operation of anaerobic digesters. Based on the biochemistry and microbiology of the anaerobic process for process analysis, control, and design, literature on kinetics of degradation was unavailable on chemical synthesis-based wastewater using first-order, Monod, Contois, and Grau kinetic models. Some researchers have investigated different wastewater treatment performance and its compounds using these models [5–9]. For biokinetic modeling of the reactor, Monod, Contois, Grau, and

Y. A. Öktem (✉)
Department of Environmental Engineering, Istanbul University-Cerrahpaşa, Istanbul, Turkey
e-mail: oktemy@istanbul.edu.tr

N. Balkaya, S. Guneysu (eds.), *Recycling and Reuse Approaches for Better Sustainability*, Environmental Science and Engineering,
https://doi.org/10.1007/978-3-319-95888-0_19

first-order kinetic models were used. These models have shown high correlations at studies. Although microbial kinetics of various types of industrial wastewaters has been studied, there is still lack of information in the literature about kinetics of chemical synthesis-based pharmaceutical wastewaters. In this study, these four types of kinetic models for describing the process kinetics of an acidogenic reactor chemical synthesis-based pharmaceutical wastewater were compared.

Materials and Methods

The reactor was made of Plexiglas with an active volume of 5 L. Characteristics of the wastewater are given in Table 1.

Inoculation sludge was obtained from an anaerobic reactor in alcohol industry. Analytical methods were used to determine reactor performance, and routine analyses were carried out in steady-state conditions. During the operation of the reactor, pH, temperature, gas production, and VFA were monitored daily. COD and TS/TVS analyses were carried out once every other day. All the analyses such as COD, solids, pH, etc. were conducted in accordance with the standards methods [10].

Reactor Operation

Acidification degree was interpreted according to different SRTs, HRTs, and OLRs, processing the chemical synthesis-based pharmaceutical wastewater.

Table 1 Wastewater characteristics

Parameters	Concentration (mg/L)
COD	40,000–60,000
TKN	800–900
PO_4-P	3–6
Suspended solids (SS)	800–1000
Volatile suspended solids (VSS)	500–700
Alkalinity (as $CaCO_3$)	900–1000
pH[a]	7–8

[a]Unitless

Results and Discussion

Performance of Acidogenic Reactor

Optimum operating conditions for acidification of wastewater were found to be an HRT of 12 h, pH range of 5.5 ± 0.1, and OLR of 13 kg COD/m^3 day. At this point, maximum acidification of 44% was achieved.

Determination of Microbial Kinetic Constants

For a completely stirred reactor without biomass recycle, the rate of accumulation of microorganisms can be shown as:

$$\frac{dx}{dt}V = QX_0 - QX + \mu XV_r - K_d XV_r \tag{1}$$

where dx/dt is the rate of change of microorganism concentration and is measured in terms of volatile suspended solids (VSS), V_r is the reactor volume (liter), X_0 and X are the concentrations of biomass in the influent and the reactor/effluent (mgVSS/L), and μ and K_d are the specific growth and decay rates (1/day), respectively.

Monod Model

If assumed that the concentration of microorganisms in the influent can be neglected, then at steady-state, $dx/dt = 0$. Equation (1) can be written as follows:

$$\mu = \frac{1}{HRT} + K_d \tag{2}$$

The relationship between the specific growth rate and the rate-limiting substrate concentration can be expressed by the Monod equation [11]:

$$\mu = \frac{\mu_m \times S}{(K_s + S)} - K_d \tag{3}$$

where: S: Effluent concentration (mg COD/L), μ_m: The maximum specific growth rate (1/day), K_S: Half saturation concentration (Monod constant) (mg COD/L)

The general expression of Monod kinetic model is given below:

$$-\frac{dS}{dt} = \frac{\mu_{\max} XS}{Y(K_S + S)} \tag{4}$$

Y: Cell yield coefficient (mg VSS/mg COD)

If Eq. (4) is integrated and then reduced, the equation can be written as:

$$\frac{\mu_m \times S}{(K_s + S)} = \frac{1}{HRT} + K_d \tag{5}$$

If Eq. (5) is linearized, the equation obtained will be:

$$\frac{HRT}{1 + HRT \times K_d} = \frac{1}{\mu_m} + \frac{K_S}{\mu_m} \cdot \frac{1}{S} \tag{6}$$

By plotting Eq. (6), the values of $\mu_{\max}$ and K_S can be calculated from the slope and the intercept of the line. Figure 1 was plotted for determining the values of $\mu_{\max}$ and K_S.

If the rate of change in the substrate concentration in the acidogenic reactor is expressed mathematically, this becomes:

$$\frac{dS}{dt} V = QS_0 - QS - V_r \frac{dS}{dt} \tag{7}$$

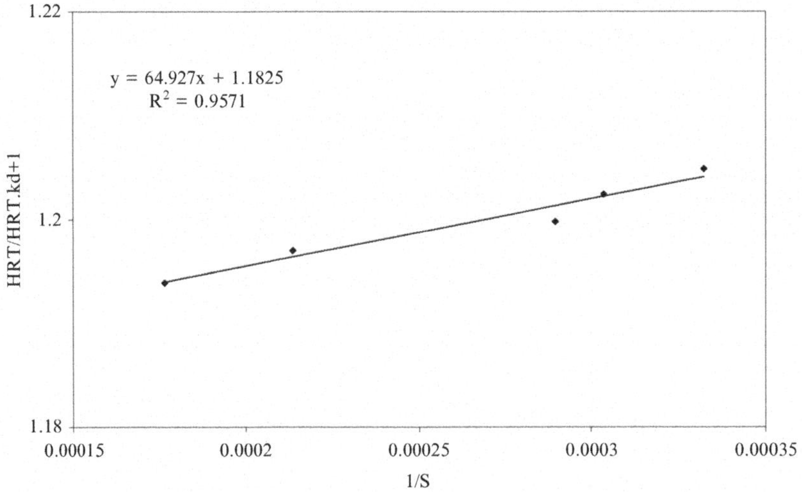

Fig. 1 Monod model $\mu_{\max}$ and K_s

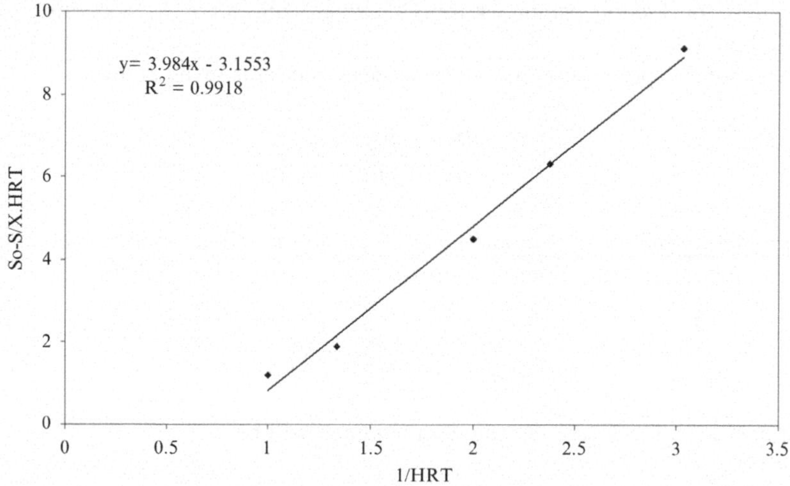

Fig. 2 Monod model Y and K_d

Then, the above equation can be arranged according to Monod equations (4) and (7) and defined as follows:

$$\frac{Q}{V}(S_0 - S) = \frac{1}{Y}\left[\frac{1}{HRT} + K_d\right] \times X \tag{8}$$

Equation (8) can be linearized, and then the equation obtained will be:

$$\frac{S_0 - S}{HRT \times X} = \frac{1}{Y} \times \frac{1}{HRT} + \frac{K_d}{Y} \tag{9}$$

where, by plotting, the values of Y and K_d obtained in Eq. (9) can be calculated from the slope and the intercept of the line. The values of Y and K_d were determined from Fig. 2.

Contois Model

By applying Contois model to the experimental data, the kinetic constants were calculated, and these obtained values of kinetic constants of the model were found to be same as the calculated values of the Monod model. The relationship between μ_m and the rate of limiting S concentration can be expressed by the Contois equation [12] as follows:

$$\mu = \frac{\mu_m \times S}{BX + S} - K_d \tag{10}$$

where B is the kinetic parameter mg COD/mg VSS, $BX = K_S$ according to the Contois kinetics

The general expression of the Monod kinetic model is given below:

$$-\frac{dS}{dt} = \frac{\mu_{\max} XS}{Y(BX + S)} \tag{11}$$

By applying the Contois model given above to Eq. (5), a kinetic evaluation can be done. Substituting the Contois expression, Eq. (10), in Eq. (5) gives:

$$\frac{\mu_m \times S}{BX + S} = \frac{1}{HRT} + K_d \tag{12}$$

If Eq. (12) is linearized, the equation obtained will be:

$$\frac{HRT}{1 + HRT \times K_d} = \frac{1}{\mu_m} + \frac{X}{S} \times \frac{B}{\mu_m} \tag{13}$$

Similarly, the values obtained in Eq. (13) can be used for plotting values of $\mu_{\max}$ and B. The value of $\mu_{\max}$ can be calculated from the intercept of the straight line, while B can be obtained from the slope line. Figure 3 was plotted to determine the kinetic values ($\mu_{\max}$ and B) according to the Contois model.

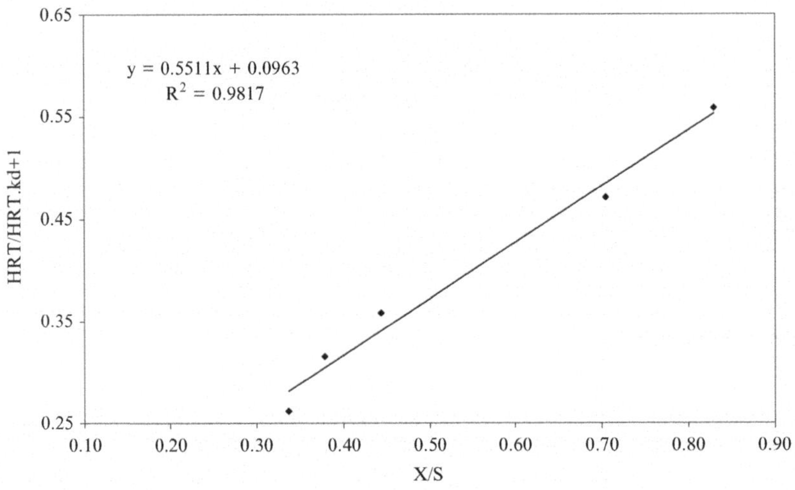

Fig. 3 Contois model

Grau Model

Grau model in Eq. (14) that was established by Grau et al. [13] is that the predicted effluent substrate concentration is function of the influent substrate concentration.

$$\mu = \frac{\mu_m \times S}{S_0} - K_d \tag{14}$$

Acidogenic reactor is a complete stirred reactor and behaves as a chemostat reactor; therefore, θ_c and θ have the same values. According to the Grau model, for determining kinetic constants, Eq. (14) is reduced and linearized, and then the obtained equation is shown below:

$$\frac{S}{S_0} = \frac{1}{\mu_m} \times \frac{1}{\theta} + \frac{K_d}{\mu_m} \tag{15}$$

The values obtained in Eq. (15) can be used for plotting values of $\mu_{\max}$ and K_d. Similarly, the value of K_d can be calculated from the intercept of the straight line, and $\mu_{\max}$ can be obtained from the slope line of Fig. 4.

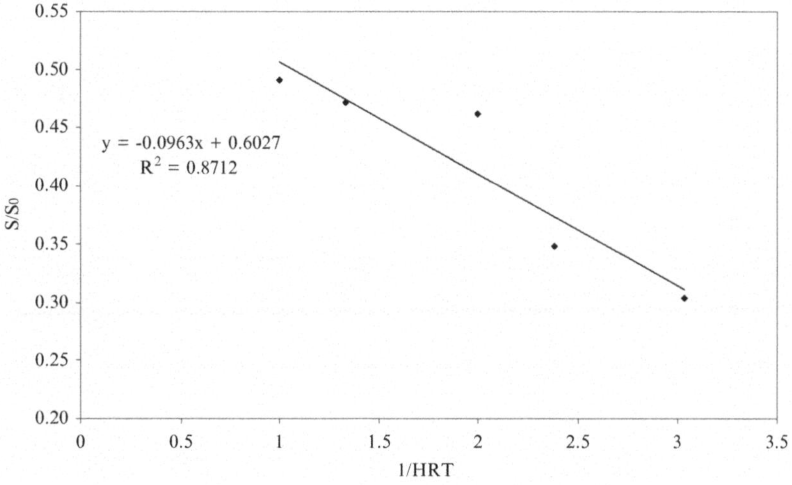

Fig. 4 Grau model

First-Order Kinetic Model

Finally, using the first-order model, kinetic constants were determined. Effluent substrate concentration (S) is calculated by the first-order model. According to θ_c and maximum substrate utilization rate k (mg COD/mg VSS day), the first-order kinetic model [14] expressions are as follows:

$$\mu = \frac{k \times S}{S_0 - S} - K_d \tag{16}$$

$$-\frac{dS}{dt} = kS \tag{17}$$

If Eq. (17) is integrated and rearranged for estimating effluent S concentration, it can be written as:

$$S = \frac{S_0}{1 + k\theta_C} \tag{18}$$

the effluent biomass concentration under steady-state condition and Eq. (16) is used to plot $1/E$ values versus μ values. The values of k and K_d are obtained in Fig. 5.

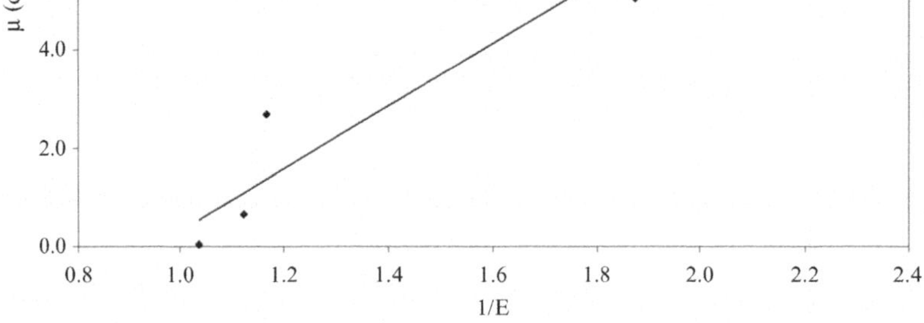

Fig. 5 First-order kinetic model

Table 2 Kinetic constants for acidifier

Substrate	Y mg VSS/mg COD	k_d 1/day	k_d 1/h	μ_m 1/day	μ_m 1/h	k 1/day	B mg COD/mg VSS
Confectionary [15]	0.31	1.56	0.065	64.8	2.7	–	–
Glucose [16]	0.17	6.1	0.254	30	1.25	–	–
Glucose [17]	0.625	0.41	0.017	–	–	–	–
Carbohydrate [18]	0.14–0.17	6.1	0.254	7.2–30	0.3–1.25	1.33 –70.6	–
Ice cream ww [19]	0.21	0.013	0.001	0.92	0.039	–	0.481
Lactose [20]	0.23	–	–	–			
Beef extract [21]		1.74	0.072	32.6	1.358		
Monod (this study)	0.25	0.79	0.033	0.84	0.035	3.37	
Contois				10.38	0.433		5.723
Grau		0.60	0.025	0.096	0.004		
First order		6.05	0.25			6.37	

Overall Evaluation of Applied Kinetic Models

By the application of the Monod kinetic model to the experimental data, μ_{max} and K_S were found to be 0.845 1/day and 55 mg COD/L, respectively, at high correlation ($R^2 = 0.96$). Y and K_d were found to be 0.25 mg VSS/mg COD and 0.79 1/day, respectively, at the highest correlation ($R^2 = 0.99$) in this study. Substrate utilization rate was calculated as 3.37 mg COD/mg VSS day. In the application of the Contois model, the Contois kinetic constant (B) was found to be 5.72 mg COD/mg VSS for each hydraulic retention time. In the application of the Grau model, K_d and μ_{max} were determined as 0.60 1/day and 0.096 1/day, respectively, at low correlation ($R^2 = 0.87$). Lastly, according the first-order kinetic model, k and K_d were found to be 6.05 mg COD/mg VSS day and 6.37 1/day, respectively. All the kinetic constants in this study found in accordance to the four different biokinetic models for best describing the acidogenic reactor are given in Table 2 as compression in the literature studies. The Monod model was found to be more assembled than the other kinetic models according to (Y) 0.25 mg VSS/mg COD and (K_d) 0.79 1/day, respectively. On account of the wastewater structure, these values were higher than the given literature values. A big part of the wastewater includes dissolved organics (approximately 70%) and is reduced to VFAs. Chemical synthesis-based pharmaceutical wastewater also contains large amounts of simple form alcohol, such as isopropyl alcohol, ethanol, etc. Since the hydrolyze phase of this wastewater takes a very short time, Y coefficient is slightly higher than the literature values [15–21]. On the other hand, (μ_{max}) 0.43 1/day was calculated by the Contois model and this value to more appropriate than the Monod model. The reason was using the equation proposed by the Contois, the specific growth rate is considered as a function of the growth-limiting nutrient in both input and effluent S concentration by using an empirical constant, which was related to the microbial population [11]. The

Monod and the Contois models are more aptly than the other kinetic models, especially when it comes to treating chemical synthesis-based pharmaceutical wastewater.

Acknowledgments The author gratefully acknowledges the support of the Scientific Research Projects of Istanbul Technical University and would like to thank them for the financial support given to the project with grant numbers 1536.

References

1. Benitez J, Rodriguez A, Malauer R (1995) Stabilization and dewatering of wastewater using hollow fiber membranes. Water Res 29:2281–2286
2. Kilroy AC, Grey NF (1992) The toxic of four organic solvents commonly used in the pharmaceutical industry of activated sludge. Water Res 26:887–892
3. Oktem YA (2004) Comparison of system performances of single and two phase anaerobic digestion treating a chemical synthesis-based pharmaceutical wastewater. Dissertation, Istanbul Technical University
4. Tomei MC, Braguglia CM, Mininni G (2008) Anaerobic degradation kinetics of particulate organic matter in untreated and sonicated sewage sludge: role of the inoculum. Bioresour Technol 99:6119–6126
5. Tiwari MK, Guha S (2013) Kinetics of the biodegradation pathway of endosulfan in the aerobic and anaerobic environments. Chemosphere 93:567–573
6. Wei Q, Zhang W, Guo J, Wu S, Tan T, Wang F, Dong R (2014) Performance and kinetic evaluation of a semi-continuously fed anaerobic digester treating food waste: effect of trace elements on the digester recovery and stability. Chemosphere 117:477–485
7. Deng L, Yang H, Liu G, Zheng D, Chen Z, Liu Y, Pu X, Song L, Wang Z, Lei Y (2014) Kinetics of temperature effects and its significance to the heating strategy for anaerobic digestion of swine wastewater. Appl Energy 134:349–355
8. Wu S, Yassine MH, Suidan MT, Venosa AD (2015) Anaerobic biodegradation of soybean biodiesel and diesel blends under methanogenic conditions. Water Res 87:395–402
9. Li Y, Jin Y, Li J, Li H, Yu Z, Nie Y (2017) Effects of thermal pretreatment on degradation kinetics of organics during kitchen waste anaerobic digestion. Energy 118:377–386
10. APHA (1995) Standard methods for the examination of water and wastewater, 19th edn. APHA, Washington
11. Monod J (1950) La technique de culture continue. Théorie et applications. Ann Inst Pasteur 79:390
12. Contois DE (1957) Kinetics of microbial growth. Theory and applications. Ph.D. Thesis, University of California, Los Angeles
13. Grau P, Dohanyas M, Chudoba J (1975) Kinetic of multicomponent substrate by activated sludge. Water Res 9:637–642
14. Lagergren S (1898) Zur theorie der sogenannten adsorption gelöster stoffe. Kungl Svenska Vetenskapsakademiens Handlingar 24(4):1–39
15. WC H, Thayanith K, Forster CF (2002) A kinetic study of the anaerobic digestion of ice-cream wastewater. Process Biochem 37:965–971
16. Massey ML, Pohland FG (1978) Phase separation of anaerobic stabilization biokinetic controls. J Water Pollut Control Fed 50:2204–2222

17. Ghosh S, Pohland FG (1974) Kinetics of substrate assimilation and product formation in anaerobic digestion. J Water Pollut Control Fed 46:748–758
18. Mazijat A, Mitsunori Y, Mitsunori W, Michimasa N, Jun'ichiro M (1997) Hydrogen gas production from glucose and its microbial kinetics in anaerobic systems. Water Sci Technol 36:279–286
19. Pavlostathis SG, Giroldo-Gomez E (1991) Kinetics of anaerobic treatment. Water Sci Technol 24(8):35–59
20. Fang HPH, Yu HQ (2001) Acidification of lactose in wastewater. J Environ Eng 127:825–831
21. Dinopoulou G, Sterritt RM, Lester JN (1988) Anaerobic acidogenesis of a complex wastewater: kinetics of growth, inhibition, and product formation. Biotechnol Bioeng 31:969–978

Evaluation of Heterogeneous Photocatalytic Treatment for Reuse of a Hotel's Greywater

Caner Kepoğlu, Can Burak Özkal, and Süreyya Meriç

Abstract Reuse of greywater for non-potable purpose such as irrigation, toilet flushing and groundwater recharge has been evaluated and applied widespread after advanced treatment processes. Specifically advanced oxidation processes present alternative option to integrate with biological treatment as prior or post-treatment to remove recalcitrant organic matters of emerging concern in greywater. This study focuses on the efficiency assessment of heterogeneous photocatalysis (PC) for oxidative degradation of real greywater samples collected at a hotel in Antalya, Turkey, by means of dissolved organic carbon (DOC) and UV_{254} absorbance removals that correspond the mineralization and degradation rates, respectively. Increasing TiO_2 dose from 0.1 to 1 g/L enhanced DOC removal rates. Reduction of suspended solids from greywater using 0.45 microfilters increased DOC removal from 27 to 35% during PC treatment.

Keywords Greywater · Degradation · Mineralization · Photocatalysis · Reuse

Introduction

Water scarcity has pushed research into eco-friendly and low-energy cost-efficient remediation of water resources among which greywater (GW) has also been considered [1, 2]. However, several emerging classes of micro-pollutants exist in wastewater and greywater matrices, and conventional treatment systems are not sufficient in terms of total removal of these compounds [5, 6]. Among the technologies known to treat greywater, water filtration technologies are high-energy demand due to necessary driving forces to push water through membranes at high fluxes [1–6]. Advanced oxidation processes (AOPs) have demonstrated to degrade various micropollutants [7]. Hydroxyl radicals generated by AOPs follow unselective multistep oxidative pathways however; complex environmental matrices play the radical

C. Kepoğlu · C. B. Özkal · S. Meriç (✉)
Department of Environmental Engineering, Namik Kemal University, Tekirdağ, Turkey
e-mail: cbozkal@nku.edu.tr; smeric@nku.edu.tr

N. Balkaya, S. Guneysu (eds.), *Recycling and Reuse Approaches for Better Sustainability*, Environmental Science and Engineering,
https://doi.org/10.1007/978-3-319-95888-0_20

scavenging role and limit their efficiency [8]. Photocatalysis (PC) is a highly selective, rapid and validated green technology that is able to react with various micropollutants. Factors that have inhibiting effect on PC process can be listed as (1) transformation of some complex organic fractions to the surface of TiO_2 saturating its structures and (2) the presence of hydroxyl radical scavenging anions in solution matrix [9]. The influence of TiO_2 concentration (0.5, 1, 2 and 3 g/L) on heterogeneous PC oxidation of GW (initial DOC: 29 mg/L) that originated from a hotel was investigated using a medium pressure mercury lamp (150 W) [10]. DOC removal increased by shifting the concentration of catalyst but limited to 2 g/L TiO_2 concentrations (greater than 65% after 150 min) [10, 11]. As the number of the literature studies are still few to reach a conclusion for PC treatment of real GW from different origin [8], this study aimed to investigate the efficiency of heterogeneous PC treatment of real GW samples (namely, raw, precipitated and filtered samples) using TiO_2 nanoparticles.

Materials and Methods

Samples and Conservation

Real GW samples were collected from a 5-star hotel located in Antalya City, Turkey. Only 176 rooms of a total number of 310 rooms facilitated the GW collection system. The GW that originated from baths and sinks are treated in an ultrafiltration-based membrane bioreactor (MBR) and reused for greenfield (30,000 m^2 cress) irrigation after a UV disinfection (planned to be optionally used in the extreme conditions such as overflow during intense season and was out of order during sampling) followed by a final chlorination [12]. Average GW originated was 100 L/capita-day according to a previous sampling (May–September 2013: unpublished data). All samples were transferred to the laboratory in cooled boxes, and they were kept cooled at +4 °C during experiments without adding any conservative chemical.

Chemicals and Synthetic Greywater Solutions

All chemicals used during experiments were of analytical grade. TiO_2 (CAS #13463–67-7, zero charge point of 6.8, specific surface area (BET) of 35–65 m^2/g and 21 nm primary particle size (TEM), molecular weight of 79.87, ≥99.5% trace metal basis (Aeroxide® P25)) was purchased from Sigma-Aldrich (Merck).

Photocatalytic Experiments

Initial adsorption is also a crucial step before PC process. According to optimized adsorption-desorption phase of organic matter, a prior adsorption-desorption time was taken into account during experiments, while irradiation was kept switch off.

The PC experiments were performed in a batch reactor system (100 mL water volume) using a multi-lamp photoreactor (14 lamps, Philips-8 W: 4.94 mW/cm^2) in 350 nm wavelength [10–14]. Photocatalyst doses varied from 0.5 to 2 g/L concerning the previous studies performed on GW treatment [11–14].

The samples were ultrasound irradiated for 10 min for effective mixture of TiO_2 using the MRC Ultrasonic (US) bath. Effects of pretreatment, precipitation for 24 h (PGW) and filtration using 0.45 μm (FGW) were also determined. Finally effluent was droplet filtered using Millex HA MF-Millipore mixed cellulose (REF: SLHA033SS/LOT: R2KA49112) cut-off filter (0.45 μm) to separate catalysts in the treated samples. The performance of the PC reactor was determined by means of UV and DOC removals. The PC reactor was operated at room temperature.

Analysis

The incident of photon flux was analysed by Universal Photometer (MRC, Israel). Degradation of real and synthetic GW samples was analysed by UV absorbance (Shimadzu, UV-1800), while mineralization was analysed by TOC or DOC measurement (Shimadzu, 6KVA model). The pH, conductivity and salinity were analysed using a multi-parameter pH metre (Hach Lange). The characteristics of real GW samples were analysed according to standard methods [15].

Results and Discussion

Characterization of Real Greywater Samples (RGW)

Table 1 shows the compositions of the RGW samples collected from the influent and effluent of the existing MBR plant at the hotel. According to Table 1, the characteristics are close to moderate compared to the literature values.

Photocatalytic Experiments

Adsorption-desorption profiles are presented in a way to represent initial conditions just before the PC process get started (Fig. 1). Following adsorption-desorption period, PC degradation of colloids takes place and results in reduction of COD

Table 1 Characteristics of RGW samples in comparison with the literature studies

Parameter	Unit	RGW[a]	France (B–S) [16]	UK (B, S, W) [1]	Austria (B) [17]	TR[a] [18]	MBR eff.	Cl_2[b]	Jordan[c] (LI) [19, 20]	Greece[c] [21]
Conductivity	(mS)	499	468	–	–	401–495	511	435		
TDS	(ppm)	250	–	–	–	–	256	218		
Salinity	(ppt)	0.23	–	–	–	–	0.24	0.20		
pH	–	7.00	7.58	6.6–7.3	6.4–8.1	7.1–7.2	7.1	7.02	6–9	
TSS	(mg/L)	20	125	29	48–120	48–55	5.1	3	<50	10
COD	(mg/L)	135	399	–	–	245	<30	<30	<120	
Soluble COD[d]	(mg/L)	105	136	86–575	–	177–277	<30	<30		
TOC	(mg/L)	41.30	50.6	12–56	–					
DOC	(mg/L)	25.46	–	–	–					
TOC	(mg/L)	82.58	–	–	–					
NH_3-N	(mg/L)	0.90	–	0.7–1	0.1–15	1.2–1.3	0.06	0.06		
T. Coliform	CFU/100 mL	No data		4×10^5	$500–2.4 \times 10^7$	1.36×10^4	–	–	–	
F. Coliforms	CFU/100 mL	No data	3.42×10^5		$170–3.3 \times 10^3$	3.57×10^3 to 1.1×10^4	–	–	–	5
E. coli	CFU/100 mL	No data	4.76×10^5				–	–	<10	

COD chemical oxygen demand, *TOC* total organic carbon, *DOC* dissolved organic carbon, *LI* landscape irrigation, *B* bath, *S* shower, *W* washbasin, *Cl_2* chlorinated effluent.

[a]Moderate greywater.

[b]This study.

[c]GW reuse standards for specific applications in different countries.

[d]COD measured in the filtrate from 0.45 μm filter.

[e]*RGW* real greywater in this study.

[f]Determined as BOD_5.

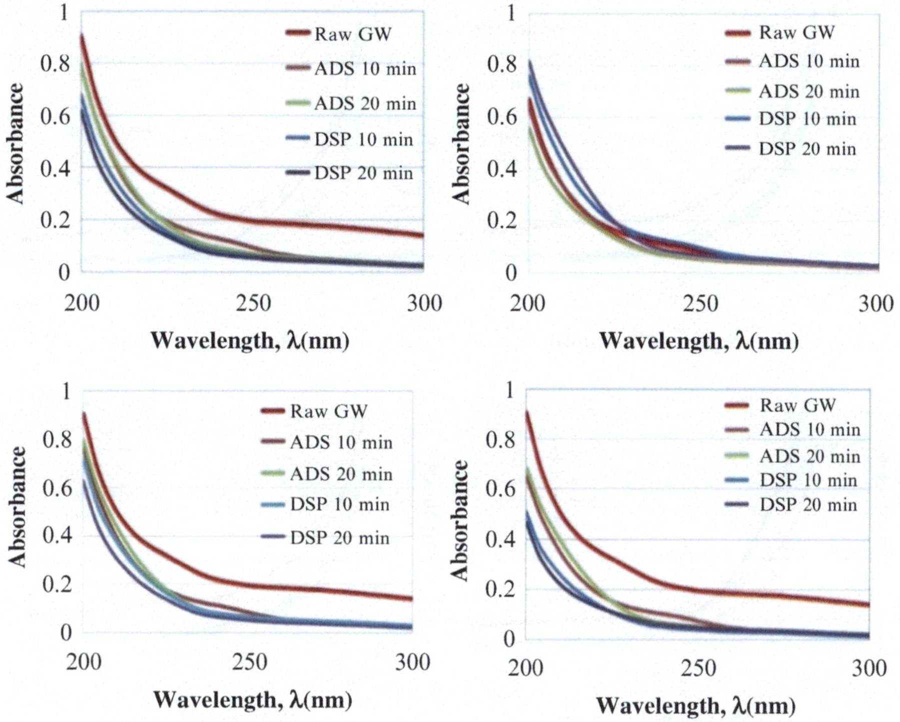

Fig. 1 Comparison of adsorption-desorption profiles of RGW using 0.1 (**a**), 0.2 (**b**), 0.5 (**c**) and 1 (**d**) g/L TiO_2 concentrations at neutral pH (*ADS* adsorption, *DSP* Desorption)

and TOC inside the system, while a simultaneous release of dissolved COD takes place [10–22].

US irradiation seemed to increase initial UV_{254} values for RGW (data not shown) due to degraded colloids that lead to release of dissolved COD into the solution matrix [23]. Optimum photocatalyst dosage was initially determined based on DOC removal rates.

Contrary to increasing trend of DOC removal at higher TiO_2 dosages, UV_{254} removal rates remained stable between 0.2 and 0.5 g/L initial TiO_2 concentrations [10]. Increasing TiO_2 dose (0.1–1 g/L) resulted in enhanced mineralization rates (DOC removal: 10–28%) for the raw GW sample [7–24].

PC degradation of organics and the mineralization processes proceed (not reaching a stable phase) throughout the 120 min of PC process time after pre-filtration (Fig. 2a). It is clearly evidenced that 0.45 μm filtration is able to retain prevailing bacteria-DOM interactions (Fig. 2b) [9]. The highest PC removal efficiency was achieved at 0.5 g/L TiO_2 initial conditions (greater than 30% DOC) on the PGW sample in which colloid-based competition for the active surface sites is negligible [7].

Regarding DOC removal rates at 30–120 min irradiation times, both sources of GW seem to play an analogous role on PC mechanism (Fig. 3). The xenobiotic

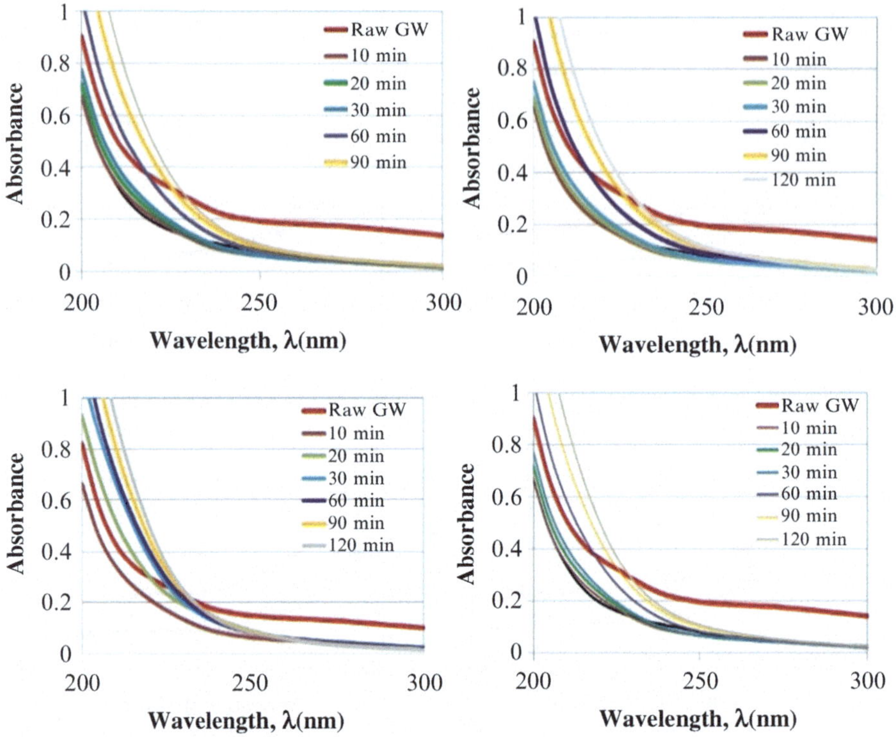

Fig. 2 Photocatalytic degradation of RGW using 0.1 (**a**), 0.2 (**b**), 0.5 (**c**), and 1 (**d**) g/l TiO_2 concentrations at neutral pH and 4.94 mW/cm^2 level of UV-A energy

Fig. 3 Photocatalytic degradation of ultrasonicated real greywater (URGW) at 0.1 (a), 0.2 (b), 0.5 (c) and 1 (d) g/L TiO_2 concentrations at neutral pH and 4.94 mW/cm^2 level of UV-A energy

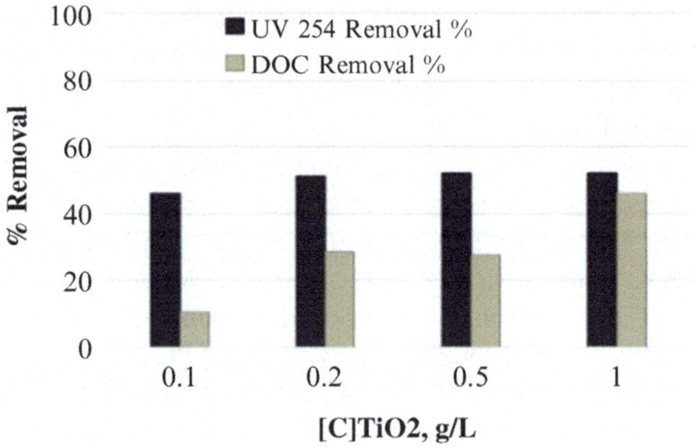

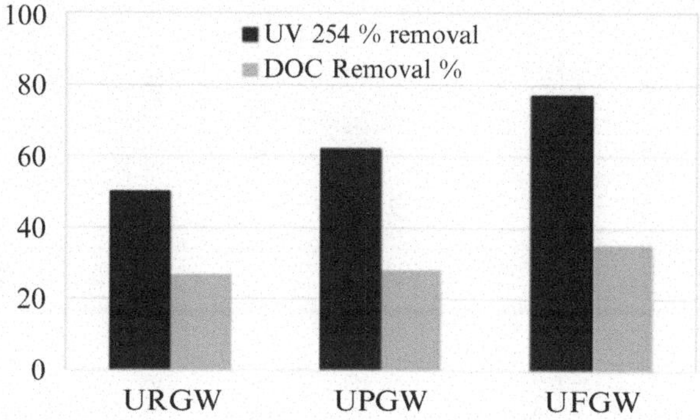

Fig. 4 Comparison of UV_{254} and DOC removal rates of GW sources at neutral pH, 4.94 mW/cm^2 level of UV-A energy, US 10 min and PC 120 min (*URGW* ultra-sonicated greywater, *UPGW* ultra-sonicated and precipitated greywater, *UFGW* ultra-sonicated and filtrated greywater)

components such as sodium laureth sulphate, benzophenone-4, benzyl salicylate, citronellal and polyphosphates were reported to cause inhibitive effects, while polyethylene glycol and derivatives were reported to enhance PC degradation. Those components have been theoretically retained in the 0.45 μm cellulose acetate filters and resulted with an overall enhancement of PC activity (35% of DOC removal efficiency), while PC mineralization performance remained stable for raw and precipitated samples [25].

The order of DOC removals in the PC-treated URGW, UPGW and UFGW samples was 27, 28 and 35%, respectively, at natural pH (Fig. 4). The RGW sample displayed a stronger inhibitive effect on the PC mechanism with regard to pretreated samples in well accordance with literature findings [25].

Table 2 compares the results obtained in different studies regarding the PC treatment of GW samples. According to Table 2, PC itself proved a consistent removal efficiency in our study after 120 min of irradiation. From the results of this study and the literature summarized in Table 2, it is possible to conclude that PC is to be an adequate option as an alternative to MBR treatment. However, it has apparently concluded from different study findings that a pretreatment step is a prerequisite for PC treatment based on energy and treatment efficiency aspects.

Contrary to general view, MBR treatment of GW has potential to pose several threats. Limitations in filter pore size and molecular size fraction distribution of DOM result in reaction of residual organics with final chlorination (Fig. 5). That's why AOPs are more pronounced to ensure safe effluents in terms of DOM and bacteria and/or other recalcitrant pollutants compared to many physical, chemical or biological processes alone or combined [26].

Briefly speaking, DOM is not the only constraint in PC reactions, but also pH, anion concentration and alkalinity of biologically treated GW characteristics have potential to cause alterations in the process performance [16].

Several researchers have proposed that chlorination or alternative disinfection options are necessary to meet the requirements of reuse standards shown in Table 1. Accordingly, as the PC process itself may fail to meet level of disinfection or the risk of photoactivation of bacteria, the ideal treatment option would necessitate a combination of PC process with a final chlorination (disinfection) [26].

Table 2 Comparison of photocatalytic process conditions for treatment of GW samples

Sample type	Initial TOC (mg/L)	Initial DOC (mg/L)	Experimental conditions	Reactor conditions	% TOC removal after a certain process period (min)						Ref.
					30	120	150	210	300	360	
Simulated GW considering appropriate quantities of the commercial personal care products calculations based on the DOC analysis[a]	–	93	pH 3; TiO_2 P25 0.5 g/L; UV-A (9 W)	Bench scale: 0.5 L Pyrex cell	–	>40	–	>50			[10]
			TiO_2 P25 0.5 g/L + H_2O_2 0.5 g/L; UV-A (9 W); pH: 6.5		–	–	–	<60		–	
			pH 3; TiO_2 P25 0.5 g/L; solar UV energy	Pilot scale: 15 L fountain type					>50		
			TiO_2 P25 0.5 g/L + H_2O_2 0.5 g/L; pH: 6.5, solar UV energy						>35		
Real GW sample[b]	24.6	–	pH 3; UV-A + 0.07 g/L TiO_2	Pilot scale: 31 L operated in recirculation mode	–	–	–	–		57	[27]
Hotel GW sample[c]	–	28.90	UV-A + 0.5 g/L TiO_2	125 W HPK mercury lamp, pH: 6.90		>15	–	40	–		[28]
			UV-A + 1 g/L TiO_2	Lab scale: 1 L cylindrical glass photoreactor		>15		55			
			UV-A + 2 g/L TiO_2			<20		>60			
			UV-A + 3 g/L TiO_2			<20		>60			
URGW	41.3	25.46	pH 7; 4.92 mW/cm^2 UV-A + 0.5 g/L TiO_2	Lab scale: 100 mL	–	27	–	–		–	This study
UPGW	25.5	–			–	28	–	–		–	
UFGW	36.4	–			–	35	–	–		–	

[a] TiO_2, Evonik P25 catalyst (TiO_2 P25, 70% anatase and 30% rutile with a BET of 55 ± 15 m^2/g.

[b] Initial pH, 7.3, main constituents anionic surfactants (sodium laureth sulphate, sodium cocoamphoacetate, sodium lauryl sulphate, ammonium laureth sulphate), cationic surfactants (cocamide MEA), non-ionic surfactants (lauryl glucoside, cetyl alcohol), fragrance, antimicrobial agents. TiO_2: Aeroxide P25, 21 nm, 50 ± 15 m^2/g BET.

[c] Turbidity >50 NTU; COD, 53.8 mg/L, Cl^- 38.9 mg/L; PO_4^{3-}, no data; SO_4^{2-}, 19.3 mg/L; anionic surfactants, 1.76 mg/L; $[C]_{E.coli}$, 19–50 CFU/mL; TiO_2, Aeroxide® P 25.

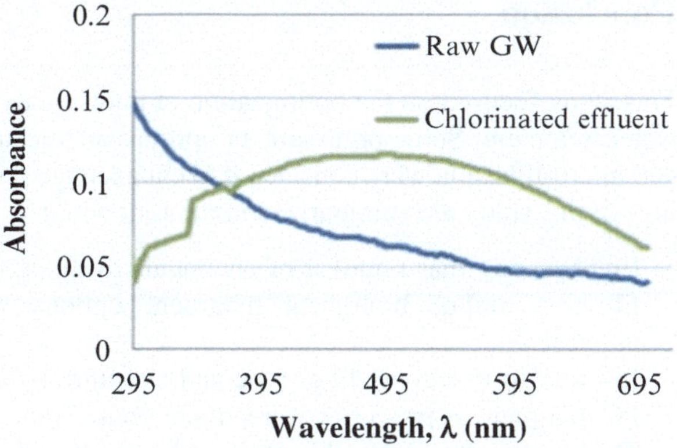

Fig. 5 Comparison of absorbance evolution in RGW and chlorinated effluent samples

Table 3 Proposed PC reactor conditions for GW treatment compared to existing MBR treatment unit

Terms of reactor	Conditions
Photocatalyst dose	>0.5 g/L
Pretreatment	Physical/ultrasound [10]
Adsorption at dark conditions	>10 min
Final chlorination	Yes (NOM—Bacteria cell interaction may inhibit PC disinfection performance [14–22]
Process (retention) time	3–4 h (would vary according to UV energy and GW characteristics as well as reuse purpose/legislations)
UV irradiation conditions	Simulated UV-A equivalent or higher than 30 W/m^2, should meet the average 2 mW/cm^2 UV-A energy requirement that is theoretically max 3.5% of solar energy [14]
Surface area	20–1000 m^2/m^3 of GW 350 m^2/m^3 for biologically treated GW [29]
Mixing and flow conditions	Turbulent flow in order to ensure perfect mixing of TiO_2 along the flow direction [30]
Materials	Plexiglass, quartz glass or a material that provides equal transparency to visible UV [14]

Given the fact that PC process itself has been proven to be an adequate option for mineralization of various organics present in greywater, fundamentals, design criteria, basic limitations and instructions for the PC process could be set to design an ideal photo-reactor to control of the microbiological aspects as well [28].

Considering the literature and the findings obtained in this study (see Table 2), the PC reactor conditions are possible to propose as seen in Table 3.

Conclusion

This study focused on the optimization of photocatalytic reactor for a hotel's greywater treatment. Some pretreatment options such as ultra-sonication, precipitation and micro-filtration were considered to enhance the process performance. The findings of this study are summarized in the following:

- Emphasizing that removal of chemicals of emerging concern in greywater by filtration and/or biological treatment options would be very costly and unfeasible.
- In order to overcome filter-type and structure-based limitations, integration of PC treatment as pre- or post-treatment seems to be a promising approach.
- The results obtained in this study indicated that the use of PC as sole treatment with pre-filtration was satisfying to comply with the reuse COD limits.
- It is worthy to note that organic load and photocatalyst dose are determining factors on PC mineralization rates.

Acknowledgements This study was inspired from EU COST Actions ES 1202 and ES 1205. The authors would like to thank the grant of Namik Kemal University Research Funding Office to support the MSc Thesis (NKU.BAP.00.17.YL.14.03) and the water treatment utility at the hotel for sampling collaboration.

References

1. Abu Ghunmi L, Zeeman G, Fayyad M, van Lier JB (2010) Grey water treatment in a series anaerobic—aerobic system for irrigation. Bioresour Technol 101:41–50
2. Ghaitidak DM, Yadav KD (2015) Reuse of greywater: effect of coagulant. Desalin Water Treat 54:2410–2421
3. Cirja M, Ivashechkin P, Schäffer A, Corvini PFX (2008) Factors affecting the removal of organic micropollutants from wastewater in conventional treatment plants (CTP) and membrane bioreactors (MBR). Rev Environ Sci Biotechnol 7:61–78
4. Donner E, Eriksson E, Revitt DM, Scholes L, Lützhøft HCH, Ledin A (2010) Presence and fate of priority substances in domestic greywater treatment and reuse systems. Sci Total Environ 408:2444–2451
5. Li F, Wichmann K, Otterpohl R (2009) Review of the technological approaches for grey water treatment and reuses. Sci Total Environ 407:3439–3449
6. Scheumann R, Merz C, Atasoy E, Murat S, Baban A, El Hamouri B, Kraume M (2007) Greywater treatment with membrane bioreactors. In: Proceedings: comparison of three different reactors operated with synthetic and real greywater
7. Rizzo L, Selcuk H, Nikolaou AD, Meriç Pagano S, Belgiorno V (2014) A comparative evaluation of ozonation and heterogeneous photocatalytic oxidation processes for reuse of secondary treated urban wastewater. Desalin Water Treat 52:1414–1421
8. Robert D, Malato S (2002) Solar photocatalysis: a clean process for water detoxification. Sci Total Environ 291:85–97
9. Abdullah M, Low GKC, Matthews RW (1990) Effects of common inorganic anions on rates of photocatalytic oxidation of organic carbon over illuminated titanium dioxide. J Phys Chem 94:6820–6825

10. Sanchez M, Rivero MJ, Ortiz I (2010) Photocatalytic oxidation of grey water over titanium dioxide suspensions. Desalination 262:141–146
11. Zhu X, Nanny MA, Butler EC (2008) Photocatalytic oxidation of aqueous ammonia in model gray waters. Water Res 42:2736–2744
12. Kepoğlu C (2014) Removal of personal care products and grey water treatment by ozone and photocatalysis, M.Sc. Thesis, Namik Kemal University, Institute of Applied Sciences, June (in Turkish)
13. Ozkal CB, Koruyucu A, Meric S (2016) Heterogeneous photocatalytic degradation, mineralization and detoxification of ampicillin under varying pH and incident photon flux conditions. Desalin Water Treat 57:18391–18397
14. Spasiano D, Marotta R, Malato S, Fernandez-Ibanez P, Di Somma I (2015) Solar photocatalysis: materials, reactors, some commercial, and pre-industrialized app. A comprehensive approach. Appl Catal B Environ 170–171:90–123
15. American Public Health Association/American Water Works Association/Water Environment Federation (1998) Standard methods for the examination of water and wastewater, 20th edn. American Public Health Association/American Water Works Association/Water Environment Federation, Washington
16. Chaillou K, Gérente C, Andrès Y, Wolbert D (2011) Bathroom greywater characterization and potential treatments for reuse. Water Air Soil Pollut 215:31–42
17. Christova-Boal D, Eden RE, McFarlane S (1996) An investigation into greywater reuse for urban residential properties. Desalination 106:391–397
18. Atasoy E, Murat S, Baban A, Tiris M (2007) Membrane bioreactor (MBR) treatment of segregated household wastewater for reuse. Clean Soil Air Water 35:465–472
19. McCornick PG, Hijazi A, Sheikh B (2004) From wastewater reuse to water reclamation: progression of water reuse standards in Jordan. In: Wastewater use in irrigated agriculture: confronting the livelihood and environmental realities. Cabi, Cambridge, pp 153–162
20. Halalsheh M, Dalahmeh S, Sayed M, Suleiman W, Shareef M, Mansour M, Safi M (2008) Grey water characteristics and treatment options for rural areas in Jordan. Bioresour Technol 99:6635–6641
21. Andreadakis A, Gavalaki E, Mamais D, Tzimas A (2001) Wastewater reuse criteria in Greece. Paper presented at the 7th Conference on Environmental Science and Technology, Ermoupolis, Syros Island, Greece
22. Birben NC, Uyguner-Demirel CS, Bekbolet M (2016) Photocatalytic removal of microbiological consortium and organic matter in greywater. Catalysts 6:91
23. Hernández Leal L, Temmink H, Zeeman G, Buisman CJN (2010) Comparison of three systems for biological greywater treatment. Water 2:155–169
24. Chong MN, Cho YJ, Poh PE, Jin B (2015) Evaluation of titanium dioxide photocatalytic technology for the treatment of reactive black 5 dye in synthetic and real greywater effluents. J Clean Prod 89:196–202
25. Lopez L, Panther BC, Turney TW (2015) Contaminant effects on the photo-oxidation of greywater over Ti film catalysts. J Water Process Eng 7:46–53
26. Meric S, Özkal CB (2016) *E.coli* in urban water. In: Eslamian S (ed) Urban water reuse handbook. CRC Press, New York, pp 83–99
27. Tsoumachidou S, Velegraki T, Antoniadis A, Poulios I (2017) Greywater as a sustainable water source: a photocatalytic treatment technology under artificial and solar illumination. J Environ Manag 195:232–241
28. Boyjoo Y, Ang M, Pareek V (2012) Photocatalytic treatment of shower water using a pilot scale reactor. Int J Photoenergy 3:14–22
29. Malato S, Blanco J, Vidal A, Richter C (2002) Photocatalysis with solar energy at a pilot-plant scale: an overview. Appl Catal B Environ 37:1–15
30. Fiorentino A, Ferro G, Alferez MC, Polo-López MI, Fernández-Ibañez P, Rizzo L (2015) Inactivation and regrowth of multidrug resistant bacteria in urban wastewater after disinfection by solar-driven and chlorination processes. J Photochem Photobiol B Biol 148:43–50

An Overview of Results and Conclusions from ITU Segregated Streams/ECOSAN Research Group Experience on Gray Water Management

Bilsen Beler-Baykal

Abstract Gray water is a segregated domestic wastewater stream, which may be returned to almost any point in the water cycle after pertinent treatment. It is a reliable and renewable source of water. The experience of ITU's segregated streams and ECOSAN research group has revealed/confirmed some issues/facts about gray water management based on Turkish case studies. Characterization of gray water and/or its sub-streams to be recycled and reused is a key factor in determining final uses as well as selecting the most appropriate treatment schemes and feasible layouts. Pathogens may be as significant a parameter as the predominantly mentioned organic matter. Carefully performed water balances regarding the amount which is to be supplied by reclaimed gray water and the amount of gray water that may possibly be produced within an area are critical in the success and feasibility of the process. Proximity of the gray water source to the location of end use is yet another significant determinant. Economic analysis for reclaimed gray water use for toilet flushing in the Turkish megacity Istanbul has shown that the suggestion is feasible with payback periods of less than 5 years with optimal number of units in developments. Social acceptance surveys indicate that there is a considerable approval for the use of reclaimed gray water. All in all, the idea of recycling and reuse of reclaimed gray water makes a lot of sense, and the conclusions drawn in terms of technical/environmental, economic, and social tiers of sustainability have provided motivating results.

Keywords Domestic wastewater management · Stream segregation · Gray water · Water balances · Location and proximity · Economic feasibility · Social acceptance · Domestic wastewater management

B. Beler-Baykal (✉)
Department of Environmental Engineering, Istanbul Technical University, Istanbul, Turkey
e-mail: baykalb@itu.edu.tr

N. Balkaya, S. Guneysu (eds.), *Recycling and Reuse Approaches for Better Sustainability*, Environmental Science and Engineering,
https://doi.org/10.1007/978-3-319-95888-0_21

Introduction

Management of domestic wastewater after segregation into different fractions , as described in Chap. 2, which are to be processed separately to be revaluated for further end uses is one of the recent promising options with significant contributions toward sustainability of the environment and limited sources. Gray water is one of those segregated domestic wastewater streams which consists of all wastewater coming from various washing activities in the household. It is practically the entire domestic wastewater except that part which comes from the toilet bowl (black water). Gray water will be produced regardless of the type of domestic wastewater separation either in two components or in three components.

Since in terms of pollutants gray water is a milder stream as compared to conventional domestic wastewater as well as other segregated streams, it may be returned to almost any point in the water cycle after proper treatment and may be used as an alternative source of water. As it will indispensably be produced as a result of human activity wherever people live, it is a reliable and renewable source. Substitution of pristine water with reclaimed gray water in line with the "fit for purpose use" is a promising practice for the sustainability of water resources and will especially benefit areas under water stress or water scarcity.

Gray water constitutes the highest portion by volume among possible segregated streams as it amounts up to 3/4 of domestic wastewater and contains organic matter as the primary pollutant. Pathogens are also significant as will be discussed later.

The most preferred end uses seem to be toilet flushing and irrigation at this time although a line of different end uses have been suggested like car washing, firefighting, industrial reuse including cooling water, groundwater recharge, etc.

While Australia and Germany [1–3] are among the leaders of gray water recycling, efforts of gray water reclamation and reuse in Turkey are rather new; however, interest in the subject matter is increasing in time. This section focuses upon conclusions drawn from Turkish case studies and aims to present a quick overview of gray water management first and then to share some of the gray water-related work undertaken at Istanbul Technical University by the ITU segregated streams/ECOSAN research group, addressing this can be directed to various[4–6] and surveys conducted as a requirement of a doctoral level course entitled "Segregated collection and management of domestic wastewater and ECOSAN" together with other projects/activities since 2010. Within the scope of this section, specific emphasis will be placed upon the significance of characterization of gray water and its consequences, significant parameters in revaluation of gray water, examples and costs of gray water treatment, social acceptance, and applicability.

An Overview of Gray Water and Its Management

Gray water may be produced both in two-component segregation of domestic wastewater as black water and gray water and in three-component segregation (or in more widely used terms ECOSAN – ECOlogical SANitation segregation), where separation involves gray water together with yellow water (source separated human urine) and brown water (separately collected human feces).

Being separated from black water (or alternatively from yellow and brown water), gray water is practically free of most of the nutrients and over half of organic matter contained in domestic wastewater. Since gray water is equivalent to ¾ of domestic wastewater by volume, constituents therein occur in more dilute concentrations as compared to the conventional one and other streams. The level of pathogens is expected to be relatively low as it leaves toilet wastewater out. As such, gray water is a milder stream in terms of pollution potential as compared to conventional domestic wastewater as well as other streams that may be generated as a consequence of stream segregation.

In terms of nutrients, as gray water is separated from yellow water, its nutrient content is low with typical values as 3% of domestic wastewater for nitrogen and 10% for phosphorus as mass percentages. Hence, gray water is not critical in terms of the nutrient problem. It contains about 40% of organic matter that needs to be treated before it can be directed to various end uses [7, 8]. This can be done using a line of treatment schemes, which frequently employs biological processes. Additionally, pathogens are significant from the perspective of public health. As will be addressed in subsequent sections, characterization of the gray water to be handled is a crucial issue in the selection of the most suitable treatment plants and the most efficient route for recycling and reuse.

Although collection of the entire gray water in a mixed mode is a viable option, further segregation into segments as light (or weak) and strong (or dark) gray water is also possible. While light gray water comes from bathtubs, showers, and washbasins, strong gray water contains wastewater from kitchen sinks, dishwashers, and washing machines. Examples of typical gray water characteristics from the literature are presented in Table 1 [3]. A survey of the table reveals that while combined

Table 1 Examples of typical average gray water characteristics [3]

		Combined	Light	
	Unit	Household[a]	Household[b]	Student residence hall[c]
COD	mg/l	245	112	96
sCOD	mg/l	177	29	
BOD	mg/l	90	78	46
sBOD	mg/l			31
TSS	mg/l	48	37	39
VSS	mg/l	39		
TKN	mg/l	9	5.4[d]	4.6
TP	mg/l	7.3	0.2	0.9
Turbidity	NTU		49.5	26
Total coliforms	cfu/100 ml	1.36E+04		2.20E+08
Fecal coliforms	cfu/100 ml	3.57E+03		
E. coli	cfu/100 ml		3.00E+04	3.90E+05
Enterococcus	cfu/100 ml		8.50E+03	2.50E+03

[a][9]
[b][10]
[c][11]
[d]As total nitrogen

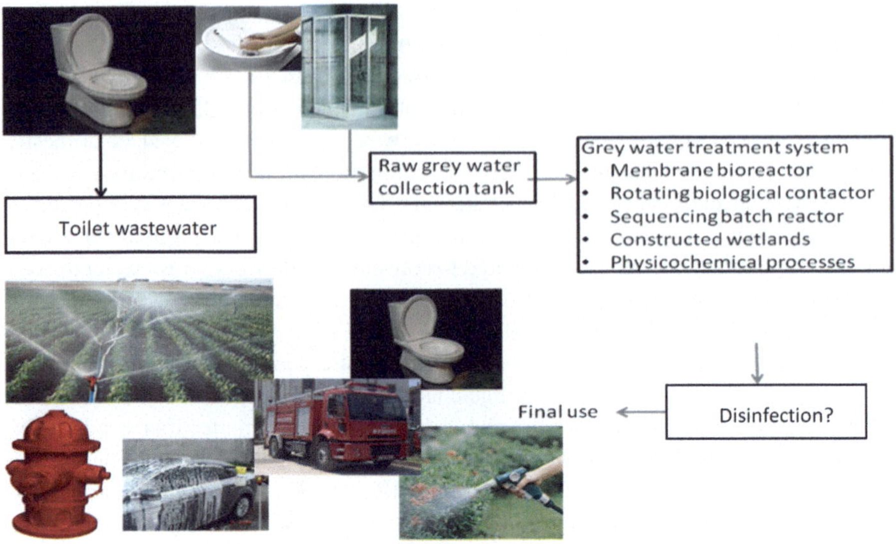

Fig. 1 Separate collection and management of gray water [14]

gray water is comparable to weak domestic wastewater or better in general, light gray water is by far milder as compared to even weak domestic wastewater with organic matter as the most outstanding pollutant and pathogens should not be overlooked.

It is interesting that the typical water use ending up in light gray water production makes up about 30% of daily water use while toilet flushes use 25–30% [12]. This makes light gray water to be collected from bathtubs, showers, and washbasins just the right amount to be directed into toilet flushes. Hence reclaiming gray water for toilet flushing is a wise practice, which will typically save about ¼ of the daily drinking water quality pristine water and eliminate "wasting" of that high a quality of water for an unworthy purpose [13].

An illustration of a typical scheme for gray water management is presented in Fig. 1 [14]. Following separation from toilet wastes, gray water is lead into a collection tank, which is spared for raw gray water only. Thereafter, raw gray water is fed in an appropriate treatment system, mostly a biological one, although depending upon its characteristics, physicochemical treatment may also be an option. With biological systems, a line of options are possible, including high-tech compact systems like membrane bioreactors (MBR) and rotating biological contactors as well as low-tech land-demanding ones like constructed wetlands. Although disinfection will be more critical for non-MBR treatment systems, all schemes with the inclusion of MBRs seem to include a disinfection unit for hygienic safety for elimination of possible health risks.

As it will inevitably be produced wherever people live, gray water may be used as a dependable and renewable alternative source of water following pertinent treatment and may be returned to almost any point in the water cycle. Currently its use as a non-potable water source has found a larger acceptance as compared to potable water, which is also possible after more stringent treatment. As such, gray water recycling and reuse is a practice which will benefit mankind through aiding sustainability of water resources and as an important tool especially for water stressed and water scarce areas.

Highlights of ITU Segregated Streams/ECOSAN Research Group Experience

Segregated streams and ECOSAN research group of Istanbul Technical University (ITU) has been focusing upon reclamation, recycling, and reuse of gray water since 2010. Investigations and results of the work by the group have revealed/confirmed some issues/facts about gray water management. Observations and conclusions within that realm include:

- Characterization of gray water and/or its sub-streams to be recycled and reused is a key factor in determining final uses as well as selecting the most appropriate and feasible treatment schemes.
- Pathogens may be as significant a parameter as organic matter.
- Carefully made water balances to relate the quantity of the demand which is to be supplied by reclaimed gray water and the quantity to be collected from possible segregation options of gray water within an area are critical in the success and feasibility of the process.
- Proximity of the gray water source to the location of end uses is another significant determinant.
- Economic analysis for reclaimed gray water use for toilet flushing in Istanbul has shown that it is feasible with payback periods of less than 5 years.
- Social acceptance surveys have provided motivating results, and they indicate that there is a considerable approval for the use of reclaimed gray water.

Characterization and Significant Parameters

As different origins and segments show significant variations, quality of the specific gray water to be handled is an important issue in making decisions about the treatment scheme as well as the preparation of the most appropriate management plan. Monitoring through rather long spans of time, preferably for a year, is recommended for dependable characterization. Table 2 presents averages from such monitoring programs to show that even though only light gray water was observed, characteristics could differ a lot [10]. The table reveals further that, even though using exactly the same origin, bathtubs, showers, and hand washbasins in the case of the 5-star hotel and the student residence hall, contents may have a difference which exceeds

Table 2 Results of physicochemical analyses in various light gray water from Istanbul [15]

Parameters (mg/L)	Five-star hotel[a]	Student residence hall[a]	University building[b]
COD	120	217	89
sCOD	40	85	39
BOD	32	76	54
TSS	87	145	12
VSS	47	96	4
TKN	6.99	6.76	–
NH_3-N	2.58	0.62	0.14
TP	2.29	1.78	–
PO_4-P	1.05	0.75	0

[a]Bathtubs/showers/hand washbasins
[b]Hand washbasins

100% as exemplified by the organic content. In both cases, reclaimed gray water quality was highly satisfactory. In the case of the five star hotel which featured MBR treatment for example, year round monitoring of the influent and effluent streams revealed that it was mostly the membrane rather than the biological treatment which was predominant in the removal of organics and actually the bio section could be eliminated. This suggests that for this specific light gray water which was low in organic concentrations and had a low BOD with considerable solids of organic origin, a physicochemical treatment unit employing membrane filtration only would be sufficient. Hence, the "biological" segment of the MBR treatment unit was in fact obsolete due to the low organic content which was mostly in particulate form and that actually it was the physical segment, the ultrafiltration membrane, which performed the treatment [15]. Lab-scale experiments have also shown that whenever gray water quality permits, physicochemical systems may be used as an option [16].

Organic matter is cited by far as the most predominant pollutant in gray water in the literature which has to be treated before reuse. Although not emphasized as clearly and strongly as organic matter in the literature, control of pathogens is essential in safeguarding hygienic well-being [17].

Costs and Payback Periods

Economic considerations constitute one of the main concerns especially for new approaches. Initial costs, operating costs, and payback periods for MBR treatment in the Turkish megacity Istanbul were investigated for households and hotels. The results presented in Figs. 2 and 3 reveal that 200 units or more is a fair size for best outcomes in terms of economics both for households and for hotels with similar investment costs of about 200 and operating costs as 10 Euros per housing unit or hotel guestroom both. Payback periods on the other hand are about 2 years for households and 1 year for hotels with the conditions prevailing in Istanbul. Moreover

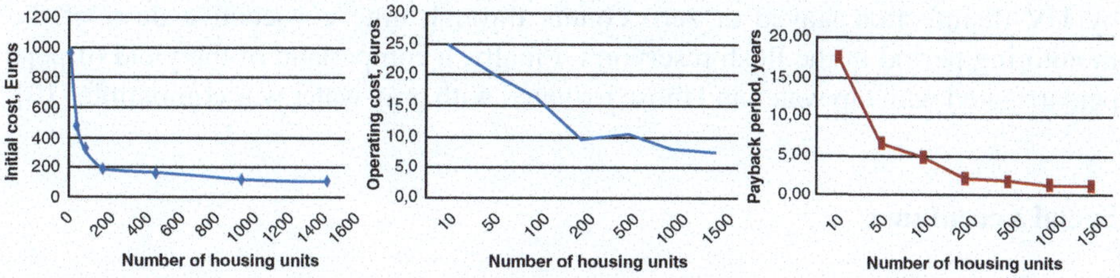

Fig. 2 Costs and payback periods for MBR treatment in households in Istanbul [14]

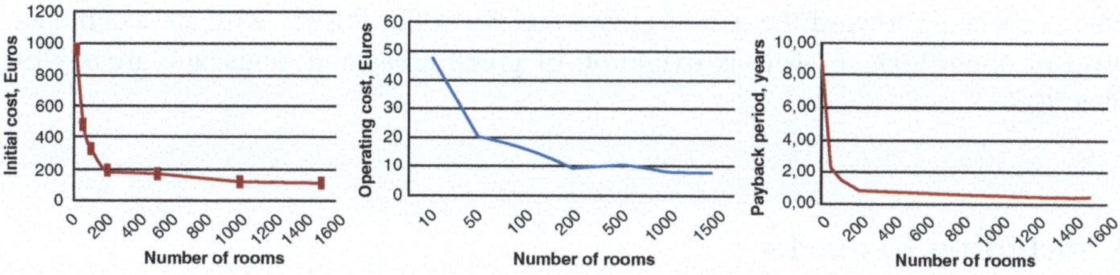

Fig. 3 Costs and payback periods for MBR treatment in Istanbul hotels [14]

if an additional piping cost of 25–100% is anticipated, the payback period, in all circumstances, would be less than 5 years in any case, which is an acceptable one [14]. Acceptable economic results for gray water recycling were also confirmed by Efe [6].

Applicability

An analysis of gray water recycling regarding various end uses and their combinations including toilet flushing, irrigation, car washes, urban cleaning, and industrial use as cooling water based upon real data from a Turkish neighborhood near Istanbul was presented by Efe [6] to show the significance of water balances. Matching quantities between the demand for an end use and what can be collected in a defined area is one of the fundamental steps for gray water management. Since with such an application, a wastewater stream, i.e., gray water, would be an input for water supply, water balances and incorporation of water/wastewater cycles as well as a holistic look are specifically important.

Another critical feature of gray water management is the proximity of the source of gray water to the final users as transportation may play a determining role in the applicability and feasibility of this practice [6].

Additionally, it was demonstrated by Giresunlu and Beler-Baykal [18] that once treated properly, there was no growth of pathogenic indicators in toilet flushes and the zero counts recorded in successfully reclaimed gray water in an MBR followed by UV disinfection stayed at zero counts through nine consecutive days of the monitoring period in the flush reservoirs. Finally, a comparison of the yield of peppers irrigated with tap water and those irrigated with gray water was compatible [19].

Social Acceptance

Social behavior and acceptance of consumers is one of the key elements to turn implementations into success stories, especially new and non-conventional ones. Surveys to assess public acceptance regarding gray water reuse were undertaken to indicate motivating results [20]. The results revealed that there was a high acceptance especially toward the use of gray water for toilet flushes with an acceptance level of about 90% as well as irrigation of green areas and landscape areas with over 80%.

Concluding Remarks

Gray water is a reliable, renewable, and sustainable source of water, the use of which should not be postponed for the welfare of mankind and prolonged use of limited water resources. A good starting point would be conversion of toilet flushes from drinking quality water to reclaimed gray water which would save about 25% of pristine water which may be done easily especially in the case of new buildings and in periods of rebuilding.

References

1. Nolde E (1999) Gray water reuse systems for toilet flushing in multi_storey buildings – over ten years experience in Berlin. Urban Water 1:275–284
2. Nolde E (2003) Graywater recycling systems in Germany—results, experiences and guidelines. Water Sci Technol 51(10):203–210
3. Giresunlu E, Soyer E, Beler Baykal B (2016) Physicochemical treatment of light gray water: an experimental study using membrane filtration. In: 13th IWA specialized conference on small water and wastewater systems and 5th IWA specialized conference on resources-oriented sanitation, Athens, Greece, 14–16 Sept 2016
4. Oructut N (2013) Gray water management and an appraisal of urban gray water practice (Gri su yönetimi ve kentsel ağırlıklı uygulamaların değerlendirilmesi). Master's Thesis, Istanbul Technical University, Environmental Sciences and Engineering Program, 2013 (in Turkish with an extended abstract in English)
5. Giresunlu E (2015) Characterization and analysis of weak urban gray water. Master's Thesis, Environmental Sciences and Engineering Program, Istanbul Technical University
6. Efe H (2017) An analysis on gray water source separation as an alternative for conventional wastewater management using the case of Kocaeli Yenikent neighborhood. Master's Thesis, Environmental Sciences, Engineering and Management Program, Istanbul Technical University

7. Otterpohl R, Braun U, Oldenburg M (2003) Innovative technologies for decentralized water-wastewater and biowaste management in urban and peri-urban areas. Water Sci Technol 48(11): 23–31
8. Beler Baykal B (2015) Stream segregation in household use: a review of gray water as an alternative source of water and yellow water as an alternative source of fertilizers. Water Qual Expo Health 7(1):27–37
9. Atasoy E, Murat S, Baban A, Tiris M (2007) Membrane bioreactor (MBR) treatment of segregated household wastewater for reuse. Clean Soil Air Water 35(5):465–472
10. Chaillou K, Gérente C, Andrés Y, Wolbert D (2011) Bathroom graywater characterization and potential treatments for reuse. Water Air Soil Pollut 215(1):31–42
11. Birks R, Hills S (2007) Characterization of indicator organisms and pathogens in domestic graywater for recycling. Environ Monit Assess 129(1):61–69
12. Beler Baykal B, Giresunlu E (2013) Gray water as an alternative source of (flush) water in a mega city of a water stressed country. In: IWA conference on alternative water resources management and integrated technologies for sustainable urban water system, Qingdao, China, 25–28 June 2013
13. Beler Baykal B, Baykal BA (2017) Past, present and future of toilet flushes – an appraisal through water scarcity, water quality and "fit for purpose" use. In: 2nd IWA regional conference on water, wastewater and environment: past, present and future of the world's water resources, Cesme/Izmir, Turkey, 14–22 Mar 2017
14. Beler-Baykal B, Oructut N (2017) Gray water as reliable/renewable source of water for cities of the future and an appraisal based on Istanbul, Turkey. In: IWA cities of the future conference: embrace the water, Gothenburg, Sweden, 12–14 June 2017
15. Giresunlu E, Beler Baykal B (2016) A case study of the conversion of gray water to a flush water source in a Turkish student residence hall. Water Sci Technol Water Supply 16(6):1659–1667
16. Giresunlu E, Beler Baykal B, Afacan E (2016) A comparative study on light gray water from buildings with different functions. In: 8th international conference on environmental science and technology, American Academy of Sciences 2016, Houston, 6–10 June 2016
17. Giresunlu E, Beler Baykal B (2015) An appraisal of microbiological indicators for gray water reuse as a priority issue. In: CEST 2015 14th international conference on environmental science and technology, Rhodes, Greece, 3–5 Sept 2015
18. Giresunlu E, Beler Baykal B (2015) Tracking pathogenic indicators in the gray water cycle for reuse in flushes. In: IWA alternative water resources conference, conference programme and proceedings, Nanjing, China, 26–28 May 2015, p 18–22
19. Dogan G, Giresunlu E, Beler-Baykal B (2017) A preliminary investigation on the use of segregated wastewater streams as fertilizer and irrigational water upon pepper, New York, USA, 5–9 Aug 2017
20. Beler Baykal B, Efe H, Afacan E, Arslan E (2017) A preliminary survey on public attitude for the reuse of gray water as an alternative source of water in Turkey. In: IWA S2Small2017 conference on small water and wastewater systems and resources oriented sanitation (14th IWA specialized conference on small water and wastewater systems (SWWS) together with the 6th IWA specialized conference on resources-oriented sanitation (ROS) and the 3rd international conference terra preta sanitation and decentralized wastewater system, Nantes, France, 22–26 Oct 2017

Wet Peroxidation of Olive Oil Mill Wastewater for Sludge Minimization

Sinan Guneysu and Semiha Arayıcı

Abstract Advanced oxidation process treatment method has been adapted to olive oil mill wastewater (OMW) for treatability in conventional plants by the aim of sludge minimization. Lots of olive oil mills have facilities in the Aegean Region, Turkey. Wastewater, used in this study, was collected from these factories. Because of the advanced and expensive treatment methods of this wastewater, regulations and limitations are ignored. 95 °C temperature and 10 bar pressure conditions were used with H_2O_2 as an oxidant in wet oxidation method. Over 96% of oil-grease and 48% of chemical oxygen demand (COD) removal were achieved using wet air peroxidation in conditions with 2.5 mL H_2O_2/100 mL in 1 h. The influence of hydrogen peroxide with temperature and pressure makes the oil in wastewater soluble and in that way can be treatable in conventional plants.

Keywords Olive oil mill wastewater · Wet oxidation · Hydrogen peroxide · Waste minimization

Introduction

Olive oil mill wastewater (OMW) is an effluent of the olive oil production process. It is known that OMW is a great problem in the Mediterranean countries, which farm olive and produce olive oil. Turkey is the fourth greatest producer in these countries which has a production of about 1.64 million tons/year of oil and 2.1 million tons/year of total olive production. Turkey has 174,594 million olive trees for farming and produced 1,640,000 tons of olive oil in the year 2017 [1]. 1,968,000 m^3 wastewater is produced by the production of olive oil in decanter system. Also, 18,105 tons of waste oil and 20,075 tons of solids in wastewater are discharged in a year.

S. Guneysu (✉) · S. Arayıcı
Department of Environmental Engineering, Istanbul University-Cerrahpaşa, Istanbul, Turkey
e-mail: guneysu@istanbul.edu.tr; semara@istanbul.edu.tr

N. Balkaya, S. Guneysu (eds.), *Recycling and Reuse Approaches for Better Sustainability*, Environmental Science and Engineering,
https://doi.org/10.1007/978-3-319-95888-0_22

By this means, the characterization of olive oil mill industry wastewater parameters ranges between 37 and 318 g/L (chemical oxygen demand), 15 and 135 g/L (biological oxygen demand), 6 and 14.5 g/L (oil-grease), 6 and 69 g/L (suspended soils), and 4.6 and 5.18 (pH) [2].

Different treatment methods on OMW were tested, but most of them have a poor usage or expensive. Certain treatment methods that are used on OMW are diluting [3], settling [4], flotation [5], centrifugation [6], filtration [7, 8], thermal processes [9], evaporation [10], pyrolysis [11], neutralization [6], flocculation [12], chemical oxidation processes [13], advanced oxidation processes [14], Fenton processes [15], electrocoagulation [16], and aerobic and anaerobic treatment processes [17].

In recent years, chemical oxidation or Fenton processes, which are advanced treatment methods, have been developed for use in OMW treatment [4]. In Turkey, OMW has a limited treatment, so that only settling or land filling methods apply widely by municipalities [2]. However, these physical treatment methods make secondary pollution, like air pollution [18], groundwater pollution, and soil pollution problems. Also many problems concerning the high toxicity, oil content, air emissions, and biodegradability of the OMW appear in the settling tank. Therefore, a pretreatment is needed to remove the toxicity and oil content of OMW.

Wet peroxidation process increases the influence of hydrogen peroxide by high temperature and pressure. So, much more hydroxyl radicals can be produced with fewer hydrogen peroxide, and this helps degradation of the oil content of the OMW. By degrading the oil content of the wastewater, it can be easily treated with domestic wastewater in municipality treatment plants.

Simply, photolysis reactions of H_2O_2 and kinetics are given below [19].

Starter

$$H_2O_2 \xrightarrow{hv} 2\bullet OH \tag{1}$$

Process

$$H_2O_2 + HO\bullet \xrightarrow{k_2} HO_2\bullet + H_2O \quad k_2 = 2,7\cdot 10^7 1/Ms \tag{2}$$

$$H_2O_2 + HO_2\bullet \xrightarrow{k_3} HO\bullet + H_2O + O_2 \quad k_3 = 3,71/Ms \tag{3}$$

Termination

$$HO\bullet + HO\bullet \xrightarrow{k_4} H_2O_2 \quad k_4 = 1,1\cdot 10^{10} 1/Ms \tag{4}$$

$$HO_2\bullet + HO_2\bullet \xrightarrow{k_5} H_2O + O_2 \quad k_5 = 1,2\cdot 10^7 1/Ms \tag{5}$$

$$HO\bullet + HO_2\bullet \xrightarrow{k_6} H_2O + O_2 \quad k_6 = 7,5\cdot 10^9 1/Ms \tag{6}$$

Oxidation

$$\text{Waste} + \text{HO}\bullet \rightarrow \text{Products}\ \left(CO_2 + H_2O + HO_2\bullet + \text{ inorganic compounds}\right) \quad (7)$$

The first, second, and fourth reactions are identified as hydroxyl radical formations.

Olive Oil Production Method

Nowadays, four methods are used for olive oil production [20]:

- Traditional press method
- Three-phase process method
- Two-phase process method
- Virgin olive oil production method

Traditional Press Method

This method uses the hydraulic pressure to produce oil. In press process, firstly the olives are washed with warm water and then kneaded and crushed. After that, this olive paste is pressed for separating oil and wastewater in decantors [21, 22]. In press method, olive oil mill wastewater has a characterization of 90–100 g/L BOD, 90–130 g/L COD, and pH range of 4.5–5.0 [20].

Three-Phase Process Method

At the end of this process, there are three phases: oil, wastewater, and bagasse (oil waste). Since this process uses a lot of process water, excess amount of wastewater is formed [23].

Two-Phase Process Method

In this system, no process water is needed during the production, and two phases are formed as oil and bagasse after the process. The system is very attractive ecologically because the liquid phase does not form, and a large part of the wastewater comes out with bagasse. The resulting solid phase contains 50–60% water and 2–3% oil [23].

Table 1 Material and energy balance of olive production methods (l/ton olive)

Process type	Input	Input amounts	Waste	Waste amounts	COD (g/L)
Traditional press method	Washing water Energy	0.1–0.12 m^3 40–63 kWh	Oil Bagasse (25% water + 6% oil) OMW (88% water)	200 kg 400 kg 600 L	90–130
Three-phase decantor	Washing water Cleaning water Oil cleaning water Energy	0.1–0.12 m^3 0.5–1 m^3 10 L 90–117 kWh	Oil Bagasse (50% water + 4% oil) OMW (94% water + 1% oil)	200 kg 500–600 kg 1,000–1,200 L	60–90
Two-phase decantor	Washing water Energy	0.001 m^3 <90 to 117 kWh	Oil Bagasse (60% water + 3% oil)	200 kg 800–950 kg	10–15

Virgin Olive Oil Production Method

Adhesion of oils to all surfaces forms the basis of the infiltration process. A metal layer with a recessed surface used for this process adhesion is immersed in olive paste, and the cavities on the layer are filled with oil. In this method, a large number of metal sheets are used, and the system used is known as the "Sinoles system." A machine with a surface area of 6000 m^2 consisting of 5120 plates can process 350 kg of dough in 7–8 min. However, it is stated that the method cannot be used alone but can be integrated with press or centrifugation methods [20].

Table 1 summarizes the process water, wastewater, bagasse, and energy requirements of olive oil production types.

Materials and Methods

In this study, 1–2.5 mL of hydrogen peroxide was mixed with 100 mL of OMW and reacted at 95 °C temperature and 10 bar pressure in 1 h.

OMW Samples

The wastewater sample used in this study was directly collected from the effluent of the process plants in the Aegean Region in the production season (usually November and December). The samples were quickly analyzed and stored at +4 °C [24].

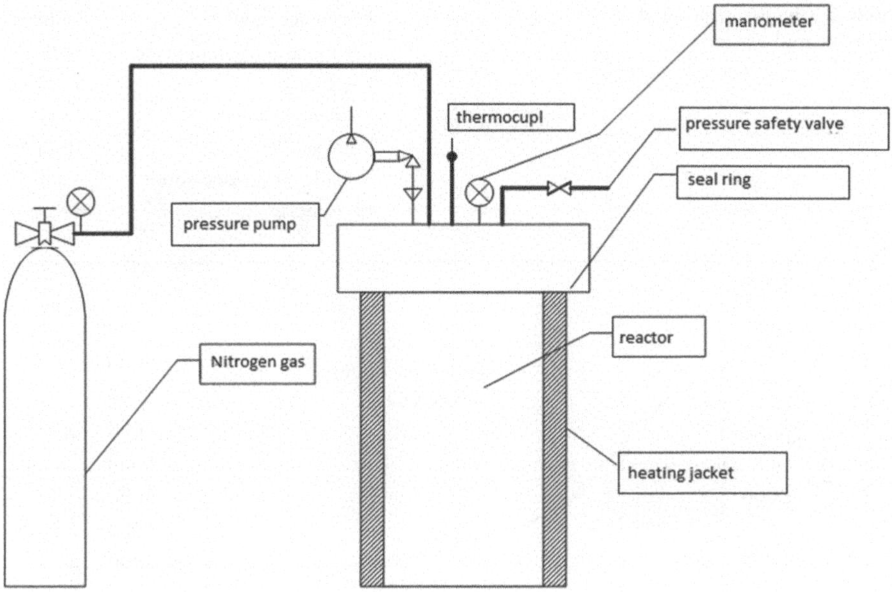

Fig. 1 Schematic drawing of wet peroxidation reactor [25]

Wet Peroxidation Reactor

The reactor was designed for working at high temperature and pressure safely. It is constructed from stainless steel and works up to 300 bar pressure perfectly. Furthermore, the oxidant agent we used in this study was not harmful to the reactor. The reactor was equipped with a gas inlet valve for increasing the pressure, a hole for injecting the oxidant agent addition, and a safety pressure valve for preventing from the dangers of high pressure. Figure 1 shows the schematic drawing of wet peroxidation reactor.

Results

Turkey-Aegean Region OMW was selected for this study and characterized as given in Table 2.

As seen from Table 1, high content of COD and oil-grease content make the conventional treatment impossible.

It has been obtained from the experiments that the oil-grease removal efficiency was about 97%, whereas COD removal efficiency was 48% by the wet peroxidation process with 2.5 mL H_2O_2/100 mL OMW. The removal of the oil-grease parameter was significant as the main performance criteria in this study. Table 2 shows the COD removal by the hydrogen peroxide amount used in the process. As mentioned

Table 2 Characterization of raw olive mill waste

Parameter (mg/L)	Value
Chemical oxygen demand	88,064
Oil-grease	9,217
Suspended solids	5,020
Total solids	10,212
Volatile suspended solids	4,480
Phosphorus	84

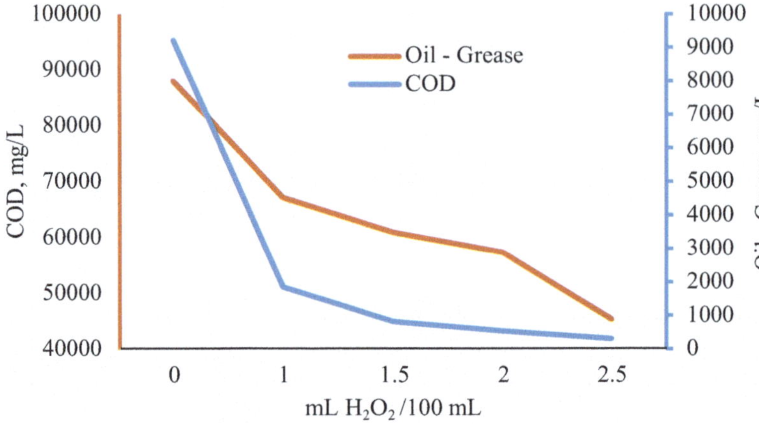

Fig. 2 Removal of COD and oil-grease by wet peroxidation

in Table 2, wet peroxidation conditions were determined as 10 bar reactor pressure, 95 °C reaction temperature, and 1 h reaction time within the 1–1.5–2–2.5 mL hydrogen peroxide additions to 100 mL of OMW.

Figure 2 indicates the removal of COD and oil-grease by wet peroxidation. The y-axis of the diagram shows COD and oil-grease values as mg/L, whereas the x-axis shows the mole of hydrogen peroxide added to the reactor per 100 mL of wastewater. Graphics show that the addition of hydrogen peroxide increases the treatment efficiency. However, the economic charges are so important that we suggest the optimum value was 2 mL H_2O_2/100 mL OMW addition, whereas COD and oil-grease removals were 57,330 and 536 mg/L, respectively.

Discussion

With the aim of the increase, the efficiency with different oxidation experiments was tried out at a temperature of 95 °C, 10 bar nitrogen gas pressure, and 1 h reaction time. Table 3 states the oxidized values of COD and oil-grease by addition of

Table 3 COD and oil-grease values after wet peroxidation

Concentration	COD (mg/L) [88,064 mg/L]	Oil-grease (mg/L) [9,217 mg/L]
1 mL H_2O_2/100 mL	67,193	1,850
1.5 mL H_2O_2/100 mL	60,928	812
2 mL H_2O_2/100 mL	57,330	536
2.5 mL H_2O_2/100 mL	45,360	302

1 – 1.5 – 2 and 2.5 mL hydrogen peroxide as an oxidant, per 100 mL of wastewater. Therefore, removal capacity of COD and oil-grease was calculated as 48.5% and 96.7%, respectively.

The higher the rate of hydrogen peroxide in addition to the wastewater, the higher the removal capacity on COD and oil-grease. We know from the literature that to remove 1 g of COD, theoretically 0.0625 mol H_2O_2 is needed [26]. Olive oil mill wastewater used in this study was characterized as 88,064 mg/L COD value, and theoretically 5 mol of H_2O_2 are need for oxidation. The optimum hydrogen peroxide dose, which is needed less than the theoretical value, can be explained as increasing the oxidant efficiency by reaction temperature and pressure. Furthermore, the removal ratio of oil-grease was calculated as 96.7%, but still 302 mg/L residual value couldn't be removed. This value is too high to be discharged according to the Water Pollution and Control Regulations [27], and a second treatment has to be done.

Conclusion

In Turkey, 3842 million m^3 wastewater is treated in plants per year, and 1138 million tons of sludge is produced from these facilities. If olive oil mill wastewater has a chance to be treated by a plant, 20,075 tons of sludge would be produced from 1968 million m^3 wastewater treated [1]. It is important and successful to perform the preoxidation for the treatment of olive oil wastewater when sludge minimization is targeted.

References

1. Turkish Istatistical Institute (2018) http://www.tuik.gov.tr/PreIstatistikTablo.do?istab_id=1073. Accessed 10 Apr 2018
2. Şengül F, Pınar C, Yıldırım T (1996) Removal and treatment of olive oil production wastewater Çanakkale Province, environmental problems, Çanakkale Province. In: Proceedings book, İzmir, Turkey, pp 67–74

3. Boari G, Mancini IM (1990) Combined treatments of urban and olive mill effluents in Apulia. Water Sci Technol 9:235–240
4. Al- Malah K, Azzam MOJ, Abu-Lail NI (2000) Olive mills effluent (OME) wastewater post-treatment using activated clay. Sep Purif Technol 2:225–234
5. Curi K, Velioglu SG, Diyanmanoglu V (1980) Treatment of olive oil production wastes. In: Treatment and disposal of liquid and solid industrial wastes: proceedings of the third Turkish German environmental engineering symposium, Istanbul, Turkey. Pergamon Press, Oxford, pp 189–205, isbn:978 008 023 9996
6. Mitrakes M, Papageorgiou G, Docoslis A, Sakellaropoulos G (1996) Evaluation of various pretreatment methods for olive oil mill wastewaters. J Water Pollut Control Fed 6:10–16
7. Velioglu SG, Curi K, Camillar SR (1987) Laboratory experiments on the physical treatment of olive oil wastewater. Int J Dev Technol 5(1):49–57
8. Bradley RM, Baruchello L (1980) Primary wastes in the olive oil industry. Effluent Water Treat J 20(4):176–177
9. Rozzi A, Malpei F (1996) Treatment and disposal of olive mill effluents. Int Biodeterior Biodegrad 38(3):135–144
10. Arjona R, Garci'a A, De Ollero Castro P (1999) The drying of alpeorujo, a waste product of the olive oil mill industry. J Food Eng 41(3):229–234
11. Di Giacomo G, Bonfitto E, Brunetti N, Del Re G, Jacoboni S (1989) Pyrolysis of exhausted olive oil husks coupled with two-stages thermal decomposition of aqueous olive oil mills effluents. In: Pyrolysis and gasification. Elsevier, New York, pp 586–590
12. Fiestas Ros De Ursinos A, Borja-Padilla R (1992) Use and treatment of olive mill wastewater: current situation and prospects in Spain. Grasas Aceites 43(2):101–106
13. Massignan L, De Leo P, Traversi D, Aveni A (1988) Use of membrane technology for processing wastewater from olive-oil plants with recovery of useful components. In: Cecille L, Toussaint JC (eds) Proceedings of symposium on future industrial processes of membrane processes. Commission of the European Communities, Elsevier Applied Science, Brussels, Belgium, pp 179–189
14. Legrini O, Olivers E, Braun AM (1993) Photochemical processes for water treatment. Chem Rev 93(2):671–698
15. Sun JH, Sun SP, Liang G, Qiao LP (2007) Degradation of azo dye Amido black 10B in aqueous solution by Fenton oxidation process. Dyes Pigments 74(3):647–652
16. Gotsi M, Kalogerakis N, Psillakis E, Samaras P, Mantzavinos D (2005) Electrochemical oxidation of olive oil mill wastewaters. Water Res 39(17):4177–4187
17. Borja R, Alba J, Banks CJ (1996) Anaerobic-digestion of wash waters derived from the purification of virgin olive oil using a hybrid reactor combining a filter and a sludge blanket. Process Biochem 31(3):219–224
18. Israilidies CJ, Mourafeti VN, Karvouni G (1997) Olive oil wastewater treatment with the use of an electrolysis system. Bioresour Technol 61(2):163–170
19. Lee Y, Lee C, Yoon J (2004) Kinetic and mechanism of DMSO (dimethylsulfoxide) degradation by UV/H_2O_2 process. Water Res 38(10):2579–2588
20. Caglayan O (2018) Production methods. Research Institute for Water and Waste Management, Germany. http://www.fiw.rwth-aachen.de/cms/index.php?id=349. Accessed 01 Apr 2018
21. Demichelli M, Bontoux L (1996) Survey on current activity on the valorization of by-products from the olive oil industry. EC-Institute for Prospective Technological Studies Doc. No. 16466 EN, 53
22. Şengül F (1991) Characterization and treatment of industrial wastewaters. Dokuz Eylül University, Faculty of Engineering Press, İzmir, Turkey, Part 8
23. Masghouni M, Hassairi M (2000) Energy applications of olive-oil industry by-products: —1. The exhaust foot cake. Biomass Bioenergy 18(3):257–262

24. APHA-AWWA-WEF (ed) (2001) Standard methods for the examination of water and waste-water, 20th edn. APHA-AWWA-WEF, New York
25. Guneysu S (2009) Research of olive mill wastewater with different methods. PhD Thesis, Istanbul University, Institute of Sciences, pp 63–64
26. Genc N, Yonsel S, Dagasan L, Onar AN (2002) Wet oxidation: a pre-treatment procedure for sludge. Waste Manag 22(6):611–616
27. Ministry of Forest and Water (2018) http://www.ormansu.gov.tr. Accessed 25 Mar 2018

Biosorption of Dye from Aqueous Solutions by a Waste Lignocellulosic Material

Nilgün Balkaya

Abstract In the present study, the biosorption of Astrazon Blue FGRL (AB) which is one of the cationic dyes most commonly used in nylon and acrylic textiles from aqueous solution was studied onto tea waste (tea dust discharged after using), a waste lignocellulosic material. The effects of different parameters including biosorbent dosage, initial pH, contact time, initial dye concentration, and temperature were studied. Tea waste was characterized by Brunauer-Emmett-Teller (BET) surface area, FTIR, and SEM. The experimental equilibrium data were fitted to the Langmuir and Freundlich isotherms. The Freundlich isotherm model fitted to the experimental data better than the Langmuir isotherm. The maximum biosorption capacity, $q_{\max}$, was found to be 263.16 mg/g. The experimental data were discussed in detail comparing with some other low-cost adsorbents reported for AB removal in the previous literature, considering $q_{\max}$, adsorbent surface area, experimental conditions, isotherm models, and thermodynamics of the AB adsorption. The thermodynamic data indicated that AB biosorption was feasible but nonspontaneous, endothermic, and a chemisorption reaction.

Keywords Astrazon Blue FGRL · Tea waste · Biosorption · Aqueous solution · Thermodynamic · Equilibrium

N. Balkaya (✉)
Department of Environmental Engineering, Istanbul University-Cerrahpaşa, Istanbul, Turkey
e-mail: nbalkaya@istanbul.edu.tr

N. Balkaya, S. Guneysu (eds.), *Recycling and Reuse Approaches for Better Sustainability*, Environmental Science and Engineering,
https://doi.org/10.1007/978-3-319-95888-0_23

Introduction

Discharge of colored wastewaters from textile industries into natural streams causes several environmental problems which is accompanied by the necessity of removal of the dyes from wastewater before discharging into water bodies. In wastewater treatment, various methods such as biological treatment, coagulation, membrane separation, advanced oxidation, and adsorption are used to remove dye.

Adsorption is a simple, very efficient separation process. It is easily operated. Therefore, among the mentioned techniques above, it attracts extensive attention. However, the use of inexpensive adsorbents having high adsorption capacity is a very important issue in terms of adsorption technology [1]. Powdered and granular activated carbons are the most commonly used adsorbents in the treatment of industrial wastewater. However, they are not low cost. Whereas, adsorbents to be used in wastewater treatment should be low cost, easily available, and environmentally friendly and have high adsorption capacity. Agro-based waste materials/by-products, industrial by-products, and biomasses such as bacteria, yeast, and fungi do meet these requirements. Lignin and cellulose are the two major constituents found in agro-based waste materials and by-products. Besides, these materials may also contain alcohols, aldehydes, ketones, and carboxylic, phenolic, and ether groups, which are the other polar functional groups of lignin [2]. Nowadays, the agro-based waste materials/by-products, which are lignocellulosic materials, attract particular interests due to their environmentally friendly properties, inexpensiveness, and easy availability as an adsorbent.

In this study, the removal of Astrazon Blue FGRL (AB) dye by tea waste (tea dust discharged after use), which is a lignocellulosic material, was investigated. AB is a cationic dye which is widely used for dyeing acrylic and nylon. These groups of dyes have complex chemical structures. Therefore, they are not destroyed by chemical, physical, and biological treatment [3]. Even at low concentrations, the majority of the cationic dyes are harmful to human beings and toxic to microorganisms [4].

Up to now, researchers have worked on the utilization of macroalga *C. lentillifera* [5], dried biomass of baker's yeast [6], fly ash, apricot stone activated carbon [7], dried sea grape (*Caulerpa lentillifera*) [8], and sepiolite [7, 9] for Astrazon Blue FGRL (AB) removal from aqueous solution. However, according to our literature knowledge, the utilization of tea waste for adsorbing AB has not been reported.

Tea plant (*Camellia sinensis* (L) Kuntze) is grown in various countries in the world. After tea leaves were dried and cured by tea factories, it is widely consumed as a beverage. Turkey is one of the largest producers of tea in the world. Tea plant is usually grown in Rize, Turkey. In Turkey and many other parts of the world, it is one of the most popular and inexpensive beverages. The used tea after preparation of this beverage is completely disposed of as a waste. It is therefore available in huge amounts as solid wastes from houses, cafes, and restaurants in nearly all parts of the world.

In the scope of this study, the effect of various operational parameters (adsorbent dosage, contact time, initial dye concentration, pH, temperature) on AB removal was studied. The biosorption equilibrium isotherms and thermodynamic parameters of AB dye on tea waste were evaluated. The AB removal using tea waste was compared and discussed in detail with various low-cost adsorbents reported in the literature, in the light of the adsorbent surface area, experimental conditions, isotherm models, and thermodynamic and maximum adsorption capacity.

Experimental

Biosorbent

Tea dust discharged after use (tea waste) was collected from households and used without any additional pretreatment except for washing and sieving to obtain the desired adsorbent sizes. The Turkish tea waste was washed several times with boiled water and then washed with an adequate amount of distilled water until the washing water contained no color. The washed tea was dried in an oven at 70–80 °C for a few hours and then sieved to the particle size of 0.8–1 mm. The dried material was stored in dark glass bottles for further use.

Chemicals

Astrazon Blue FGRL (AB) was obtained by DyStar, from Turkey. This dye consisted of two main components which were C.I. Basic Blue 159 and C.I. Basic Blue 3 (Fig. 1). The ratio of the two components was 5:1 (w/w).

A 1000 mg/L stock solution of the AB was prepared in distilled water. The solution was diluted to the required concentration for the experiments.

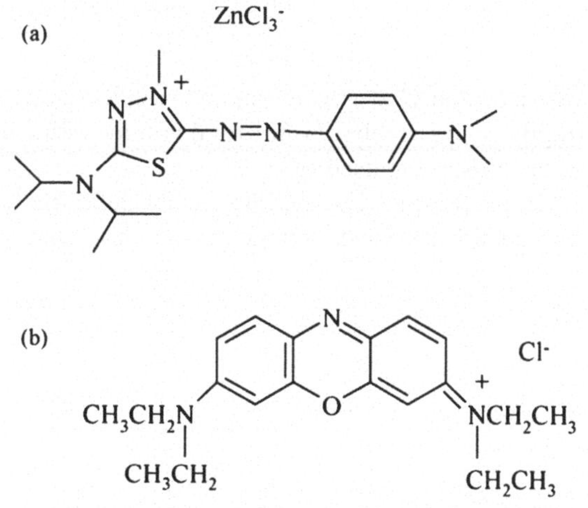

Fig. 1 Chemical structures of (**a**) C.I. Basic Blue 159 and (**b**) C.I. Basic Blue [7]

Characterization of Biosorbent

In this study, the characterization of tea waste was performed by utilizing Brunauer-Emmett-Teller (BET) surface area, Fourier transform infrared (FTIR) spectroscopy (Shimadzu model FTIR-8201 PC (1000–4000 cm^{-1}), and scanning electron microscopy (SEM) (Hitachi 2300 Scanning Electron Microscope). The analysis of Brunauer-Emmett-Teller (BET) surface area was carried out in the METU Central Laboratory-Training Centre in Ankara, Turkey.

Batch Biosorption Experiments

In the batch biosorption experiments, the tea waste was weighed and then placed in 250 mL stoppered Erlenmeyer flasks containing 50 mL of dye solution. The flasks were then placed in a water-bath shaker and agitated at a stirring rate of 200 rpm. The suspensions were filtered, and the concentrations of AB in the filtrate were analyzed using a UV/VIS spectrophotometer (Perkin Elmer UV/VIS spectrometry) by monitoring the absorbance (at 599 nm). The experiments were conducted with varying biosorbent dosages (50, 100, 200, 300, 400, 600 g/L), contact times (5, 15, 30, 45, 60, 75 min), initial AB concentrations (25, 50, 100, 200 mg/L), initial pH (2.0, 4.0, 6.0, 8.0, 10.0), and temperature (20, 30, 40, 50 °C). The desired pH of the solution was adjusted using dilute HCl and/or dilute NaOH.

The amounts of dye uptake per unit weight of the biosorbent q_e (mg/g) and dye removals (%) were calculated by the following equations, respectively:

$$q_e(\mathrm{mg/g}) = (C_0 - C_e)\frac{V}{M} \tag{1}$$

$$\text{Dye removal}(\%) = \frac{(C_0 - C_e)}{C_0}100 \tag{2}$$

where C_0 and C_e are the initial and equilibrium concentrations of dye in the solution (mg/L), respectively, V is the total volume of the dye solution (L), and M is the mass of the biosorbent used (g).

To ensure the repeatability of the data, the biosorption experiments were performed at least twice, and the mean values were presented.

Theory

Equilibrium Isotherms

Langmuir model (Eq. 3) and Freundlich model (Eq. 4) are expressed in linear form as follows [10, 11]:

$$\frac{C_e}{q_e} = \frac{1}{bq_{max}} + \frac{C_e}{q_{max}} \tag{3}$$

$$\log q_e = \log K_F + \frac{1}{n} \log C_e \tag{4}$$

where C_e is the equilibrium dye concentration (mg/L), q_e is the amount of dye uptake per unit weight of the biosorbent (mg/g), q_{max} is the maximum biosorption capacity of the biosorbent (mg/g), b is the Langmuir isotherm constant (L/mg), and K_F (mg/g) and $1/n$ are the constants representing the biosorption capacity and the biosorption intensity, respectively.

Thermodynamic

The thermodynamic parameters (the change in free energy (ΔG), enthalpy (ΔH), and entropy (ΔS)) were calculated using the following equations:

$$\Delta G = -RT \text{In} K_d \tag{5}$$

$$K_d = \frac{q}{C_e} \tag{6}$$

$$\text{In} K_d = \frac{\Delta S}{R} - \frac{\Delta H}{RT} \tag{7}$$

where K_d is the distribution coefficient for the adsorption; ΔH, ΔS, and ΔG are the changes in enthalpy, entropy, and Gibb's free energy, respectively; R is the gas constant; T is the absolute temperature; q is the equilibrium concentration of dye cations sorbed onto biosorbent (mg/L); and C_e is the equilibrium dye concentration (mg/L).

Results and Discussion

Characterization of Tea Waste

Specific Surface Area

The Brunauer-Emmett-Teller (BET) surface area of the tea waste as a natural material with plant origin was determined as 0.871 m^2/g. It can be said that the BET surface area of the tea waste is significantly smaller than that of the activated carbon. It is known that the surface area of several activated carbons used for wastewater treatment is about 1000 m^2/g [12].

Lignocellulosic adsorbents (natural raw materials mainly with plant origin) such as agro-based waste materials usually exhibit low surface area. Physicochemical modifications of these materials can enlarge surface area, type of adsorbing sites, porosity, etc. Thus, sorptive capacity can be improved. However, it is noted that the modification process may compensate for the cost of additional processing [13]. Therefore, tea waste was not subjected to any pretreatment.

Fourier Transform Infrared (FTIR) Analyses

In this study, the chemical characterizations were studied by Fourier transform infrared (FTIR) spectroscopy in order to identify the functional groups that might have participated in the biosorption and also indicate the surface site(s) on which biosorption has taken place. The FTIR spectra for tea waste before and after dye biosorption are shown in Fig. 2. The FTIR spectral characteristics of the tea waste are shown in Table 1. The *spectral data obtained* from *FTIR analyses in the present study* were found to be similar to the ones (bonded OH groups, aliphatic C–H group, secondary amine group, etc.) obtained from some other studies on heavy metal or dye biosorption by tea (tea dust discharged after use, waste tea leaves from tea factory, etc.).

From the FTIR spectra, various functional groups were detected on the surface of the biosorbent sample before and after biosorption. As can be seen from Fig. 2, the spectra display the number of adsorption peaks. Figure 2 also shows that the comparison of the spectra of the tea waste unloaded and loaded with dye showed changes in the absorption intensities of various peaks.

According to the spectra, it can be said that the biosorbent exhibits a complex nature. Vien-Lin et al. [14] revealed that the troughs due to the bonded OH groups can be observed in the range of 3340–3380 cm^{-1}. Fazal and Rafique [15] reported that the trough was observed at 3404.47 cm^{-1} (–OH, –NH). In our study, the bonded OH groups were seen at 3310.6 cm^{-1} before adsorption. The band changed after biosorption (3298.1 cm^{-1}). The considerable shift to the lower wave number suggested that chemical interactions between the dye cations and the hydroxyl (OH) groups occurred on the tea waste surface [15].

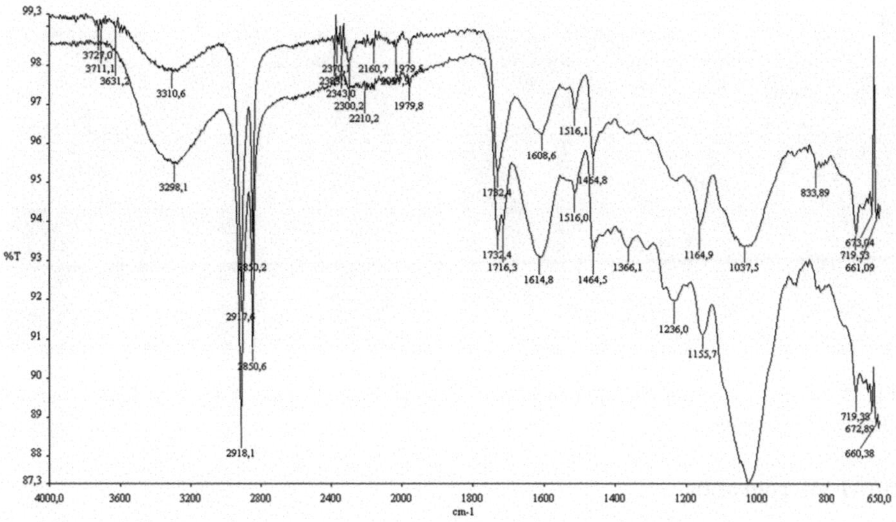

Fig. 2 The FTIR spectra for the tea waste before and after AB biosorption

Table 1 IR absorption bands and involvement of possible groups in biosorption of AB by the tea waste

Frequencies (cm^{-1})		
Before biosorption	After biosorption	Assignment
3310.6	3298.1	Bonded –OH groups
2850.2	2850.6	Aliphatic C–H groups
1732.4	1732.4	Carbonyl stretch of unionized carboxylate
1608.6	1614.81	Symmetric bending of CH_3
1516.1	1516.0	Secondary amine group
1464.8	1464.5	Symmetric bending of CH_3
1164.9	1155.7	C–O stretching of ether groups
661.09	661.38	–CN– Stretching

The bands observed at about 2850.2 and 2850.6 cm^{-1} before and after biosorption could be assigned to the aliphatic C–H group [16, 17]. A shoulder at wave number 1732.4 cm^{-1} was observed. It might be due to the carbonyl stretch of unionized carboxylate [18]. As stated by Malkoc and Nuhoglu [17], the peaks observed at 1511 and 1546 cm^{-1} correspond to the secondary amine group. We observed this peak at 1516 cm^{-1}. Symmetric bending of CH_3 was observed to shift to 1464 cm^{-1}.

Auta and Hameed [19] revealed that 1034.20 cm^{-1} band width had some molecules containing sulfur/oxygen bonds (S=O). We observed this peak at 1037.5 cm^{-1}. It was reported that the peaks observed at 1148 and 619 cm^{-1} could be assigned to C–O stretching of ether groups and –CN stretching, respectively [20]. In our study, the C–O stretching of ether groups was seen at 1164.9 and 1155.7 cm^{-1}. –CN stretching was seen at 661.09 and 661.38 cm^{-1}.

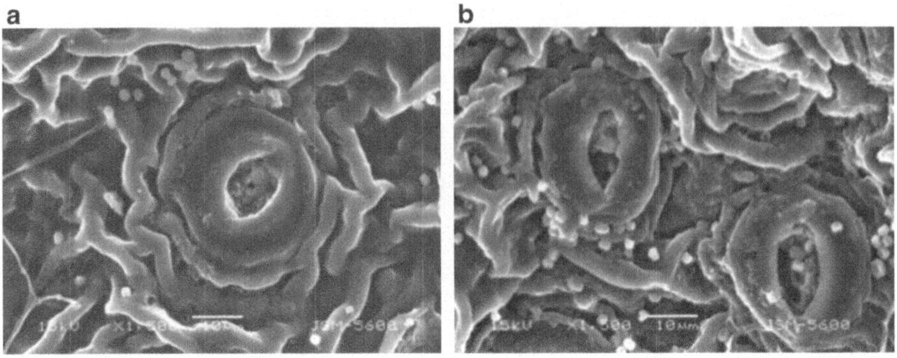

Fig. 3 Scanning electron micrographs of the tea waste (magnification: 1500×) (**a**) before and (**b**) after dye biosorption

SEM Micrographs

The SEM micrographs of the tea waste before and after dye biosorption are given in Fig. 3, respectively. It can be seen that the tea waste mainly consists of fibers with open stomata. Similar finding for green tea residue has been reported by Yang and Cui [21]. As can be seen from Fig. 3, a significant difference does not exist in the surface morphology of the tea waste before and after dye biosorption.

The Effect of Biosorbent Dosage

The effect of biosorbent dosage on AB removal is shown in Fig. 4. The percentage of dye removal increased when the dosage of the tea waste per liter of solution was increased. This trend is a result of the increased adsorbent surface area and the availability of more adsorption sites arising from the increased dosage of the adsorbent [18]. On the contrary, the dye uptake (q) decreased with an increase in the biosorbent dosage. As stated by Jiang et al. [22], this may be due to the presence of more surface area for a fixed amount of dye. Similar findings for AB adsorption onto macroalga *Caulerpa lentillifera* [5], cationic dye methylene blue adsorptions onto tea waste [18], and sugar beet pulp [23] have been reported.

The Effect of pH

The effect of initial pH on the biosorption of AB onto tea waste was investigated by varying the initial pH from 2 to 10, under a constant initial dye concentration of 50 mg/L, adsorption time of 30 min, biosorbent size of 0.8–1.0 mm, and a tea waste dosage of 5.0 g/L. The results of the pH studies are shown in Fig. 5. As known, the

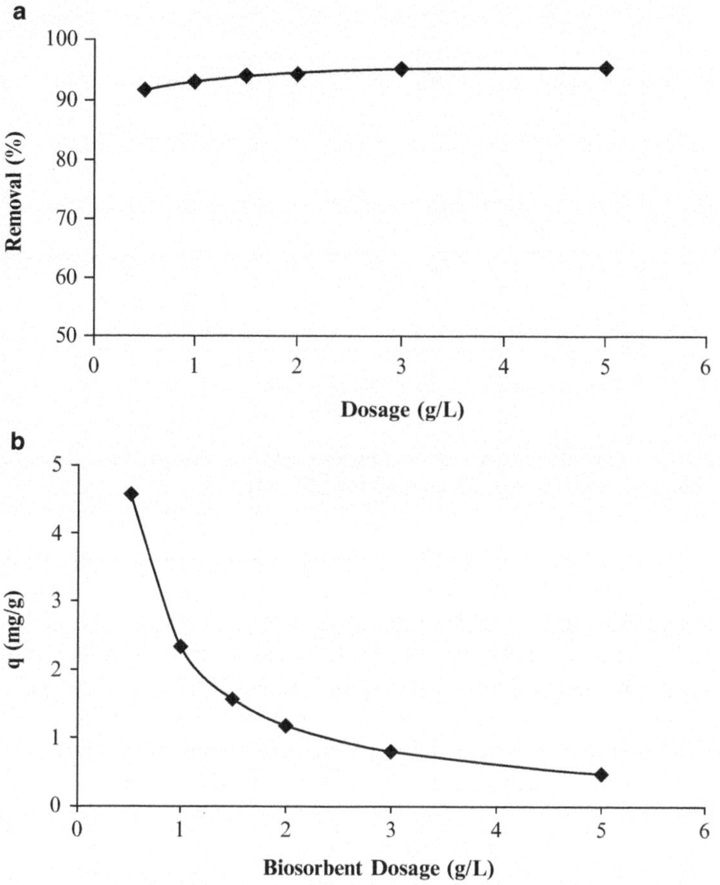

Fig. 4 The effect of biosorbent dosage on AB removal onto the tea waste: (**a**) dye removal (%) (**b**) dye uptake, q (mg/g) (C_0, 50 mg/L; biosorbent particle size, 0.8–1 mm; pH, 5.0; contact time, 30 min)

solution pH affects the surface binding sites of the adsorbent and the degree of ionization of the dye [24].

As can be seen from Fig. 5, the dye uptake increased as the initial pH increased from pH 2 to 4 and then slightly decreased at pH beyond 5. AB is a cationic dye, which exists in aqueous solutions in the form of positively charged ions. As stated by various researches [24, 25], dye removal is inhibited at low pH (less than 4.0). This can be explained by the high concentration of protons in the solution which leads to positive charge density on dye binding sites. As described by Uddin et al. [18], when the pH increases, the surface of the tea waste may become negatively charged as a result of the adsorption of OH^-, and the carboxyl groups of the tea waste may become deprotonated. Thus, negatively charged biosorption sites are produced. Accordingly, the biosorption process is favorably preferred by means of the electrostatic forces of attraction.

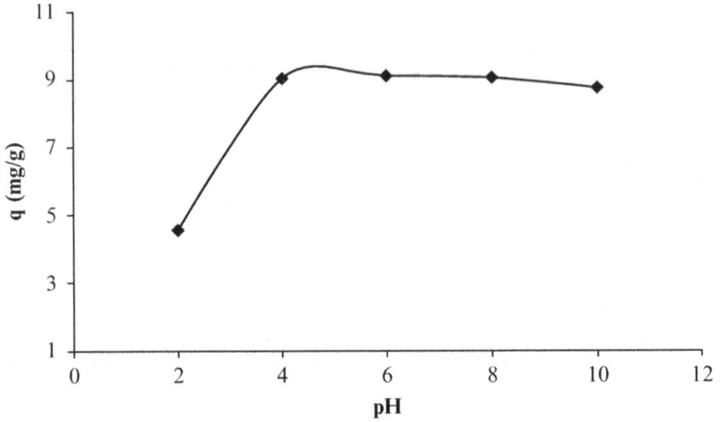

Fig. 5 The effect of initial pH on AB removal by tea waste (C_0, 50 mg/L; biosorbent particle size, 0.8–1 mm; biosorbent dosage, 5.0 g/L; contact time, 30 min)

However, as can be seen from Fig. 5, the dye biosorption at alkali pH is slightly decreased. This behavior may be due to the competition of OH^- [26]. A similar trend has been observed elsewhere [26, 27].

In this study, the initial pH of 5.0 was selected as the more adequate value for the other biosorption experiments. As stated by Liu and Huang [28], the pH values of the cationic dye solutions are generally below 6.0, and instead of the alkaline mediums, the cationic dyes become stable in the acidic mediums. Moreover, the pH values of the cationic dye effluents arising from the dyeing processes range from 4.0 to 7.0.

The Effect of Initial Concentration and Contact Time

The effect of contact time and initial dye concentration on dye removal is shown in Fig. 6. As can be seen from Fig. 6, an increase in the initial dye concentration leads to an increase in the dye uptake. In other words, the more concentrated the dye solution, the higher the biosorption capacity. Presumably, this trend results from the high driving force for mass transfer in high dye concentration [29]. The resistance to dye uptake decreases as the mass transfer driving force increases [7, 30].

The results were in agreement with Karagözoglu et al. [7], Marungrueng and Pavasant [5], and Ongen et al. [9] who studied the adsorption of AB from aqueous solutions. The increased dye uptakes with increasing initial concentrations have also been observed in the adsorption of a cationic dye, methylene blue, onto tea waste [18] and rejected tea [30].

As can be seen from Fig. 6, rapid dye uptake values were obtained in the initial stages of contact time. However, as the contact time increased, biosorption became

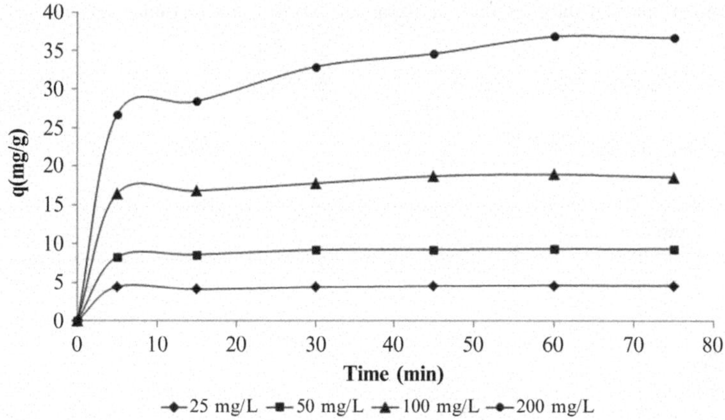

Fig. 6 The effect of contact time and initial dye concentration on AB removal by tea waste (biosorbent particle size, 0.8–1 mm; biosorbent dosage, 5.0 g/L; pH, 5.0)

progressively slower. The fast removal of dye in the initial stage and thereafter saturation gradually could also occur as a consequence of a large number of surface sites which were initially available for the biosorption of the dye molecules. However, steric repulsion could occur between the solute molecules, which consequently leads to a slower biosorption process [31]. Similar phenomena have been observed in the adsorption of AB onto sepiolite, fly ash, and apricot stone activated carbon [7].

Equilibrium Studies

In this study, the Langmuir and Freundlich isotherm models were used for the mathematical statement of the dye biosorption onto tea waste. Fitting of the Freundlich model to the experimental data was better than the Langmuir model. The constants and correlation coefficients (r) of the Langmuir and Freundlich isotherm models for AB biosorption on tea waste are presented in Table 2. The equilibrium isotherm of AB in aqueous solutions onto tea waste is shown in Fig. 7.

In the Freundlich equation, the surface is assumed to be heterogeneous. Therefore, the reason for the better fit of the Freundlich isotherm to the experimental data (*r*: 0.9565) may be a consequence of the heterogeneous distribution of the active sites on the tea waste surface [32]. The Langmuir isotherm is known with its dominant ion-exchange mechanism. On the other hand, the Freundlich isotherm exhibits adsorption-complexation reactions that occur in the biosorption process [33].

Table 2 The Freundlich and Langmuir constants for AB biosorption on tea waste

Langmuir			Freundlich		
q_{max} (mg/g)	b (L/mg)	R	K_F (mg/g)	n	R
263.16	0.006	0.9265	2.76	0.80	0.9565

R correlation coefficient

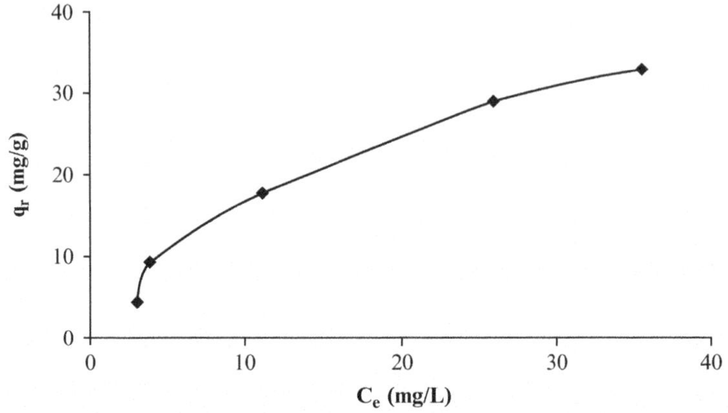

Fig. 7 Equilibrium isotherm of AB in aqueous solutions onto tea waste

The Comparison in Detail of the AB Adsorption Capacities of Various Low-Cost Adsorbents Reported in the Literature

Up to now, some kinds of biomasses (macroalga *Caulerpa lentillifera*, baker's yeast), natural materials (sepiolite), industrial waste products (fly ash), and activated carbon prepared from agricultural by-product (apricot stone activated carbon) have been used for AB removal by various researchers [5–9]. The comparison of the maximum adsorption capacity (q_{max}) and the adsorbent surface area of the tea waste tested in this study with these adsorbents reported in the literature is given in Table 3. Besides, the experimental conditions employed in these studies, isotherm models, thermodynamics, and kinetics related to AB removal by these adsorbents, are presented in this table.

Indeed, every single adsorbent has distinct physical and chemical characteristics including porosity, surface morphology, surface area, stability, and physical strength. In addition, depending on the experimental conditions, the adsorption capacities of the adsorbents vary. Thus, it is difficult to make a comparison of the adsorption performances [34]. However, as can be seen from Table 3, q_{max} of the tea waste is similar to or greater than those reported for other kinds of adsorbent. Despite the tea waste's low surface area, it exhibited a high biosorption capacity for AB. The high biosorption capacity of the tea waste may be attributed to its cellulosic structure which has a high affinity for dyes [35].

Table 3 Comparison of the various adsorbents for AB removal

Experimental conditions							
Adsorbents	ASA (m^2/g)	Time (h)	T (°C)	Isotherm models	TD	q_{max} (mg/g)	Rs
Macroalga *Caulerpa lentillifera*	–	–	18	Freundlich	Endothermic	30.67	[5]
Macroalga *Caulerpa lentillifera*	–	–	27	Freundlich	Endothermic	37.17	[5]
Macroalga *Caulerpa lentillifera*	–	–	40	Freundlich	Endothermic	48.65	[5]
Macroalga *Caulerpa lentillifera*	–	–	50	Freundlich	Endothermic	49.26	[5]
Macroalga *Caulerpa lentillifera*	–	–	70	Langmuir and Freundlich	Endothermic	46.73	[5]
Dried biomass of baker's yeast	–	4	20	Langmuir	Endothermic	69.93	[6]
Dried biomass of baker's yeast	–	4	30	Langmuir	Endothermic	75.19	[6]
Dried biomass of baker's yeast	–	4	50	Langmuir	Endothermic	96.16	[6]
Apricot stone activated carbon	566	24	30	Langmuir	Endothermic	181.50	[7]
Apricot stone activated carbon	566	24	40	Langmuir	Endothermic	188.68	[7]
Apricot stone activated carbon	566	24	50	Langmuir	Endothermic	201.61	[7]
Sepiolite	234.3	24	30	Langmuir	Endothermic	155.52	[7]
Sepiolite	234.3	24	40	Langmuir	Endothermic	190.11	[7]
Sepiolite	234.3	24	50	Langmuir	Endothermic	209.21	[7]
Fly ash	0.342	24	30	Langmuir	Endothermic	128.21	[7]
Fly ash	0.342	24	40	Langmuir	Endothermic	141.84	[7]
Fly ash	0.342	24	50	Langmuir	–	152.44	[7]
Dried sea grape (macroalga *Caulerpa lentillifera*)	5.0	–	25	Freundlich		80.70	[8]
Sepiolite (meerschaum)	377.9	1	20	Freundlich	Exothermic	312.50	[9]
Tea waste	0.87	1	20	Freundlich	Endothermic	263.16	This study

- No data available, *PSO* pseudo-second-order kinetic model, *ASA* adsorbent surface area, *Time* contact time, *T* temperature, *TD* thermodynamic, *Rs* references

When Table 3 was examined in detail, it was seen that:

1. An interesting feature of the adsorbents is that, on the contrary to what was expected, q_{max} values of most of the adsorbents with high surface area were lower than those of the adsorbents with low surface area such as the tea waste. For example, the surface areas of the apricot stone activated carbon and the sepiolite samples are 566 m^2/g [7], 234.3 [7], and 377.916 m^2/g [9], respectively. However,

the sepiolite samples have higher q_{max} values than those of the apricot stone activated carbon. In spite of the fact that the surface area of fly ash [7] is lower than that of the dried sea grape (macroalga *Caulerpa lentillifera*) [8], its q_{max} value is higher than that of the dried sea grape (macroalga *Caulerpa lentillifera*). The reason may probably be due to the effect of pore diameter as well as the varying experimental conditions. As also pointed out by Punjongharn et al. [8], the effect of the pore diameter is more important than that of the surface area. Cationic dye molecules could hardly enter the small pores of the adsorbent, and access to the binding sites inside the pores becomes difficult.

2. AB adsorption by macroalga *Caulerpa lentillifera* [5], dried sea grape (macroalga *Caulerpa lentillifera*) [8], sepiolite [9], and the tea waste (this study) is best described by the Freundlich adsorption isotherm model which may show adsorption-complexation reactions that occur in the adsorption process. The Langmuir isotherm model established the best prediction for the adsorption of AB by dried biomass of baker's yeast [6], apricot stone activated carbon [7], sepiolite [7], and fly ash [7]. As stated above, the Langmuir isotherm is known with its dominant ion-exchange mechanism.
3. Although macroalga *Caulerpa lentillifera* and sepiolite are tested for AB removal from aqueous solution by various researchers [5, 7, 9], q_{max} values found for these adsorbents by various researchers are different from each other. It can be said that, as expected, this trend is due to the different experimental conditions.
4. q_{max} and b values of all the adsorbents (apricot stone activated carbon, sepiolite, fly ash, macroalga *Caulerpa lentillifera*, and dried biomass of baker's yeast) reported in the literature except for the sepiolite tested by Ongen et al. [9] clearly increased depending on the temperature. The adsorption of AB on various adsorbents as well as the tea waste is endothermic in nature.

The Effect of Temperature

The effect of temperature on AB biosorption was studied at the dye concentration of 50 mg/L and contact time of 30 min under the optimum experimental conditions of biosorbent size of 0.8–1.0 mm, biosorbent dosage of 5.0 g/L, and at pH of 5.0, and the results are shown in Fig. 8. The dye uptake, q, on the tea waste increased with the increasing temperature (Table 4), indicating that a high temperature favors AB removal by biosorption on tea waste (Fig. 8) and the biosorption of AB on tea waste was endothermic in nature.

Although adsorption reactions are normally exothermic, the adsorption experiments for different temperatures in this study show endothermic results. It also implies that the adsorption process is dominated by the diffusion process. As stated by Li et al. [36], the diffusion is an endothermic process. Therefore, it can be said that the interparticle diffusion rate of the adsorbate molecules into the pores increased with an increase in the temperature. When the temperature increases, the molecular mobility increases, and the solution viscosity decreases. The increase in the molecular mobility and the decrease of solution viscosity enhance the diffusion rate of dye through the boundary layer and within the internal pores of the tea waste

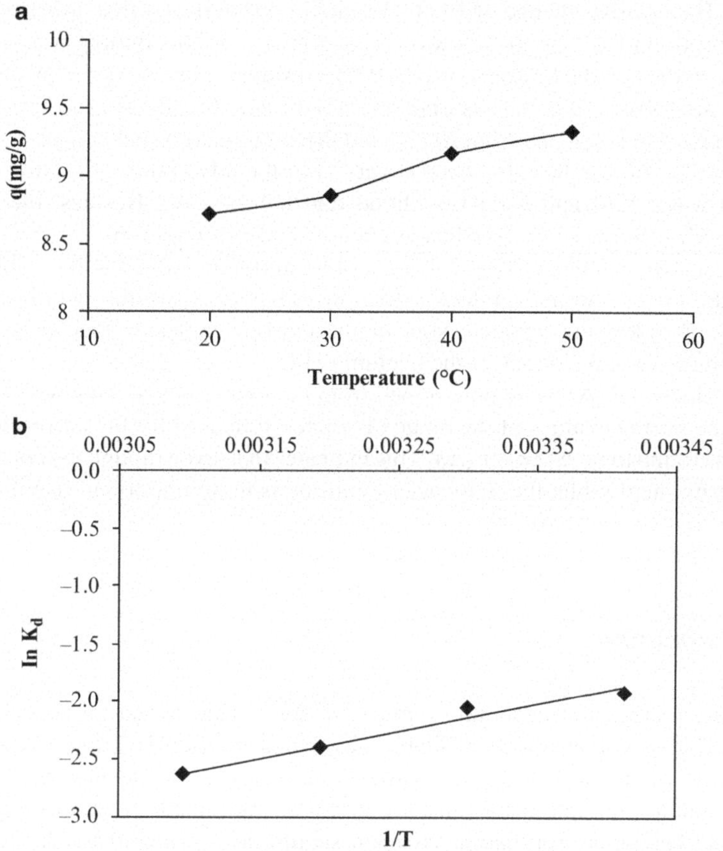

Fig. 8 (**a**) The effect of temperature on AB removal by tea waste (C_0: 50 mg/L; biosorbent particle size, 0.8–1.0 mm; biosorbent dosage, 5.0 g/L; pH, 5.0; contact time, 30 min) (**b**) The plot between ln K_d and $1/T$

Table 4 Thermodynamic parameters for the biosorption of AB by tea waste

q (mg/g)				ΔG (kJ/mol)				ΔH (kJ/mol)	ΔS (J/mol K)
293 K	303 K	313 K	323 K	293 K	303 K	313 K	323 K		
8.72	8.86	9.17	9.33	4.68	5.17	6.24	7.09	19.61	−82.48

particles [30]. Similar effects on the removal of cationic dyes were reported by other researchers for temperature [24, 28, 30].

The calculated thermodynamic parameters (ΔG, ΔH, and ΔS) are given in Table 4. The positive and weak value of ΔG indicates that the biosorption of AB dye is feasible but nonspontaneous [37]. Also, as can be seen from Table 4, the positive values of ΔG increase with increasing temperature. It can be said that this indicates the presence of an energy barrier at high temperature in biosorption [38].

The thermodynamic parameters ΔH and ΔS for the biosorption of AB by tea waste were obtained from the slope and the intercept of a plot of ln K_d versus $1/T$

(Fig. 8). The positive change of 19.609 kJ/mol in ΔH indicates that, as stated above, the biosorption of AB by the tea waste is endothermic. This finding was similar to the other studies on the biosorption of AB. For instance, positive enthalpy of adsorption was also observed in the adsorption of AB on sepiolite, fly ash and apricot stone activated carbon [7], *C. lentillifera* [5], and dried biomass of baker's yeast [6]. The positive value of ΔH has also been observed in the adsorption of methylene blue onto tea waste [26] and NaOH-modified rejected tea [39]. Besides, the positive value of ΔH (19.609 kJ/mol) indicates chemical adsorption [5, 6].

The negative value of ΔS (−82.477 J/mol K) normally indicates that a significant change does not occur in the internal structure of the tea waste during biosorption of AB [40]. Besides, the negative sign of the entropy indicates that at last things become more organized than at the beginning [4].

The values of $T_A \Delta S$ were determined from the experimental data, where T_A represents the average values of the range of temperature used for the adsorption studies. It was found to be $\Delta H < -T_A \Delta S$. This indicates that, even though the contribution of ΔH is not negligible, the influence of entropy is more important than that of the enthalpy in activation [40].

Cost Estimation

The cost-effective and economic removal of dyes from industrial wastewaters is possible if low-cost and easily available adsorbents are used. In Turkey as well as in many parts of the world, the tea waste is an easily available material from the houses, cafeterias, and restaurants as a waste. As stated by Amarasinghe and Williams [41], an appropriate mechanism should be considered for the collection and storing of the tea waste.

Hydroxyl and carboxylic groups in agricultural wastes make them amenable to easy desorption and regeneration with basic or acid solution [13]. However, it can be said that the regeneration of the tea waste is not required due to its abundant availability and low cost. It can also be disposed of after use without the need for expensive regeneration.

In the world market, activated carbon has a price that changes over a wide range (US$ 3–12 for kg) depending on the origin, quality, and quantity [42]. The cheapest commercially available activated carbon available in Turkey is US$ 1.5 for kg. As stated above, tea waste, which is a household waste, is available in huge amounts. Therefore, the total cost of tea waste will be extremely low when compared with those of the activated carbon and some other adsorbents for AB dye. For cost analysis, the expenses for collection and storing can be considered.

As stated by Baek et al. [25], the dye-loaded adsorbent can be disposed of by incineration to prevent further impact on the environment. The dye-loaded tea waste can be utilized as a fuel in the boilers/incinerators after dried. Thus, both energy recovery from the used biosorbent and the safe disposal of the biosorbed dye can be obtained by providing a greener solution.

Conclusion

In this study, the utilization of the tea waste (tea dust discharged after use), which is a green environmentally friendly adsorbent, in the removal of toxic AB dye from aqueous solution was reported.

The main conclusions can be summarized as follows:

1. When the optimum experimental conditions were determined as a function of the process parameters such as the biosorbent dosage, initial AB concentration and contact time, initial pH, and temperature, it was seen that the values of the biosorbent dosage, contact time, initial pH, and temperature were found to be 5.0 g/L, 30 min, 5.0, and endothermic, respectively.
2. The uptake of the dye, q, decreased at low initial pH (less than 4.0) and increased when the temperature of the aqueous medium increased. The effect of initial pH and temperature on AB removal using the other low-cost adsorbents (sepiolite, fly ash, apricot stone activated carbon, *C. lentillifera*, and dried biomass of baker's yeast) reported in the literature exhibited similar trends to the tea waste.
3. The Freundlich equation fitted well to the experimental data. The maximum biosorption capacity, q_{max}, was found to be 263.16 mg/g by the Langmuir isotherm.
4. It is worth mentioning that when q_{max} values obtained for biosorption of AB onto tea waste (this study) and various other low-cost adsorbents reported in the literature were compared with each other, it was seen that q_{max} values of the adsorbents with low surface area such as the tea waste were higher than those of most of the adsorbents with high surface area. It was realized that the adsorbents with high surface area do not always exhibit high q_{max} values. However, it should not be forgotten that the adsorption capacities of the adsorbents differ depending on the experimental conditions.
5. From the thermodynamic parameters (E_a, ΔG, ΔH, and ΔS), it was concluded that the biosorption process takes place by chemical adsorption (probably indicating adsorbent/dye complexation) and is endothermic in nature.
6. It can be said that the biosorption of AB onto the tea waste is a complex process. It is thought that ion-exchange, complexation, and electrostatic interactions play an important role in the whole biosorption process of the tea waste for AB removal, as revealed for metal sorption using tea industry waste by Cay et al. [33].
7. In spite of the fact that the tea waste has a significantly low surface area compared to the commercial activated carbons, the use of tea waste for biosorption of AB from wastewater streams seems to be feasible. From the economical point of view, the tea waste can be used as an alternative adsorbent to activated carbon. It has a low cost and good affinity for AB.

Further studies on AB biosorption using tea waste are currently underway. The kinetics and column studies will be the subject of the next study.

Acknowledgments This research was supported by the Research Project Unit at Istanbul University (Project No: BYP–5090).

References

1. Jia Z, Li Z, Ni T, Li S (2017) Adsorption of low-cost absorption materials based on biomass (Cortaderia selloana flower spikes) for dye removal: kinetics, isotherms and thermodynamic studies. J Mol Liq 229:285–292
2. Pagnanelli F, Mainelli S, Veglio F, Toro L (2003) Heavy metal removal by olive pomace: biosorbent characterization and equilibrium modeling. Chem Eng Sci 58:4709–4717
3. Eren E, Afsin B (2007) Investigation of a basic dye adsorption from aqueous solution onto raw and pre-treated sepiolite surfaces. Dyes Pigments 73:162–167
4. Hong J-M, Lin B, Jiang J-S, Chen B-Y, Chang C-T (2014) Synthesis of pore-expanded mesoporous materials using waste quartz sand and the adsorption effects of methylene blue. J Ind Eng Chem 20:3667–3671
5. Marungrueng K, Pavasant P (2006) Removal of basic dye (Astrazon Blue FGRL) using macroalga *Caulerpa lentillifera*. J Environ Manag 78:268–274
6. Farah JY, El-Gendy N, Farahat LA (2007) Biosorption of Astrazone Blue basic dye from an aqueous solution using dried biomass of Baker's yeast. J Hazard Mater 148:402–408
7. Karagözoglu B, Tasdemir M, Demirbas E, Kobya M (2007) The adsorption of basic dye (Astrazon Blue FGRL) from aqueous solutions onto sepiolite, fly ash and apricot shell activated carbon: kinetic and equilibrium studies. J Hazard Mater 147:297–306
8. Punjongharn P, Meevasana K, Pavasant P (2008) Influence of particle size and salinity on adsorption of basic dyes by agricultural waste: dried Seagrape (*Caulerpa lentillifera*). J Environ Sci 20:760–768
9. Ongen A, Ozcan HK, Elmaslar Ozbas E, Balkaya N (2012) Adsorption of Astrazon Blue FGRL onto sepiolite from aqueous solutions. Desalin Water Treat 40:129–136
10. Langmuir I (1918) The adsorption of gases on plane surfaces of glass, mica and platinum. J Am Chem Soc 40:1361–1403
11. Freundlich HMF (1906) Über die adsorption in lösungen. Z Phys Chem 57:385–470
12. Culp RL, Wesner GM, Culp GL (1978) Handbook of advanced wastewater treatment. Van Nostrand Reinholds Company, New York
13. Johnson TA, Jain N, Joshi HC, Prasad S (2008) Agricultural and agro-processing wastes as low cost adsorbents for metal removal from wastewater: a review. J Sci Ind Res 67:647–658
14. Vien-Lin D, Colthup NB, Fateley WG, Grasselli JC (1991) The handbook of infrared and raman characteristic frequencies of organic molecules. Academic Press, San Diego
15. Fazal A, Rafique U (2012) Acid/base treated spent black tea characterization of functional groups and lead sorptive mechanism. Int J Chem Environ Eng 3:217–224
16. Kapoor A, Viraraghavan T (1997) Heavy metal biosorption sites in *Aspergillus niger*. Bioresour Technol 61:221–227
17. Malkoc E, Nuhoglu Y (2006) Removal of Ni(II) ions from aqueous solutions using waste of tea factory: adsorption on a fixed-bed column. J Hazard Mater B135:328–336
18. Uddin MT, Islam MA, Mahmud S, Rukanuzzaman M (2009) Adsorptive removal of methylene blue by tea waste. J Hazard Mater 164:53–60
19. Auta M, Hameed BH (2011) Preparation of waste tea activated carbon using potassium acetate as an activating agent for adsorption of Acid Blue 25 dye. Chem Eng J 171:502–509
20. Malkoc E, Nuhoglu Y (2006b) Fixed bed studies for the sorption of chromium(VI) onto tea factory waste. Chem Eng Sci 61:4363–4372
21. Yang X, Cui X (2013) Adsorption characteristics of Pb(II) on alkali treated tea residue. Water Res Ind 3:1–10
22. Jiang Z, Xie J, Jiang D, Yan Z, Jing J, Liu D (2014) Enhanced adsorption of hydroxyl contained/anionic dyes on nonfunctionalized Ni@SiO_2 core-shell nanoparticles: kinetic and thermodynamic profile. Appl Surf Sci 292:301–310
23. Malekbala MR, Hosseini S, Yazdi SK, Soltani SM, Malekbala MR (2012) The study of the potential capability of sugar beet pulp on the removal efficiency of two cationic dyes. Chem Eng Res Design 90(5):704–712

24. Sánchez-Martín J, González-Velasco M, Beltrán-Heredia J, Gragera-Carvajal J, Salguero-Fernández J (2010) Novel tannin-based adsorbent in removing cationic dye (methylene blue) from aqueous solution: kinetics and equilibrium studies. J Hazard Mater 174:9–16
25. Baek MH, Ijagbemi CO, Se-Jin O, Kim DS (2010) Removal of malachite green from aqueous solution using degreased coffee bean. J Hazard Mater 176:820–828
26. Giahi M, Rakhshaee R, Bagherinia MA (2011) Removal of methylene blue by tea wastages from the synthesis waste waters. Chinese Chem Letters 22:225–228
27. Kumar KV, Sivanesan S, Ramamurthi V (2005) Adsorption of malachite green onto *Pithophora sp.*, a fresh water algae: equilibrium and kinetic modeling. Process Biochem 40:2865–2872
28. Liu MH, Huang JH (2006) Removal and recovery of cationic dyes from aqueous solutions using spherical sulfonic lignin adsorbent. J Appl Polym Sci 101:2284–2291
29. Chen H, Zhao J, Dai G, Wu J, Yan H (2010) Adsorption characteristics of Pb(II) from aqueous solution onto a natural biosorbent, fallen *Cinnamomum camphora* leaves. Desalination 262:174–182
30. Nasuha N, Hameed BH, Mohd Din AT (2010) Rejected tea as a potential low-cost adsorbent for the removal of methylene blue. J Hazard Mater 175:126–132
31. Dhodapkar R, Rao NN, Pande SP, Kaul SN (2006) Removal of basic dyes from aqueous medium using a novel polymer: Jalshakti. Bioresour Technol 97:877–885
32. Rubin E, Rodriguez P, Herrero R, Cremades J, Barbara I, Sastre de Vicente ME (2005) Removal of methylene blue from aqueous solutions using as biosorbent *Sargassum muticum*: an invasive macroalga in Europe. J Chem Technol Biotechnol 80:291–298
33. Cay S, Uyanık A, Ozaşık A (2004) Single and binary component adsorption on copper (II) and cadmium (II) from aqueous solution using tea industry waste. Sep Purif Technol 38:273–280
34. Sharma P, Kaur R, Baskar C, Chung W-J (2010) Removal of methylene blue from aqueous waste using rice husk and rice husk ash. Desalination 259:249–257
35. El Ashtoukhy ESZ (2009) *Loofa egyptiaca* as a novel adsorbent for removal of direct blue dye from aqueous solution. J Environ Manag 90:2755–2761
36. Li K, Zheng Z, Huang X, Zhao G, Feng J, Zhang J (2009) Equilibrium, kinetic and thermodynamic studies on the adsorption of 2-nitroaniline onto activated carbon prepared from cotton stalk fibre. J Hazard Mater 166:213–220
37. Gueu S, Yao B, Adouby K, Ado G (2007) Kinetics and thermodynamics study of lead adsorption on to activated carbons from coconut and seed hull of the palm tree. Int J Environ Sci Technol 4:11–17
38. Yazdani M, Mahmoodi NM, Arami M, Bahrami H (2012) Surfactant-modified feldspar: isotherm, kinetic, and thermodynamic of binary system dye removal. J Appl Polym Sci 126:340–349
39. Nasuha N, Hameed BH (2011) Adsorption of methylene blue from aqueous solution onto NaOH-modified rejected tea. Chem Eng J 166:783–786
40. Eren E (2009) Investigation of a basic dye removal from aqueous solution onto chemically modified Unye bentonite. J Hazard Mater 166:88–93
41. Amarasinghe BMWPK, Williams AR (2007) Tea waste as a low cost adsorbent for the removal of Cu and Pb from wastewater. Chem Eng J 132:299–309
42. Kavitha D, Namasivayam C (2008) Capacity of activated carbon in the removal of acid brilliant blue: determination of equilibrium and kinetic model parameters. Chem Eng J 139:453–461

Index

N. Balkaya, S. Guneysu (eds.), *Recycling and Reuse Approaches for Better Sustainability*, Environmental Science and Engineering,
https://doi.org/10.1007/978-3-319-95888-0

L

M

N

O

P

R

S

The manufacturer's authorised representative in the EU is Springer Nature Customer Service Centre GmbH, Europaplatz 3, 69115 Heidelberg, Germany. If you have any concerns regarding our products, please contact ProductSafety@springernature.com

Printed and bound by CPI Group (UK) Ltd, Croydon, CR0 4YY
05/10/2026
02228139-0013